T0366021

Recycled Ceramics in Sustainable Concrete

Emerging Materials and Technologies

Series Editor
Boris I. Kharissov

Recycled Ceramics in Sustainable Concrete
Properties and Performance
Kwok Wei Shah and Ghasan Fahim Huseien

Recycled Ceramics in Sustainable Concrete

Properties and Performance

Kwok Wei Shah and Ghasan Fahim Huseien

CRC Press is an imprint of the
Taylor & Francis Group, an informa business

First edition published 2021
by CRC Press
6000 Broken Sound Parkway NW, Suite 300, Boca Raton, FL 33487-2742

and by CRC Press
2 Park Square, Milton Park, Abingdon, Oxon, OX14 4RN

Library of Congress Cataloging-in-Publication Data

Names: Shah, Kwok Wei, author. | Huseien, Ghasan Fahim, author.
Title: Recycled ceramics in sustainable concrete : properties and performance / by Kwok Wei Shah and Ghasan Fahim Huseien.
Description: First edition. | Boca Raton, FL : CRC Press, 2021. |
Series: Emerging materials and technologies | Includes bibliographical references and index. |
Summary: "This book explores the use of novel waste materials in the construction industry as sustainable and environmentally friendly alternatives to traditional cement production technologies. It specifically focuses on using waste ceramics as a binder and aggregate replacement for concrete. Written for materials, chemical, and civil engineers as well as others who develop construction materials, this book provides readers with a thorough understanding of the merits of using waste ceramics to produce sustainable concrete"-- Provided by publisher.
Identifiers: LCCN 2020043481 (print) | LCCN 2020043482 (ebook) |
ISBN 9780367636876 (hbk) | ISBN 9781003120292 (ebk)
Subjects: LCSH: Concrete--Additives. | Ceramic materials--Recycling.
Classification: LCC TP884.A3 S525 2021 (print) | LCC TP884.A3 (ebook) |
DDC 624.1/834--dc23
LC record available at https://lccn.loc.gov/2020043481
LC ebook record available at https://lccn.loc.gov/2020043482

ISBN: 9780367636876 (hbk)
ISBN: 9781003120292 (ebk)

Typeset in Times
by Deanta Global Publishing Services, Chennai, India

Contents

Preface

Recycled Ceramics in Sustainable Concrete: Properties and Performance is comprised of 12 chapters. Each chapter discusses one of the applications of ceramic waste in the concrete industry. Recycling of ceramic waste materials, found in abundance, not only prevents deleterious environmental hazards, but can actually produce wealth by adding value through ecology.

The first chapter deals with the landfill problems of ceramic waste and the environmental impact. The environmental problems of cement, concrete durability problems, and energy problems in the cement industry are presented in this chapter. Meanwhile, the benefit of utilizing pozzolanic materials on strength development and durability performance are discussed.

The preparation stages of ceramic waste as fine and coarse aggregate and powder are presented in Chapter 2. The characteristics of ceramic waste as aggregate material are discussed including sieve analysis, shape, abrasion, and water absorption. The chemical composition and physical properties of wastes ceramic powder are presented in this chapter. Tests to determine the mineral properties, including X-ray diffraction (XRD), scanning electron microscopy (SEM), Fourier-transform infrared spectroscopy (FTIR), thermogravimetric analysis (TGA), and derivative thermogravimetric analysis (DTG), are discussed.

In the remaining chapters (3–12), the applications of ceramic waste as natural aggregate or cement replacement in normal or geopolymer concrete are discussed. Ceramic waste is used in normal concrete to reduce environmental problems, save energy, improve the strength and durability performance of the cement, and reduce the demand of natural resources. Ceramic waste is also used to produce self-compacting concrete by replacing the slag in geopolymer concrete industry. The performance of ceramic waste in aggressive environments, such as sulphuric acid, sulphate, and elevated temperatures, are discussed. Finally, using ceramic waste as repair materials and in structural applications is covered.

The versatility of this book, compared to others, lies in its timely compilation about the most significant development of the 20th century (i.e. concrete from waste materials). The authors believe no such book currently exists which compiles information about an extensive variety of wastes that could be used in the concrete industry to generate low-cost, environmentally friendly materials. Moreover, a few chapters reveal a combination of these wastes, new approaches to old materials, and unique demands related to waste materials. The availability of the book to engineers, technologists, researchers, contractors, consulting firms, and government agencies dealing with construction, the environment, the general public, etc., is very crucial.

Effective utilization of ceramic waste materials in the concrete and construction industries, whose growth seems unlimited, and mounting evidence of worldwide interest suffice the need to produce a collective anthology of a wide variety of ceramic waste materials available today.

Authors

Dr. Kwok Wei Shah is presently an assistant professor and deputy program director with the Department of Building, School of Design and Environment, National University Singapore. He is an advisory board member of the Vietnam Green Building Council and sits on its Education Committee. He lectures for REHDA GreenRE in Malaysia and is a visiting fellow at University Technology of Malaysia, UTM. He is a visiting professor at Tianjin University of Technology, China. He was appointed BCA Ambassador for 3 years and a member of the SPRING and SGBC technical committees. He served as a technical consultant for Ascendas Services Pte Ltd, and chief technical adviser for Bronx Culture Pte Ltd.

Shah's research interest is in nanotechnology and nanomaterials for green building applications. He has done outstanding research work on a novel low-cost high-volume aqueous silica-coating technique that has been granted a US patent (US 20130196057 A1). His research paper published by *Nanoscale* (DOI:10.1039/C4NR03306J; impact factor = 6.739) titled "Noble Metal Nanoparticles Coated with Silica by a Simple Process That Does Not Employ Alcohol" was highlighted by popular science websites including ScienceDaily, Phys.org, and A*STAR. Shah's research on microencapsulated phase change materials enhanced by highly thermal conductive nanowires (*Journal of Materials Chemistry A*; DOI: 10.1039/C3TA14550F; impact factor = 6.626) led to the development of the "M-KOOL" phase change cooling technology, which was featured on Phys.org, Channel News Asia, *Straits Times, Business Times, TODAY,* The Star Online, and *Lianhe Wanbao.* So far, Shah's achievements include 3 first-authored papers, 9 co-authored papers, 1 book chapter, 12 patent disclosures, and 1 commercial licensing.

Dr. Ghasan Fahim Huseien is a research fellow at the Department of Building, School of Design and Environment, National University of Singapore, Singapore. He received his PhD from the University Technology Malaysia in 2017. He is involved in applied research and development on the evaluation and application of various types of materials to repair cracks, damaged surfaces in concrete; and "the utilization of industrial and agriculture wastes (to reduce cost, energy and environmental problems) for the existing and future construction of concrete structures, both in the laboratory and in the field". He is a peer reviewer for several international journals as well as an adviser for master's and PhD students. He is a member of the Concrete Society of Malaysia and American Concrete Institute.

Huseien has more than 5 years' experience in applied research and development as well as 10 years' experience in manufacturing smart materials for sustainable building and smart cities. He has expertise in advanced sustainable construction materials covering civil engineering, environmental sciences and engineering, chemistry, earth sciences, geology, and architecture departments. He authored and co-authored 50-plus publications and technical reports, 3 books and 15 book chapters, and participated in more than 25 national and international conferences/workshops.

His past experience in projects includes the application of nanotechnology in construction and building materials, self-healing technology, and geopolymers as sustainable and eco-friendly repair materials in the construction industry.

1 Recycling of Ceramic Wastes
Emerging Research and Opportunities

1.1 INTRODUCTION

To achieve a sustainable world, reducing carbon dioxide (CO_2) emissions into the atmosphere is one of the important factors to be considered. The biggest producers of CO_2 emissions are related to the construction industries, that is cement manufacturing, steel factories and transportation. Worldwide, around 5% of the total of CO_2 emissions is contributed by the cement industry alone [1]. Therefore, partially replacing the weight of cement with pozzolanic wastes is one effective way to reduce the environmental impact of CO_2 emissions from cement production [2]. Previously, natural pozzolanic wastes have been used as construction material due to their positive effects on mechanical properties and durability of mortar [3]. However, because of stringent environmental regulations, researchers have focused on using industrial waste as a pozzolanic material [4, 5]. Nowadays huge amounts of ceramic wastes are produced by the ceramic industries which cause environmental impact and landfill problems. World production of ceramic tiles was approximately 12.4 billion square metres in 2015 [6]; Malaysia manufactured 92 million square metres of ceramic in the same year and that is increasing by 2.2% each year [7]. It is estimated that 10–30% of total ceramic production goes as waste [8]. Most of this ceramic waste cannot be recycled and later creates disposal issues [9]. The use of ceramic waste in concrete contributes to cost and energy savings, ecological balance, and conservation of natural resources [10]. The large generation of ceramic waste with little utilization encouraged researchers to use it in mortar as cement replacement [11].

Since aggregates in concrete and mortar comprise about 60–75% of the total volume, any reduction in natural aggregate consumption will have significant impact on the environment. Besides the environmental constraints of stone pits, noise, dust, vibrations, and the impact on the countryside, the consumption of a non-renewable materials tends to considerably limit their exploitation. Consequently, alternative materials such as construction and demolition wastes as well as other industrial by-products are increasingly being tested and used as sustainable natural aggregates substitutes [12]. The use of inorganic industrial waste in making concrete will lead to sustainable concrete design and a greener environment [13]. The need to develop concrete with non-conventional aggregates is highly needed for environmental as well as economic reasons. A review of earlier research showed that industrial waste and other wastes have been introduced in concrete making not only to improve the properties of concrete but also to reduce cost. The inclusion of recycled steel tyre fibres in concrete was found to avoid the opening of cracks and increase energy absorption [14]. In addition, structural lightweight concrete has been produced using palm oil shells waste [15] and demolished masonry waste [16] as aggregates. Other research revealed that an improvement in the modulus of elasticity of concrete was observed with partial replacement of crushed stone coarse aggregates with crushed vitrified soil aggregates [17].

The ceramic industry is known to generate large amounts of calcined clay waste each year. Reusing this waste in concrete could be a win–win situation. Ay and Ünal [18] reported that the use of ceramic waste as pozzolanic material in different percentage as cement replacement shows better performance when the tile waste had been used as pozzolanic material by up to 30% of weight of cement. It was also reported that ceramic wastes are potentially suitable for replacement of Portland cement without detrimental effect on the compressive strength [9]. Other research showed that concrete containing 20% ceramic waste powder as cement replacement material reached 75% of compressive strength of the control specimen at 7 days of curing and increased the durability performance of concrete [19]. Research on the utilization of crushed ceramic waste as fine aggregates was conducted by Alves et al. [20] and it was reported that the concrete showed good strength properties. A study that was conducted on the effect of crushed ceramic and basaltic pumice used as fine aggregates showed better compressive strength in comparison to the control specimen [21]. A similar finding was also reported that compressive strength was unchanged when ceramic waste was used to partially replace conventional crushed stone coarse aggregates [22, 23]. This indicated that ceramic waste could be potentially transformed into useful coarse or fine aggregates. The properties of coarse and fine aggregates of ceramic waste are well within the range of the value of concrete-making aggregates. The properties of ceramic waste coarse aggregates concrete are not significantly different from those of conventional concrete [22]. Therefore, the use of ceramic wastes as coarse aggregates in concrete has increased because it has various advantages over other cementitious materials [24].

East Asia is one of the developing regions where the construction of many infrastructures is still ongoing and, consequently, requires a high demand of concrete that will affect the use of natural aggregates as resources. Therefore, the present work focuses on the recycle and reuse of industrial waste to produce new construction material. In this research, due to its abundant availability, ceramic waste was studied as a potential material for cement, coarse and fine aggregates replacements. Nowadays, the high demand for ordinary Portland cement (OPC) partly contributes to environmental pollution and the release of carbon dioxide gasses to the atmosphere. The excessive amount of carbon dioxide gas in the atmosphere can cause many problems not only to the environment but also to reinforced concrete structures. In addition, ceramic waste is not biodegradable and consumes a lot of landfill space. Finding new ways to recycle this waste in the construction of infrastructures can be useful to preserve natural resources and the environment. In order to evaluate the true potential of ceramic waste for new applications, as cement and aggregates replacements, a comprehensive and detailed study of the fundamental properties of the material is highly needed.

1.2 ENVIRONMENTAL PROBLEMS OF CEMENT

Universally, OPC has been continually exploited as a concrete binder and different building substances. Cement manufacturing is an intensive energy consumer and significantly affects climate change. The main issues related to the cement industry are the emission of carbon dioxide to the atmosphere and energy consumption that affect environmental health and safety. The cement industry consumes a large amount of natural resources that are not renewable. It is reported that 5–6% of carbon dioxide emission produced by human activities was contributed by the manufacture of cement [25]. Due to the implementation of the Kyoto Protocol in February 2005, countries all over the world have to reduce their greenhouse gas emissions. Therefore, one of the major challenges is to reduce the CO_2 emissions from the cement manufacturing industries. Furthermore, according to the study done by Earth Institute of Colombia University, cement manufacturing as a single industry produced 5% of the total carbon dioxide emissions and cement production is growing 2.5% yearly. Around 0.8–1.3 tonnes of carbon dioxide is discharged for 1 tonne of cement manufacturing and about 0.8 tonnes of carbon dioxide is released for 1 tonne of cement clinker production [26]. In developing countries, especially China and India, the rise in cement production is causing a dramatic increase in the level of carbon dioxide as shown in Figure 1.1, which needs immediate action to find ways to reduce the usage of cement. With the increasing population, the demand of cement for construction industries is increasing and it is expected to increase until the year 2050. Therefore, by using blended cement or replacing some of the weight of cement with industrial waste that has cementitious properties can be an effective way to reduce pressure on the environment.

1.3 CONCRETE DURABILITY PROBLEMS

The serviceability of construction materials has considerable economic significance, particularly with modern infrastructures and components. For urbanization, concrete materials that are greatly exploited must meet the requisites of standard codes of practice related to strength and durability [27]. For instance, poor planning, low capacity or overload, faulty material design and structures, incorrect construction practices or unsatisfactory maintenance, and lack of engineering knowledge can often diminish the service lifespan of concrete under operation [28]. In the construction industries, fast declination of concrete structures being a major setback necessitates additional improvement. Varieties of physical, chemical, thermal and biological processes are responsible for the progressive deterioration of concrete structures during their service [29, 30]. Several studies [31–33] revealed that concrete performance

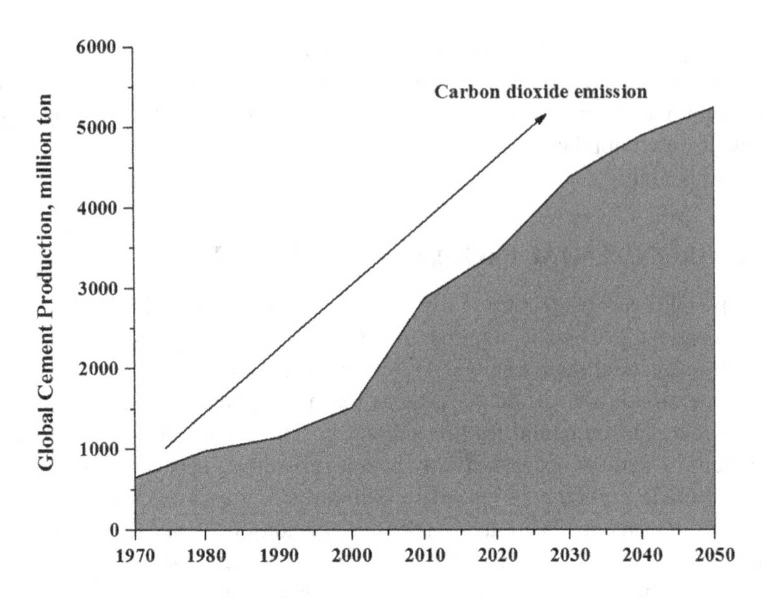

FIGURE 1.1 Global cement production.

is greatly affected by improper usage, and physical and chemical conditions of environments. It is verified that both external and internal factors involving physical, chemical, or mechanical actions are often responsible for the deterioration of the concrete structures.

Mechanical damage of concrete structure occurs due to different reasons such as impact, abrasion, cracking, erosion, cavitation, and contraction. Chemical actions that cause the declination of concrete are carbonation, reaction associated with alkali and silica, alkali, and carbonate as well as efflorescence. Moreover, outside attacks of chemicals happen primarily due to CO_2, Cl_2 and SO_4 as well as several other liquids and gases generated by the industries. Physical causes of deterioration include the effects of high temperature or differences in the thermal expansion of aggregate and of the hardened cement paste. Other reasons of deterioration is the occurrence of alternating freezing and thawing of concrete and the associated action of the deicing salts. Physical and chemical processes of deterioration often act in a synergistic way including the influence of seawater on concrete. Poor durability performance of OPC in an aggressive acidic or sulphate (especially in marine) environment is caused due to the existence of calcium complexes. These calcium complexes are very easily dissolved in an acidic atmosphere, leading to enhanced porosity and thus fast deterioration [34].

In many places worldwide, OPC structures that have existed for many decades are facing rapid deterioration [35]. The permanence of OPC is linked with the nature of concrete's constituents, where a CaO of 60–65% and the hydration product of $Ca(OH)_2$ of nearly 25% is responsible for fast structural decay. Several studies indicated that occurrence of fast reaction of $Ca(OH)_2$ in acidic surroundings allows

OPC to be deprived of water, leading to acid fusion and weakening of resistance against aggressive attacks. The intense reaction of evolved CO_2 with $Ca(OH)_2$ contribute to rapid corrosion of concretes containing OPC [35]. The safety, service life, permanence, and lifespan of the mix design of concretes are considerably influenced by crack development and subsequent erosion. These drawbacks of OPC-based concretes drove researchers to enhance properties of conventional OPC by adding pozzolanic materials, polymers, and nanomaterials so that it becomes more sustainable and endurable. The immediate consequence for affected concrete structures is the anticipated need of maintenance and execution of repairs [36]. Thus, there is a renewed interest for the development of sustainable concrete to solve all these existing shortcomings involving harsh environmental conditioning and durability.

Generally, the durability of concrete depends on a number of physical (including mechanical) and chemical causes, with the most outstanding being permeability, carbonation, leaching, aggressive chemicals (e.g. sulphate, chloride, and acid), reactive aggregates, freezing and thawing, and thermal variations. In mild environment condition, ordinary Portland cement based concrete shown excellent performance with high durability. On the other hand, in the long-term normal concrete can suffer from deterioration due to attacks from aggressive agents such as sulphate, acid, and chloride [37]. The majority of chemicals attack concrete in the form of a reaction between aggressive agents and the cement matrix, and there also may be some reactions with the aggregates such as alkali–aggregate reactions [38]. Some types of deterioration in concrete can be seen in Figure 1.2.

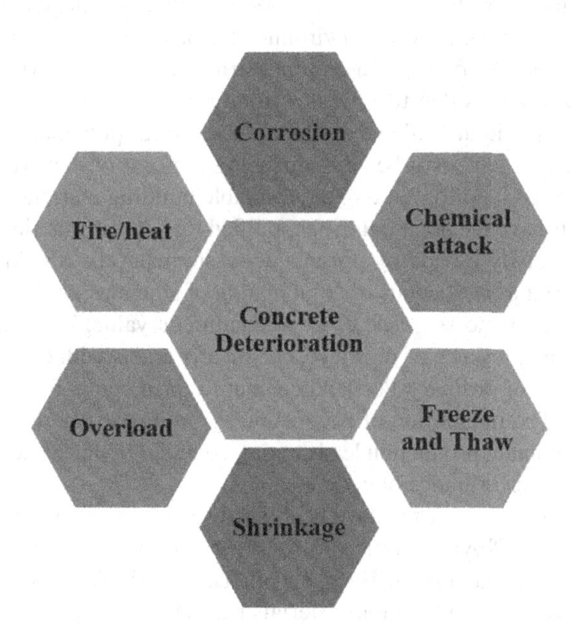

FIGURE 1.2 Concrete deterioration.

1.4 ENERGY PROBLEMS IN CEMENT INDUSTRIES

Nowadays, cement is the primary workhorse in construction sectors worldwide. Cement is processed at a high temperature (≈1450°C) in a rotary kiln by mixing limestone or chalks with clay. Next, the yielded hard nodules of clinker are crushed with a low amount of gypsum using the ball milling technique. Firing of such cement constituents at high temperatures needs substantial amounts of energy to be consumed (burning of coal or petroleum coke). Rapid decay of land setting, generation of dust during transport, creation of noise during quarrying and processing of raw materials are regarded as major environmental concerns in OPC production. According to previous publications [32, 39], OPC manufacturing devoid of considerable amounts of CO_2 release is impossible. Actually, the CO_2 emission occurs in two phases: fuel burning (to achieve very high kiln temperatures) and calcining (a chemical reaction that occurs during limestone firing). Until now, even the most efficient cement production plant emits 60% or more of CO_2 from various inevitable chemical pathways. The consumption of a considerable quantity of energy during crushing of cementitious raw materials in the clinker phase remains the foremost ecological distress in OPC industries [40, 41]. Undeniably, developing autonomic self-healing cement-based materials must be realized as a practical solution to the existing problems.

1.5 UTILIZING POZZOLANIC MATERIALS IN CONCRETE

The development of industries all over the world resulted in a huge amount of waste produced and sparked the need for environment protection due to the high demand of natural resources. The potential use of industrial wastes as construction materials led to many new findings that turn wastes from unprofitable sectors to economical sectors. Utilization of industrial and agricultural wastes as potential cement replacement in construction have two benefits: reducing the cost of construction and disposal of waste in the landfill. Among the available building material in the market, concrete products have the potential to be modified for various functions in construction projects especially by adding pozzolanic waste material. ASTM C618 defines pozzolanic material as a siliceous or combination of siliceous and aluminous material which in itself possesses little or no cementitious value, but when in a finely divided form in the presence of moisture, chemically reacts with calcium hydroxide from the hydration of ordinary Portland cement to form compounds with cementitious properties. The integration of pozzolanic by-products such as silica fume, fly ash and slag in concrete production leads to the reduction in the usage of cement and also enhances the performance of concrete.

Natural and industrial wastes have been used as construction materials, namely pulverized fuel ash (PFA), ceramic tile waste, ground blast furnace slag (GBFS), metakaolin (MK), silica fume (SF), palm oil fuel ash (POFA), and rice husk ash (RHA). Their usage as construction materials has achieved appreciable levels since they offer advantages in terms of strength and durability [42, 43]. In addition, these pozzolanic materials also have been shown to improve the properties of concrete and reduce the hydration temperature, and hence make massive construction easier to construct. Since the use of waste materials as supplementary cementing materials is growing, it is expected that modified concrete product containing pozzolanic waste will be a promising material for the construction industry in the future [44].

Generally, 10–40% of the weight of the binder can be replaced with pozzolanic materials [45]. The pozzolanic materials react with calcium hydroxide $Ca(OH)_2$ from the hydration process of Portland cement and produce secondary calcium silicate hydrate (C-S-H) [46]. Dhir et al. [47] reported an improvement in the fresh properties and compressive strength development of concrete containing pozzolanic materials. Furthermore, McCarthy and Dhir [48] reported that concrete containing pozzolanic material as cement replacement gave several benefits such as improvement in the durability performance associated with environmental and economic benefit. Ceramic wastes with a suitable fineness can be used as pozzolanic materials. It is reported that if silica content in ceramic waste is mixed with calcium hydroxide, pozzolanic reactions can take place and produced secondary C-S-H gel, thus increasing the strength and durability properties of the concrete [49, 50].

The benefit of pozzolan utilization as cement replacement material in mortar and concrete is multidimensional. Firstly, partially replacing the weight of cement with pozzolanic waste materials is economically beneficial in construction production. Secondly, it highlights the environmental issues associated with the carbon dioxide emission during cement manufacturing. The third advantage is the improvement in the durability performance of the end product, thus prolonging the service life of the structure. Furthermore, the utilization of blended cement with pozzolanic materials increased the opportunity to utilize huge amounts of industrial and agricultural wastes to produce durable construction materials.

1.6 WASTES OF CERAMIC

Ceramics is a common word used to refer to ceramic products. General ceramic products include wall tiles, floor tiles, sanitary ware, household ceramics, and technical ceramics. Ceramic tiles are manufactured by firing clay, feldspar, and quartz at high temperatures. Ceramic waste can be divided into two categories depending on the source of raw materials and the production process. The first category is all fired wastes generated by the structural ceramic factories that use only red pastes to manufacture their products, such as brick, blocks, and roof tiles. The second category is all fired waste produced in stoneware ceramics such as wall, floor tiles, and sanitary ware. The usage of white paste is more frequent and much higher in volume. Figure 1.3 shows the classification of ceramic wastes by type and production process from manufacturing factories.

The world tile production increased from 7 million m^2 in 2005 to 12.5 million m^2 in the year 2014 [7]. Figure 1.4 shows the ceramic tile production globally. In the ceramics industry, about 15–30% production goes to waste (Figure 1.5). However, ceramic waste is durable, hard and highly resistant to biological, chemical, and

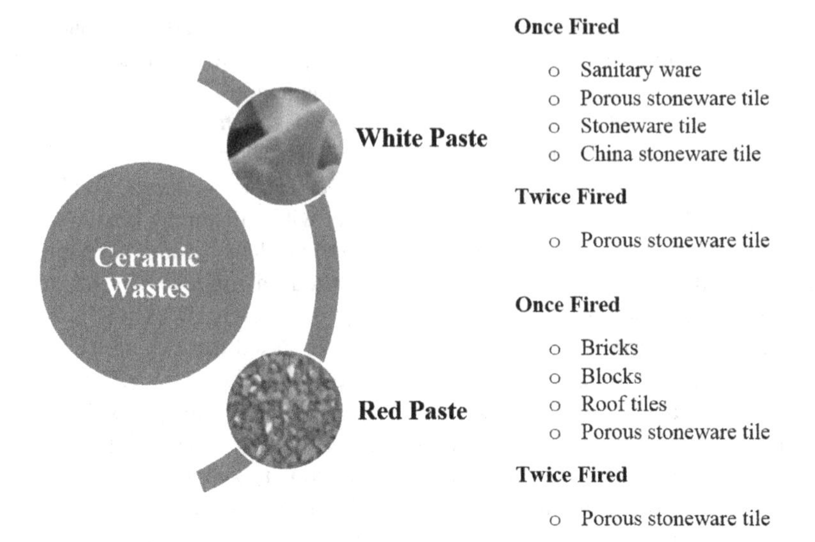

FIGURE 1.3 Classification of ceramic wastes by type and production process.

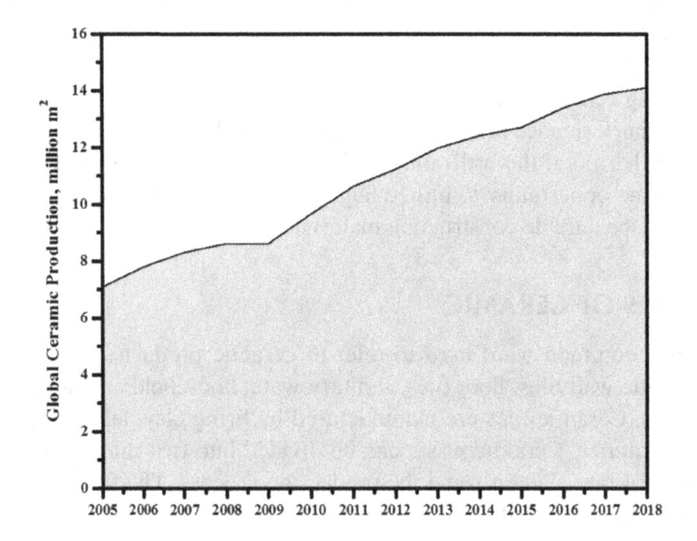

FIGURE 1.4 Global ceramic production.

physical degradation forces. Therefore, there is a need in the ceramics industry to find a way for ceramic waste disposal. Recently, many studies have focused on the use of ceramic waste powder the size of 45 μm as cement replacement in concrete or binder-based geopolymer and alkali-activated concretes. Also, crushed ceramic waste has been used as a replacement for fine and coarse aggregate natural resources.

FIGURE 1.5 Ceramic wastes.

1.7 UTILIZING CERAMIC WASTES IN CONCRETE INDUSTRY

1.7.1 Traditional Concrete

The main ingredient for producing ceramic products is clay, which contains a high amount of silicate (SiO_2). Baronio and Binda [51] reported that by heating clay minerals to 600 to 900°C and reducing the fineness (45 μm) will show highly pozzolanic properties. Calcined clays have been used as pozzolans since ancient times [52]. The heat treatments of clay minerals cause the change from crystalline structure to an amorphous phase. During the last several decades, the used of metakaolin as a pozzolanic material in mortar and concrete was widely studied. Naceri and Hamina [2] reported that waste brick can be used as cement replacement up to 10% without reducing the mechanical properties. Meanwhile, O'Farrell et al. [53] researched the effect of waste clay brick up to 30% of cement replacement on the compressive strength of mortars and found a decrease in early strength up to 28 days but an increase at later age. The effect of fine ceramics waste as partial cement replacement on the properties of concrete or mortar was studied less compared to metakaolin. However, the number of publications increased during the last few years due to the need of waste disposal [54].

Waste from ceramic roof tiles produced during ceramic manufacturing and fired at 950–1000°C were utilized in cement production. Ceramic brick, floor, and roof tile waste with a suitable fineness can be used as pozzolanic materials. It is reported that if silica content in ceramic waste is mixed with calcium hydroxide, pozzolanic reactions can take place and produce secondary C-S-H gel, thus increasing the strength and durability properties of the concrete [49, 50]. Several researchers [9, 18] studied partial replacement of cement with ceramic waste with various percentages (up to 40%). Their main focus was on the pozzolanic properties, physical properties of ceramic powder, surface area, and strength of concrete. They reported that the ceramic wastes show pozzolanic properties and also have chemical and physical properties similar to cement, thus conforming to the cement standard. Puertas et al. [55] studied cement clinkers incorporating ceramic waste as a raw material. The reactivity, and physical, chemical, and durability properties were investigated and found to be morphologically and compositionally similar in hydration behaviour when compared to conventional cement. The research was done using white ceramic wall tiles, red ceramic wall tiles, and a combination of both ceramic wall tiles at 11–14% replacement of raw materials for cement manufacturing. Heidari and Tavakoli [56] investigated the engineering properties of ceramic waste as a cement replacement in concrete production. Their results showed that by substituting 20% of cement with ceramic waste has a minor strength loss (2.4%) on the compressive strength of concrete. Similarly, Vejmelková et al. [50] reported that 20% of ceramic waste as cement replacement gave adequate mechanical properties with normal concrete, and 40% was satisfactory for chemical resistance. For concrete with grade 30, Ay and Unal [18] reported that ceramic waste powder can be used for cement replacement by up to 20% without affecting its characteristic strength.

If ceramic waste is used as a pozzolanic material in a mortar, the disposal problem of this waste can be solved, and the economic advantages can be increased for construction industries. Besides, due to the depletion of natural resources, the use of ceramic waste in large quantities will help solve the problem. Furthermore, based on previous studies, only a low volume of ceramic powder (45 µm) can be used as cement replacement in concrete due to the low reactivity. Therefore, further research is necessary to be conducted to examine the effect of a finer and higher volume ceramic powder replacement in concrete or mortar.

1.7.2 Geopolymer Concrete

Ceramic powder is a principal waste of the ceramics industry which is generated as unwanted dust during the process of dressing and polishing. It is estimated that 15–30% of ceramic wastes are produced from the total raw material used. A portion of this waste is often utilized onsite for excavation pit refilling. Ceramic waste can be used in concrete to improve strength and other durability factors. Fernandes et al. [57] reported that the waste contents at various fabrication phases in the ceramic industries can reach nearly 3–7% of its global manufacturing. This specifies that a huge amount of calcined clays are just dumped for landfilling every year. Moreover, deposition processes are becoming expensive due to the ever-growing constraints on landfilling. Therefore, industries must look for alternative solutions such as recycling such waste materials as useful products. Despite some reuses of ceramic wastes, the quantities of such wastes utilized by the construction sector is still negligible [19]. Thus, their immediate reuse in other industries appears essential. The building sector, being a user of ceramic waste, will continue to play a vital role to overcome some of the environmental issues. The concrete industry can safely use ceramic wastes without requiring any remarkable change in production and application processes. Moreover, the cost of deposition of ceramic waste in landfills can be saved together with the replacement of raw materials and natural resources, thus saving energy and protecting the environment. Some studies suggested that the construction industry can be more sustainable and beneficial if most industrial wastes can be recycled effectively as geopolymer and alkali-activated concretes [56, 58].

Samadi et al. [59] demonstrated that ceramic waste powder has a positive effect on compressive strength when ceramic waste powder replaces less than 40% of OPC the microstructure properties of mortar is enhanced. An increase in the level of replacement by more than 40% led to reduced compressive strength. Senthamarai et al. [22] reported that ceramic waste can be used as coarse aggregate in concrete. The basic trend of permeation characteristics of the ceramic waste coarse aggregate concrete is similar to those of conventional concrete. Ariffin et al. [60] investigated the effect of ceramic replaced with river sand as fine aggregate in geopolymer mortars. The results revealed that the mortar prepared with ceramic waste aggregate presented higher strength compared to conventional geopolymer mortar.

Pacheco et al. [19] showed that concrete combined with waste ceramic powder (WCP) has increased durability of performance because of its pozzolanic properties. It was realized that by replacing conventional river sand with WCP it is possible to achieve mortars with superior strength and durability performance. These WCP-substituted conventional coarse aggregate mortars are prospective but performed poorly towards water absorption. Water permeability implies that the substitution of conventional river sand by WCP is an excellent option.

1.8 CONCLUSIONS

Recently, the production of sustainable concrete using waste materials as natural resource replacements has become useful in construction industries worldwide. An exponential increase in the usage of natural resources has caused severe environmental problems. The immense benefits and usefulness of recycling waste materials technology were demonstrated in terms of sustainability, energy-saving traits, and environmental affability. An all-inclusive overview of the appropriate literature on recycling waste materials in concrete allowed us to draw the following conclusions:

i. Recycling ceramic wastes in concrete are characterized by many significant traits such as less pollution, cheap, eco-friendly, and elevated durability performance in the harsh environment. These properties make them effective sustainable materials in the construction industry.
ii. Recycling ceramic wastes reduce landfill problems and enhance the environmental performance.
iii. The use of ceramic wastes as construction materials can improve the overall performance of concrete as the wastes are available in large quantities and the overall process is cheaper than cement production.
iv. Use of ceramic wastes in concrete is advantageous in terms of improved engineering properties of cementitious materials, especially for the generation of sustainable concretes.
v. Recycling ceramic wastes in concrete will industry contributed to saving energy and costs.

REFERENCES

1. Huseien, G. F., et al., Properties of ceramic tile waste based alkali-activated mortars incorporating GBFS and fly ash. *Construction and Building Materials*, 2019. **214**: pp. 355–368.
2. Naceri, A. and M.C. Hamina, Use of waste brick as a partial replacement of cement in mortar. *Waste Management*, 2009. **29**(8): pp. 2378–2384.
3. Altwair, N. M., M. A. M. Johari, and S. F. S. Hashim, Strength activity index and microstructural characteristics of treated palm oil fuel ash. *Structure*, 2011. **5**: p. 6.
4. Frías, M., et al., Properties of calcined clay waste and its influence on blended cement behavior. *Journal of the American Ceramic Society*, 2008. **91**(4): pp. 1226–1230.
5. Sánchez de Rojas, M. I., et al., Morphology and properties in blended cements with ceramic wastes as a pozzolanic material. *Journal of the American Ceramic Society*, 2006. **89**(12): pp. 3701–3705.
6. Baraldi, L. and MECS-Machinery Economics Studies by ACIMAC, World production and consumption of ceramic tiles. *OCEANIA*, 2016. **56**: pp. 0–4.
7. Stock, D., World production and consumption of ceramic tiles. *Tile Today*, 2014. **85**(54): p. e62.
8. Patel, H., D. N. Arora, and S. R. Vaniya, Use of ceramic waste powder in cement concrete. *International Journal for Innovative Research in Science & Technology*, 2015. **2**(1): pp. 91–97.
9. Lavat, A. E., M. A. Trezza, and M. Poggi, Characterization of ceramic roof tile wastes as pozzolanic admixture. *Waste Management*, 2009. **29**(5): pp. 1666–1674.
10. Kaminskas, R., J. Mituzas, and A. Kaminskas, The effect of pozzolana on the properties of the finest fraction of separated Portland cement-Part 1. *Ceramics Silikaty*, 2006. **50**(1): p. 15.
11. Raval, A. D., I. N. Patel, and J. Pitroda, Eco-Efficient concretes: Use of ceramic powder as a partial replacement of cement. *International Journal of Innovative Technology and Exploring Engineering (IJITEE)*, 2013. **3**(2): pp. 1–4.
12. Chen, H.-J., T. Yen, and K.-H. Chen, Use of building rubbles as recycled aggregates. *Cement and Concrete Research*, 2003. **33**(1): pp. 125–132.
13. Jepsen, M. T., et al. Durability of resource saving "Green" type of concrete. In *Featured at the Proceedings of the FIB-Symposium on Concrete and Environment*, Berlin. 2001: pp. 257–265.
14. Nehdi, M. and A. Khan, Cementitious composites containing recycled tire rubber: an overview of engineering properties and potential applications. *Cement, Concrete and Aggregates*, 2001. **23**(1): pp. 3–10.
15. Alengaram, U. J., H. Mahmud, and M. Z. Jumaat. Development of lightweight concrete using industrial waste material, palm kernel shell as lightweight aggregate and its properties. In 2010 2nd International Conference on Chemical, Biological and Environmental Engineering. 2010. IEEE.
16. Padmini, A., K. Ramamurthy, and M. Mathews, Behaviour of concrete with low-strength bricks as lightweight coarse aggregate. *Magazine of Concrete Research*, 2001. **53**(6): pp. 367–375.
17. Palmquist, S. M., D. C. Jansen, and C. W. Swan, Compressive behavior of concrete with vitrified soil aggregate. *Journal of Materials in Civil Engineering*, 2001. **13**(5): pp. 389–394.
18. Ay, N. and M. Ünal, The use of waste ceramic tile in cement production. *Cement and Concrete Research*, 2000. **30**(3): pp. 497–499.
19. Pacheco-Torgal, F. and S. Jalali, Reusing ceramic wastes in concrete. *Construction and Building Materials*, 2010. **24**(5): pp. 832–838.

20. Alves, A., et al., Mechanical properties of structural concrete with fine recycled ceramic aggregates. *Construction and Building Materials*, 2014. **64**: pp. 103–113.
21. Binici, H., Effect of crushed ceramic and basaltic pumice as fine aggregates on concrete mortars properties. *Construction and Building Materials*, 2007. **21**(6): pp. 1191–1197.
22. Senthamarai, R., P. D. Manoharan, and D. Gobinath, Concrete made from ceramic industry waste: Durability properties. *Construction and Building Materials*, 2011. **25**(5): pp. 2413–2419.
23. Senthamarai, R. and P. D. Manoharan, Concrete with ceramic waste aggregate. *Cement and Concrete Composites*, 2005. **27**(9–10): pp. 910–913.
24. Zhang, Y. and T. Napier-Munn, Effects of particle size distribution, surface area and chemical composition on Portland cement strength. *Powder Technology*, 1995. **83**(3): pp. 245–252.
25. Mishra, S. and N. A. Siddiqui, A review on environmental and health impacts of cement manufacturing emissions. *International Journal of Geology, Agriculture and Environmental Sciences*, 2014. **2**(3): pp. 26–31.
26. Gartner, E., Industrially interesting approaches to "low-CO2" cements. *Cement and Concrete Research*, 2004. **34**(9): pp. 1489–1498.
27. Behfarnia, K., Studying the effect of freeze and thaw cycles on bond strength of concrete repair materials. *Asian Journal Of Civil Engineering (Building And Housing)*, 2010. **11**(2): pp. 165–172.
28. Gouny, F., et al., A geopolymer mortar for wood and earth structures. *Construction and Building Materials*, 2012. **36**: pp. 188–195.
29. Mirza, J., et al., Preferred test methods to select suitable surface repair materials in severe climates. *Construction and Building Materials*, 2014. **50**: pp. 692–698.
30. Huseien, G. F., et al., Geopolymer mortars as sustainable repair material: A comprehensive review. *Renewable and Sustainable Energy Reviews*, 2017. **80**: pp. 54–74.
31. Mueller, H. S., et al., Design, material properties and structural performance of sustainable concrete. *Procedia Engineering*, 2017. **171**: pp. 22–32.
32. Huseien, G. F., et al., Effects of POFA replaced with FA on durability properties of GBFS included alkali activated mortars. *Construction and Building Materials*, 2018. **175**: pp. 174–186.
33. Qureshi, T., A. Kanellopoulos, and A. Al-Tabbaa, Autogenous self-healing of cement with expansive minerals-II: Impact of age and the role of optimised expansive minerals in healing performance. *Construction and Building Materials*, 2019. **194**: pp. 266–275.
34. Chindaprasirt, P. and U. Rattanasak, Improvement of durability of cement pipe with high calcium fly ash geopolymer covering. *Construction and Building Materials*, 2016. **112**: pp. 956–961.
35. Hossain, M., et al., Durability of mortar and concrete containing alkali-activated binder with pozzolans: A review. *Construction and Building Materials*, 2015. **93**: pp. 95–109.
36. Norhasri, M. M., M. Hamidah, and A. M. Fadzil, Applications of using nano material in concrete: A review. *Construction and Building Materials*, 2017. **133**: pp. 91–97.
37. Chen, Y., et al., Resistance of concrete against combined attack of chloride and sulfate under drying–wetting cycles. *Construction and Building Materials*, 2016. **106**: pp. 650–658.
38. Swamy, R., Alkali-aggregate reaction: The bogeyman of concrete. *Special Publication*, 1994. **144**: pp. 105–140.
39. Gartner, E. M. and D. E. Macphee. A physico-chemical basis for novel cementitious binders. *Cement and Concrete Research*, 2011. **41**(7): pp. 736–749.
40. Li, J., S. T. Ng, and M. Skitmore, Review of low-carbon refurbishment solutions for residential buildings with particular reference to multi-story buildings in Hong Kong. *Renewable and Sustainable Energy Reviews*, 2017. **73**: pp. 393–407.

41. Wang, P., et al., Life cycle assessment of magnetized fly-ash compound fertilizer production: A case study in China. *Renewable and Sustainable Energy Reviews*, 2017. **73**: pp. 706–713.
42. Ismail, M., et al. Early strength characteristics of palm oil fuel ash and metakaolin blended geopolymer mortar. In *Advanced Materials Research*. 2013: Trans Tech Publ.
43. Ramadhansyah, P. J., et al., Thermal analysis and pozzolanic index of rice husk ash at different grinding time. *Procedia Engineering*. 2012. **31**: pp. 1–9.
44. Tangchirapat, W., C. Jaturapitakkul, and P. Chindaprasirt, Use of palm oil fuel ash as a supplementary cementitious material for producing high-strength concrete. *Construction and Building Materials*, 2009. **23**(7): pp. 2641–2646.
45. Neville, A. M., *Properties of Concrete*. Vol. 4. 1995: Longman.
46. Sabir, B., S. Wild, and J. Bai, Metakaolin and calcined clays as pozzolans for concrete: A review. *Cement and Concrete Composites*, 2001. **23**(6): pp. 441–454.
47. Dhir, R., et al., Contribution of PFA to concrete workability and strength development. *Cement and Concrete Research*, 1988. **18**(2): pp. 277–289.
48. McCarthy, M. and R. Dhir, Development of high volume fly ash cements for use in concrete construction. Fuel, 2005. **84**(11): pp. 1423–1432.
49. Toledo Filho, R., et al., Potential for use of crushed waste calcined-clay brick as a supplementary cementitious material in Brazil. *Cement and Concrete Research*, 2007. **37**(9): pp. 1357–1365.
50. Vejmelková, E., et al., Properties of high performance concrete containing fine-ground ceramics as supplementary cementitious material. *Cement and Concrete Composites*, 2012. **34**(1): pp. 55–61.
51. Baronio, G. and L. Binda, Study of the pozzolanicity of some bricks and clays. *Construction and Building Materials*, 1997. **11**(1): pp. 41–46.
52. Mehta, P. Mechanism by which condensed silica fume improves concrete strength. In Proceedings of international workshop on condensed silica fume in concrete, Ottawa, 1987: pp. 1–17.
53. O'Farrell, M., B. Sabir, and S. Wild, Strength and chemical resistance of mortars containing brick manufacturing clays subjected to different treatments. *Cement and Concrete Composites*, 2006. **28**(9): pp. 790–799.
54. Cheng, Y., et al., Test research on effects of ceramic polishing powder on carbonation and sulphate-corrosion resistance of concrete. *Construction and Building Materials*, 2014. **55**: pp. 440–446.
55. Puertas, F., et al., Ceramic wastes as alternative raw materials for Portland cement clinker production. *Cement and Concrete Composites*, 2008. **30**(9): pp. 798–805.
56. Heidari, A. and D. Tavakoli, A study of the mechanical properties of ground ceramic powder concrete incorporating nano-SiO2 particles. *Construction and Building Materials*, 2013. **38**: pp. 255–264.
57. Fernandes, M., A. Sousa, and A. Dias, Environmental impacts and emissions trading-ceramic industry: A case study. In Coimbra: Technological Centre of Ceramics and Glass, Portuguese Association of Ceramic Industry, *APICER,* (In Portuguese), 2004: pp. 1–10.
58. Limbachiya, M., M. S. Meddah, and Y. Ouchagour, Use of recycled concrete aggregate in fly-ash concrete. *Construction and Building Materials*, 2012. **27**(1): pp. 439–449.
59. Samadi, M., et al., Properties of mortar containing ceramic powder waste as cement replacement. *Jurnal Teknologi*, 2015. **77**(12): pp. 93–97.
60. Ariffin, M. A., et al., Effect of ceramic aggregate on high strength multi blended ash geopolymer mortar. *Jurnal Teknologi*, 2015. **77**(16): pp. 33–36.

2 Chemical, Physical, and Mineral Properties of Ceramic Wastes

2.1 INTRODUCTION

Nowadays, the attention towards consuming substitute materials, such as solid wastes in the construction industry, has grown continuously. According to Huseien and Shah [1], the consumption of waste materials are in line with essential environmental approaches: prevention of wastes, reutilising of waste materials, reducing landfill area, energy resuming from wastes, and saving natural **resources.** While alternative materials, such as wastes, are used in many applications, it is necessary to address their technical features, financial aspects, and environmental effects. According to Mohammadhosseini et al. [2], in the construction industry the idea of sustainability and green production encourages the use of various waste materials to replace raw materials such as supplementary cementing materials and aggregates. This, therefore, leads to eco-friendly construction by reducing the cost of construction that is associated with disposing waste materials.

Tile ceramic materials are widely utilized worldwide such that large quantities of wastes are inevitably produced by tile manufacturing industries and construction sectors. A part of these wastes is dumped in landfills [3, 4]. Notably, ceramic wastes are durable, hard, and highly resistant to biological, chemical, and physical degradations. Thus these materials cannot be recycled via existing processes [5]. In addition, approximately 30% of products in the ceramic industry are regarded as waste [6]. Reusing these wastes in concrete could be a win–win situation, for example by solving the ceramic industry waste problem and at the same time leading to a more sustainable concrete [7, 8]. Yearly, the global production of ceramic tiles is more than 10 million square metres [9]. It has been estimated in a survey that about 15–30% of production goes as waste in the ceramics industry [6], which is not reused in any form at present and hence piling up every year. This percentage of waste is increasing as more ceramic is damaged during storage, transportation and construction stages as well as house renovations. Moreover, the deposition processes are becoming expensive due to the ever-growing constraints on landfilling. In that case, industries must look for alternative solutions such as recycling such waste materials as useful products.

Yet, the development of different green concrete as environmental friendly construction materials by using ceramic wastes is attractive to researchers and people who work in the concrete industry. It's well known that concrete performance is highly influenced by the chemical composition, physical and mineral properties of ceramic wastes as partial replacement of cement and fine and coarse aggregates. Consequently, the present study intends to assess ceramic wastes properties such as the silica and aluminium content, particle size and shape as well as mineral characterizations. In this chapter, the preparation stages of ceramic wastes as coarse and fine aggregates, and micro- and nano-size powder is explained.

2.2 PREPARATION STAGES

2.2.1 Fine and Coarse Aggregates

Ceramic wastes are not only non-biodegradable but also consume much space in landfills. Thus, finding a new way to recycle this waste and subsequently using it in the construction of infrastructures can be useful to preserve natural resources and the environment. Previous research revealed that ceramic wastes have pozzolanic properties. To use ceramic wastes as fine or coarse aggregates, the ceramic wastes are collected from the construction industry or ceramic factors. The received materials are first crushed using a crushing machine, then sieved with 20 mm–75 μm mesh to isolate the large particles. The sieved ceramic is classified into two types depending on particle size and sieve stage results. The ceramic particles between 4.75 mm and 75 μm are adopted as fine aggregate. Ceramic particle sizes more than 4.74 mm and less than 20 mm are classified as coarse aggregate. Next, the sieved ceramic is collected and used in the mixing process as fine and coarse natural aggregate replacement. The procedure of treatment of the ceramic wastes is displayed in Figure 2.1.

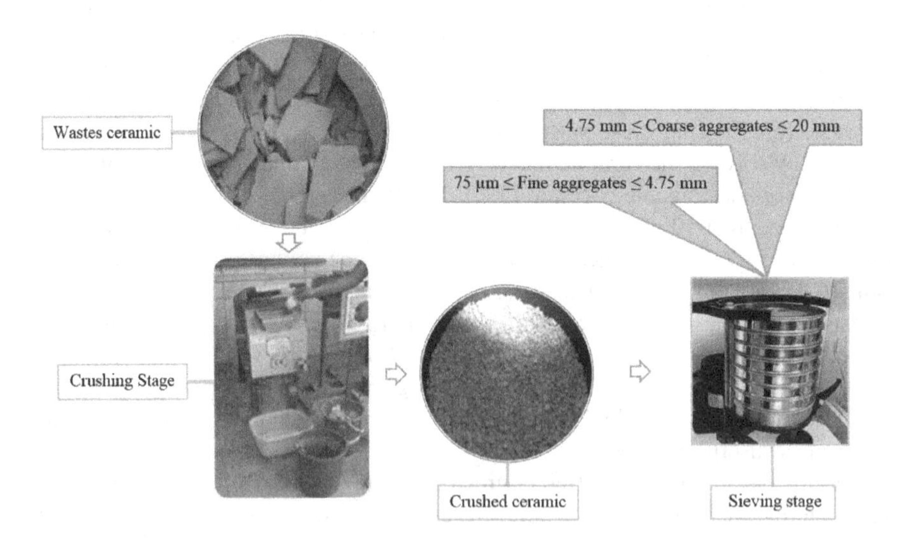

FIGURE 2.1 Preparation stages of fine and coarse ceramic aggregates.

2.2.2 Micro Ceramic Powder

To prepare micro ceramic powder, the crushed ceramic obtained from the crushing stage (Figure 2.1) is sieved with 600 μm mesh to isolate the large particles. After the sieving stage, the collected powder is ground using the Los Angeles abrasion grinding machine (Figure 2.2). The sieved ceramic is ground for 6 hours and the finesse of particles is checked after every hour interval (Figure 2.3). However, the grinding time depends on the machine's speed and the size and number of steel balls. Next, the powder is collected and used in the mixing process. The procedure of the treatment of ceramic waste is displayed in Figure 2.4.

2.2.3 Nano Ceramic Powder

To prepare nano-size ceramic powder, the collected ceramic waste from the factory is first crushed into small pieces and ground into a specific powder size using the

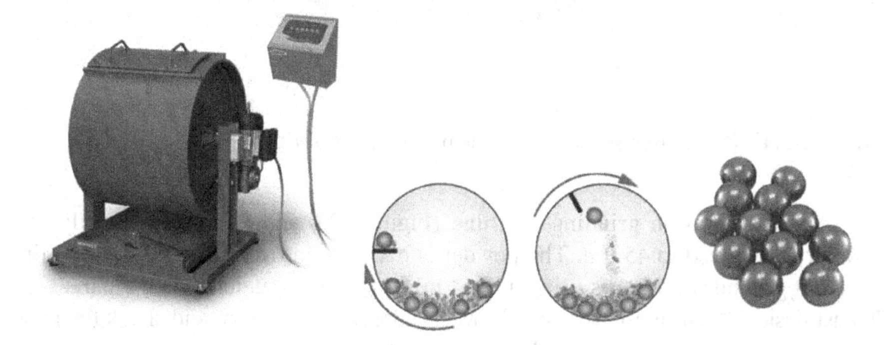

FIGURE 2.2 Los Angeles abrasion machine revealing the grinding process.

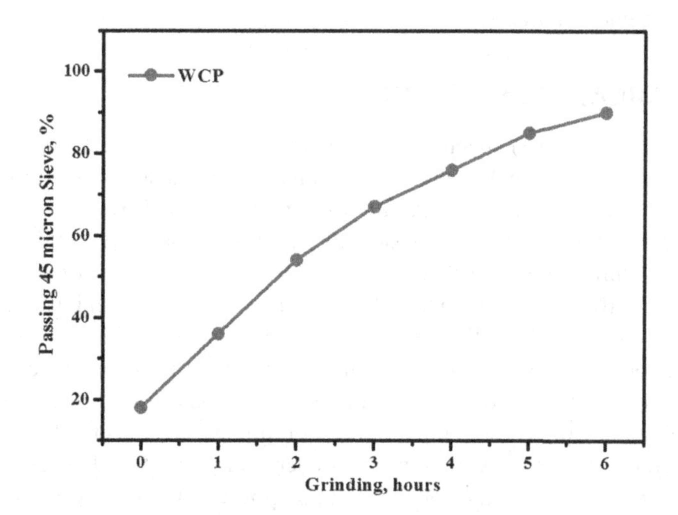

FIGURE 2.3 Time-dependent grinding behaviour of WCP.

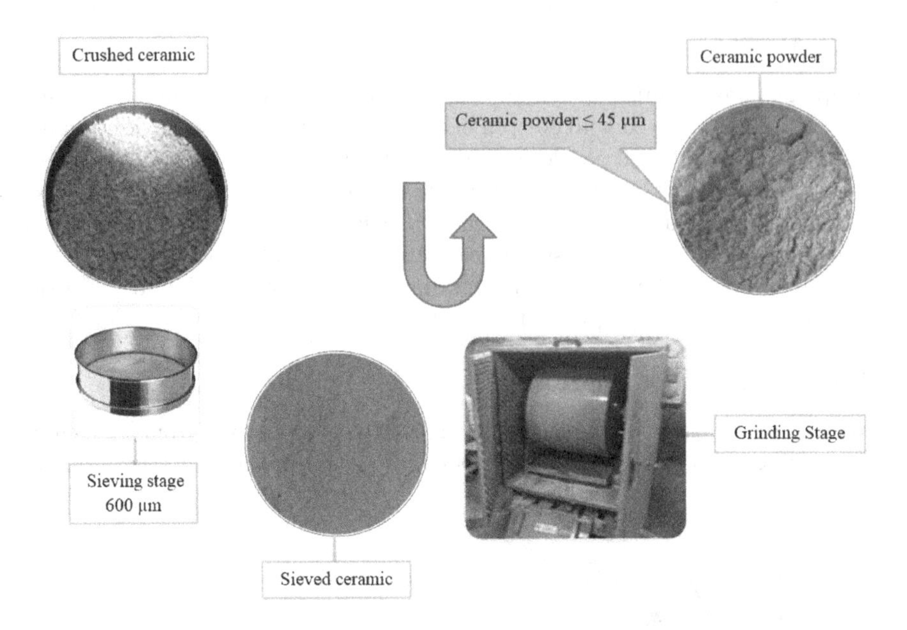

FIGURE 2.4 Preparation stages of ceramic powder (micro size).

Los Angeles abrasion grinding machine (Figures 2.1 and 2.2). Subsequently, the powder was sieved to 45 µm. The powder is dried in an oven at 110°C for 45 minutes and ground for 5 hours using a bowl mill machine to determine the nano size. Transmission electron microscopy (TEM) is adopted to measure and check the nano particle size. Next, the nano powder is collected and used in the preparation of paste, mortar, and concrete specimens. The procedure of treatment of the ceramic nano particle is presented in Figure 2.5.

2.3 CHEMICAL COMPOSITION

X-ray fluorescence (XRF) spectroscopy was used to detect the chemical compositions of waste ceramic powder (WCP) as summarized in Table 2.1. The main oxides were silica (52 to 74 weight %) and aluminium (12.2 to 19.9 wt%). The chemical composition of ceramic shows a considerably high amount of silica (SiO_2), but the lime (CaO) composition is extremely low in comparison with ordinary Portland cement (OPC) [10]. Since this material is a by-product of the combustion process, the chemical composition in this material varies from batch to batch is collected at different periods of time. The difference in the amount of the chemical components in ceramic powder is due to the material source, ceramic tile manufacturing process, and efficiency (time and temperature). Ceramic waste collected from different ceramic factories will have variation on its chemical properties. However, silica is still the major chemical composition in ceramic powder. The chemical composition of different ceramic powders used in previous research are shown in Table 2.1.

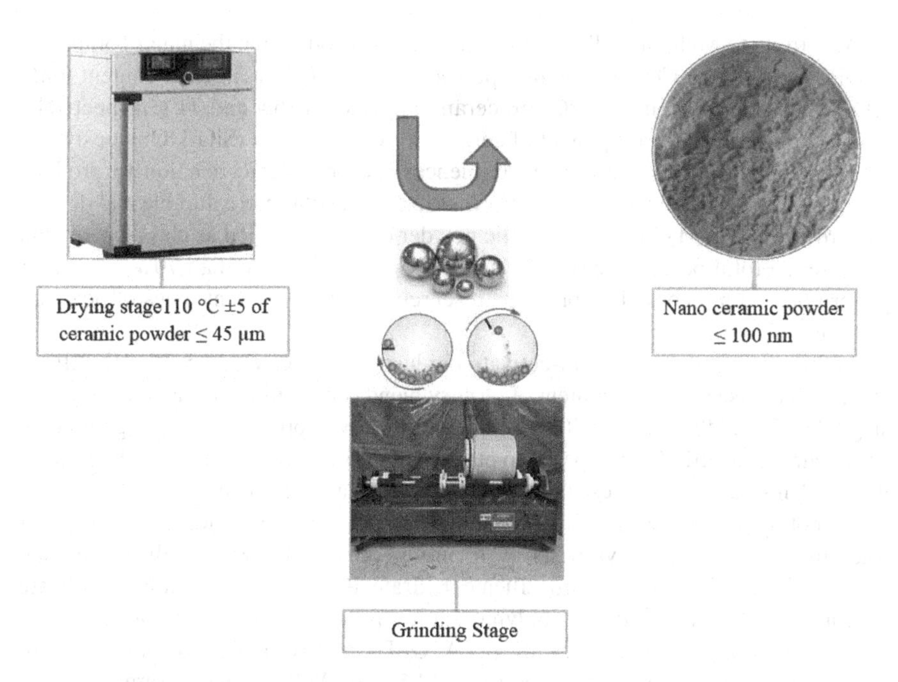

FIGURE 2.5 Preparation stages of ceramic powder (nano size).

TABLE 2.1
Chemical Composition of Ceramic Powder from Previous Research

	Elements (Weight %)								
Reference	SiO_2	Al_2O_3	Fe_2O_3	CaO	MgO	K_2O	Na_2O	TiO_2	LOI
[10]	63.3	18.3	4.3	4.5	0.7	2.2	0.8	0.6	1.6
[11]	59.5	18.1	4.1	3.7	2.4	0.9	—	—	8.0
[12]	67.0	19.9	6.3	0.1	1.4	3.5	0.2	—	0.5
[13]	52.0	15.1	6.0	4.0	4.9	4.5	0.6	0.7	9.2
[14]	63.7	19.3	7.1	1.7	1.4	1.3	3.8	0.9	0.5
[15]	66.2	14.2	5.4	6.1	2.3	2.1	0.7	—	—
[7]	58.0	18.0	1.0	8.3	0.6	1.2	0.2	0.8	—
[16]	63.4	13.9	5.4	8.2	—	2.4	0.9	0.7	1.1
[17]	69.0	16.0	0.7	0.6	4.1	1.8	3.2	0.2	5.2
[18]	74.1	17.8	3.6	1.1	—	2.7	—	0.5	0.1
[2]	74.1	17.8	3.5	1.2	—	2.6	—	0.4	0.1
[19]	72.6	12.2	0.6	0.1	0.9	0.1	13.5	—	0.1

According to Mohammadhosseini et al. [2], it was found that the major compound in OPC is CaO (68.3%), but ceramic powder shows very low in CaO content (only 1.1%). The silica content of OPC and ceramic powder is 16.4 and 74.1, respectively, which shows the main component of ceramic powder is silica (SiO_2). Obviously, the presence of a higher silica content influences the pozzolanic reaction to produce extra calcium silicate hydrate gels, thus making the mortar more durable and denser. According to ASTM C618 this ceramic powder can be classified as class F pozzolan because the total percentage of SiO_2 + Al_2O_3 + Fe_2O_3 is more than 70%. A similar pattern has been recorded by previous researchers using different types of pozzolanic materials [20].

In the geopolymer and alkali-activated industry, Huseien et al. [5, 19, 21] utilized WCP to increase the silica content of ternary blends containing ground blast furnace slag (GBFS) and fly ash (FA). It's well known the GBFS presents a very high content of calcium oxide (51.8%) compared to other materials and that leads to many durability problems such as low resistance to acid attacks and elevated temperatures. It is worth noting that the level of silicate, aluminium and calcium oxide play significant roles in alkali-activated synthesis by forming sodium aluminium silicate hydrate (N-A-S-H), calcium aluminium silicate hydrate (C-A-S-H) and calcium silicate hydrate (C-S-H) gels in the geopolymerization process. The content of potassium oxide (K_2O) was found to be very low in FA, GBFS, and WCP. The content of sodium oxide (Na_2O) was observed at a high ratio (13.5%) in WCP chemical composite. It is known that both K_2O and Na_2O can strongly affect the activation of alkaline and the geopolymerization process. The loss on ignition (LOI) contents were observed to be very low in WCP, which was consistent with the ASTM C618 standard.

2.4 PHYSICAL PROPERTIES

2.4.1 Fine and Coarse Aggregates

Aggregate is one of the important ingredients of mortar and concrete. A well-graded aggregate will improve the workability of the mix and minimize the voids. Fine aggregates used in this study consist of sand and ceramic aggregates, which are characterized based on their physical properties and grading from sieve analysis before being incorporated in the mix.

Medina et al. [28] observed that the recycled ceramic waste aggregates showed more irregular shape than natural aggregates, which provides a higher bond between ceramic waste aggregates and the cement paste. Figure 2.7 shows the shape and surface of ceramic waste aggregates as reported by previous research. Ceramic waste aggregates have a rough surface and angular shape due to the crushing process.

Brito et al. [23] stated that coarse ceramic waste aggregates have a high water absorption of 12%. They reported that the high water absorption limited the use of this type of aggregate in concrete production, due to loss in mechanical strength, workability, or durability. However, other researchers reported that the water absorption of ceramic aggregates is higher than 12% and showed a significantly higher value than normal aggregates [22, 25, 28]. This is probably due to the high porosity of crushed ceramic aggregates. Table 2.2 shows the physical properties of ceramic aggregates reported by other researchers. Researchers who used ceramic tile waste from red clay as fine aggregate replacement reported a water absorption of 7% for ceramic waste aggregates. They explained that this water absorption was due to the physical characteristic of ceramic waste aggregates used in the study.

According to Samadi et al. [32], Table 2.3 shows the result of physical properties of the fine aggregates. The size for sand and ceramic aggregates used was between 75 μm and 4.75 mm. Therefore, there is no sand or ceramic passing the 75 μm sieve size which was required according to ASTM C33. The bulk density for sand and ceramic aggregates were 1624 kg/m^3 and 1450 kg/m^3, respectively. The bulk density of ceramic aggregates used by researchers was in the range of 1100 to 1600 kg/m^3 depending on the type of ceramic material [6, 33]. The specific gravity for sand and ceramic aggregates, determined according to ASTM C128, was 2.62 and 2.38,

TABLE 2.2
Physical Properties of Ceramic Aggregates Reported by Previous Research

Reference	Type of Ceramic Waste	Water Absorption (%)	Aggregates Size (mm)	Density (kg/m^3)
[22]	Crushed brick	14.8	5	2050
[23]	Construction industry	12	9.5	2029
[6]	Industrial waste	0.7	20	2450
[24]	Crushed brick	14	4.75	2232
[25]	Hollow bricks wall	16.3	16	2160
[26]	Industrial waste	1.3	4.75	2310
[27]	Electrical insulator	0.7	30	2400
[28]	Sanitary ware	0.6	12.5	2390
[29]	Tile waste from red clay	7	4.75	2350
[30]	Crushed brick	4.8	4	2140
[31]	Sanitary ware	0.2	10	2969
[32]	White tile wastes	1.3	2.36	2380

TABLE 2.3
Ceramic Physical Properties Compared to River Sand

Property	River Sand	Ceramic Wastes	Relevant Standard and Reference
Mass passing 4.75 mm sieve (%)	100	100	ASTM C33
Mass passing 74 μm sieve (%)	0	0	ASTM C33
Bulk density (kg/m^3)	1624	1450	ASTM D1895
Specific gravity	2.62	2.38	—
Water absorption (%)	1.8	1.3	—

respectively. Similar findings were reported that the range of specific gravity of sand and ceramic aggregates was between 2.3 and 2.7 [27, 29]. In addition, the water absorption at 24 hours for sand and ceramic fine aggregates was 1.8% and 1.3%, respectively.

The result of the sieve analysis of sand and ceramic fine aggregates shown in Figure 2.6 was determined in accordance to ASTM C33. As shown in the figure, the particles' distribution of aggregates of the tested sample is between the upper and lower limits, which satisfied the ASTM standard. Therefore, the ceramic aggregates can be used as fine aggregates to replace sand in the mortar mix. This is because the grading of fine aggregates that fall within the ASTM C33-16 grading limits usually produce robust and stable mortar. A well-graded aggregate usually reduces the demand for water, thereby improving the packing density, robustness and workability of the mortar [34].

2.4.2 Micro Powder

The process for producing ceramic powder in micro and nano size is shown in Figures 2.4 and 2.5, respectively. The ceramic waste was crushed and ground until the same size of cement. Since the colour of ceramic powder depends on the actual colour of the ceramic tile, the colours reported are different for each study. Heidari and Tavakoli [35] reported the use of fine ceramic powder the size of <75 µm, 0.2% moisture content and density of 2570 kg/m^3. This was similar to previously reported research using ceramic powder of <75 µm size [7, 10]. Few researchers have studied the effect of finer ceramic powder on concrete as cement replacement [12, 13]. Meanwhile, Huang et al. [36] studied the use of the ceramic powder sized less than 300 µm as filler in concrete.

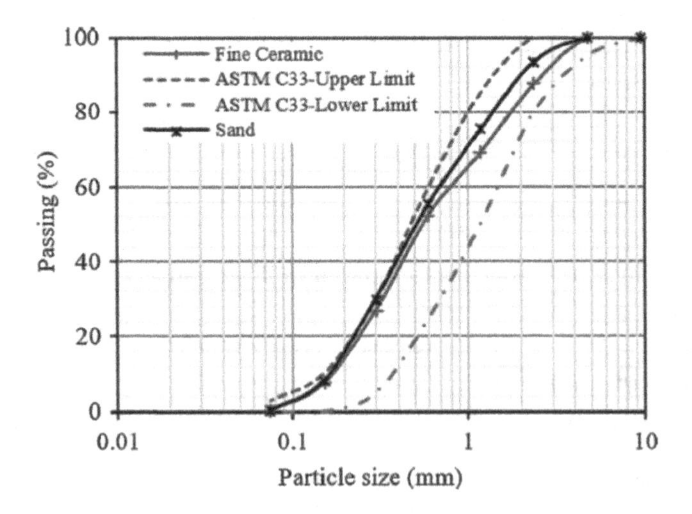

FIGURE 2.6 Sieve analysis of ceramic wastes as fine aggregates.

The colour of ceramic powder is light grey while the cement is dark grey. The physical properties of ceramic powder used is shown in Table 2.4. The specific gravity of OPC and ceramic powder is 3.15 and 2.35, respectively. Similar findings were obtained from previous research where the specific gravity of ceramic powder was in the range of 2.3 to 2.45 [11, 37]. The percentage passing through a 45 μm (No. 325) wet sieve for OPC and ceramic powder are 90.0% and 99.9%, respectively. This test was done in accordance to ASTM C430-15. Thus, ceramic powder has a smaller particle size in comparison with OPC.

The strength activity of WCP compared to OPC at 7 and 28 days of age is illustrated in Figure 2.7. The strength activity index is an indirect method for measuring and evaluating the pozzolanic property of material. The strength activity index of ceramic powder was calculated based on the control specimen (OPC). The strength activity index for 7 days of ceramic powder is 81% of OPC, which satisfied ASTM C311. At the age of 28 days, ceramic powder exhibits 95% of strength activity index

TABLE 2.4
Physical Properties of Ceramic Powder Compared to OPC

Property	WCP	OPC	Limit	Standard
Colour	Light grey	Dark grey	—	—
Specific gravity	2.35	3.15	—	ASTM C33
% passing through 45 μm	99	90	≥66	ASTM C618
Strength activity index, 7 days (%)	81	—	≥75	ASTM C618
Strength activity index, 28 days (%)	95	—	≥75	ASTM C618

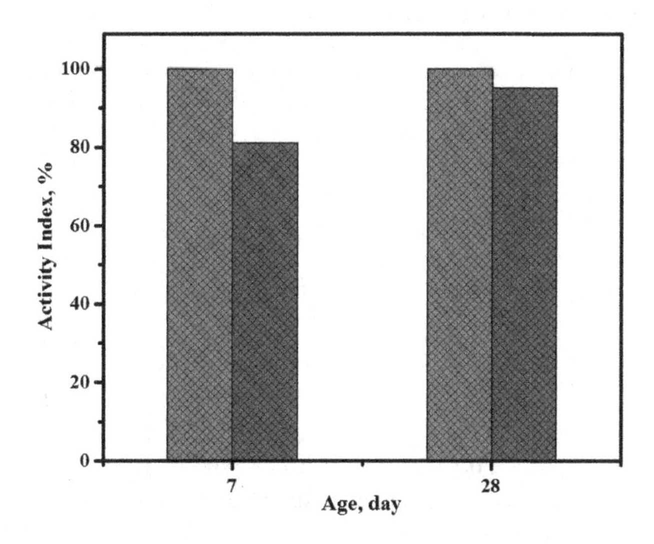

FIGURE 2.7 Strength activity index of ceramic powder.

of the OPC which is caused by the pozzolanic reaction. The activity index at 7 days is slightly lower than the control mortar. This may be due to the delayed pozzolanic reaction of $Ca(OH)_2$ with reactive silica in ceramic powder, which is similar to findings by previous researchers [20].

Figure 2.8 demonstrates the particle size distributions of WCP compared to OPC obtained using a particle size analyzer. The median particle for WCP and OPC was estimated to be 35 and 45 μm, respectively. Furthermore, the particle size of 99.9% of WCP was less than 45 μm and 90% of OPC passed 45 μm. The fineness of WCP was more than OPC. Thus, these results fulfilled the pozzolanic requirement of ASTM C618 of 66% passing 45 μm [38].

2.4.3 Nano Powder

The integration of nanotechnology into research on concrete is achieved via two major pathways known widely as nanoscience and nano-engineering or nano-modification of concrete. Employing sophisticated profiling methods and modelling at an atomic or molecular level, nanoscience analyzes the nano- and microscale structure of cement-based materials to gain insight into the impact of that structure on macroscale properties and performance [39–41]. Meanwhile, nano-engineering employs methods of nanoscale structure manipulation for the production of novel customized cementitious composites with multiple functionalities, high mechanical performance and effective durability, as well as various new properties, including low electrical resistivity, self-sensing capacity, ability to clean and repair themselves, high ductility, and crack self-control. Nano-engineering of concrete involves regulation of how the material behaves and the addition of new properties through the integration of nanoparticles and nanotubes or via molecule grafting on cement particles, phases, aggregates, and additives to achieve adjustable surface functionality aligned with particular interactions between interfaces.

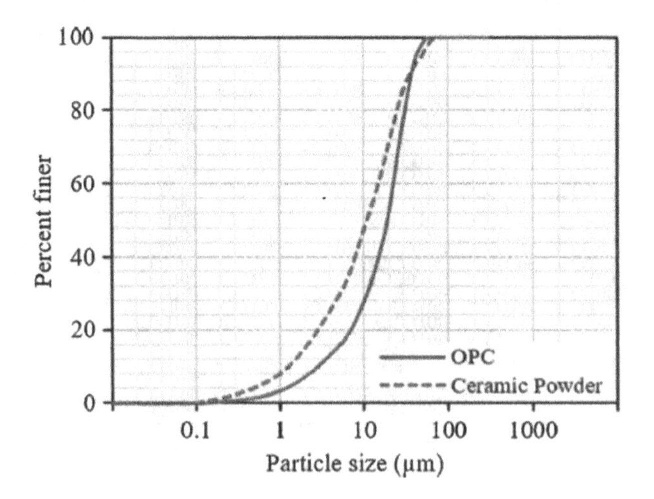

FIGURE 2.8 Particle size distribution of constituent materials.

Using materials with a particle size less than 500 nm in concrete production as admixture or part cement replacement called nanoconcrete. It was shown that the strength of normal concrete tends to enhance with the inclusion of nanoparticles. The bulk properties and packing model structure of concrete can remarkably be improved via the incorporation of nanoparticles. Nanoparticles act as excellent filling agents through the refinement of intersection zones in cementitious materials and production of high-density concrete. The manipulation or modification of these nanoparticles in the cement matrix can render new-fangled nanostructures [18, 42, 43].

2.5 MINERAL PROPERTIES

2.5.1 X-Ray Diffraction Pattern

Figure 2.9 shows the X-ray diffraction (XRD) pattern of ceramic powder. The XRD pattern of WCP revealed pronounced diffraction peaks around $2\theta = 16–30°$, which were attributed to the crystalline silica and alumina compounds. It was found that the WCP consists of quartz (SiO_2). Nonetheless, the occurrence of other crystalline peaks was ascribed to the presence of crystalline quartz and mullite phases. Silicate minerals in non-crystalline or an amorphous state are highly reactive with $Ca(OH)_2$ produced from the hydration of cement to form additional calcium silicate hydrate (C-S-H) gels. The additional C-S-H gel formed contributes towards the reduction of $Ca(OH)_2$ amounts, hence resulting in increments in strength and producing durable mortar or concrete [44]. These results suggest that the WCP is a semi-amorphous material.

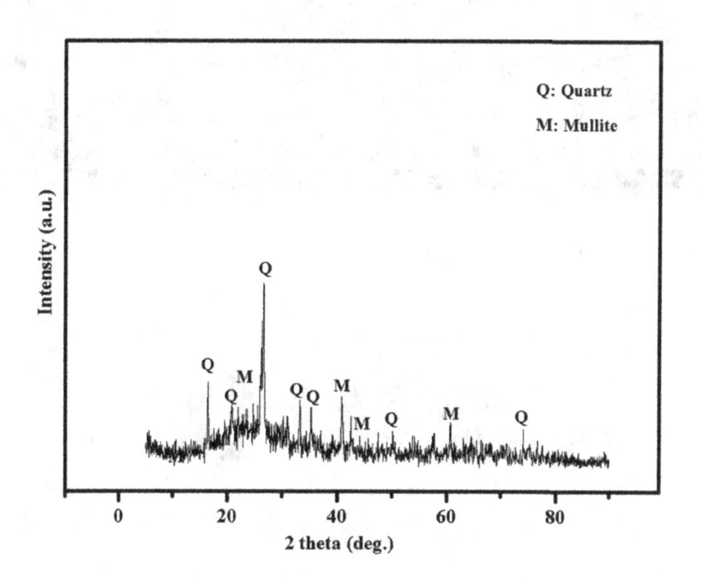

FIGURE 2.9 X-ray diffraction of ceramic powder.

2.5.2 Scanning Electron Microscopy

The morphological structure of WCP was examined using the scanning electron microscopy (SEM) technique. SEM images of WCP are presented in Figure 2.10. It is evident that WCP consisted of irregular and angular particles with rough surface. WCP consists of crushed and irregular shapes, which is similar to OPC but with finer particles size. This indicates that the particle size of WCP is smaller than OPC. Therefore, the surface area of the ceramic powder is larger than that of OPC. A larger surface area will give more reactivity of the pozzolanic material [45, 46].

2.5.3 Transmission Electron Microscopy

Transmission electron microscopy (TEM) was adopted to measure the WCP nano size. From Figure 2.11, the image size of ceramic powder particles is between 150 and 250 nm, which is less than 500 nm. The fineness of pozzolanic material is known to play an important role on the activity index. By increasing the fineness, the activity index of pozzolans increases [47]. Figure 2.10 illustrates the TEM images of the prepared WCP nano size, with evident ellipsoidal nanoparticles. The large nanoparticles in the micrograph were attributed to the agglomeration of nanoparticles because of the high surface energy and strong surface tension of ultrafine nanoparticles. The fine particle size resulted in a large surface area, which enhanced

FIGURE 2.10 SEM image of ceramic powder.

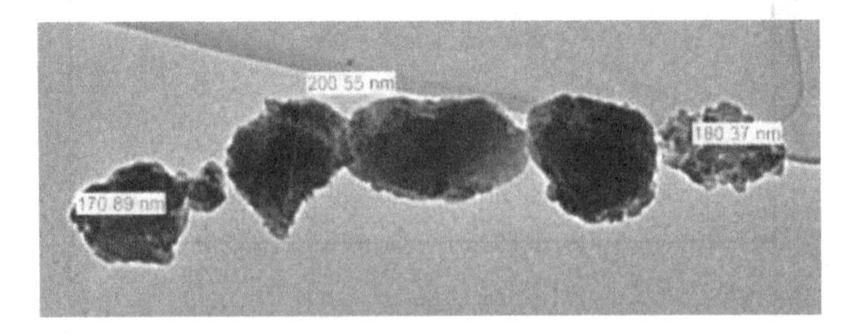

FIGURE 2.11 TEM image of ceramic powder nano size.

the catalytic activity of the nanoparticle. The average nanocrystallite size of approximately 200 nm is consistent with the nanosilica constituted structure reported in in some studies [3, 48]. Furthermore, nanoceramics with this structure exhibit distinct characteristics, such as super plasticity [49].

2.5.4 Fourier-Transformed Infrared Spectroscopy

Fourier-transformed infrared spectroscopy (FTIR) analysis provides a detailed signature of chemical bonding vibrations of the constituents and allows a comprehensive structural analysis. Transformation of WCP from a crystalline to amorphous state with temperature was examined by recording their FTIR spectra in terms of wavenumber (cm^{-1}), which is equivalent to the frequency or energy of bonding vibrations (Figure 2.12). The stretching modes are sensitive to the Si–Al composition of the framework and may shift to a lower frequency with an increasing number of tetrahedral aluminium atoms [50]. Sitarz and Handke [51] have reported the occurrences of broad bands at 500–650 cm^{-1}, indicating the existence of silicate and aluminosilicate glass phases. It also possessed short-range structural order in the form of rings of tetrahedral or octahedral. The spectral bands at 460 cm^{-1} were related to Al–O/Si–O in plane and bending modes (this silica content will give pozzolanic reaction with the hydration product in the cement paste), 730 cm^{-1} was assigned to octahedral site modes of Al, 820 cm^{-1} was allocated to tetrahedral Al–O stretching [52], and 1400 cm^{-1} was endorsed to asymmetric stretching vibrations of Al–O/Si–O bonds [53]. WCP presented broad bands with medium intensity (C=O) between 1460 and 1600 cm^{-1}. The bands appeared in the regions of 3500 cm^{-1} was assigned to bending vibrations of (C=O – H) and stretching vibration of (–OH), respectively.

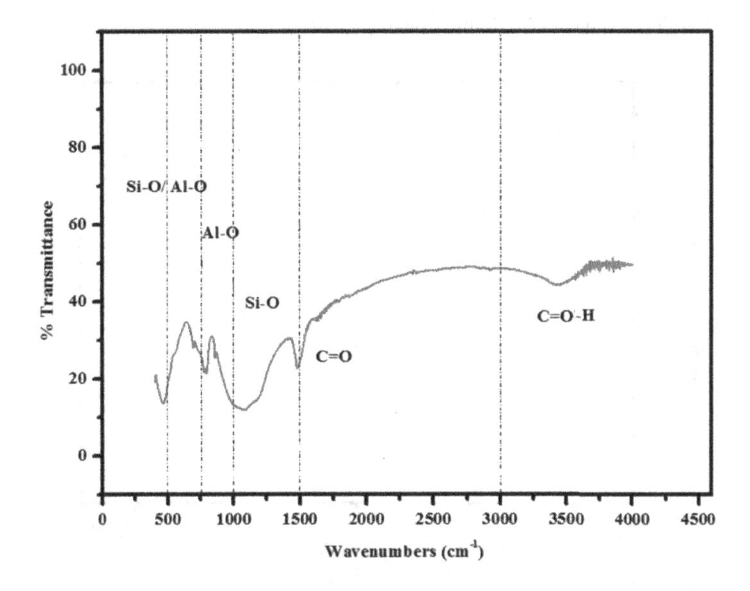

FIGURE 2.12 FTIR spectra of ceramic powder.

Hydroxyl ions being characteristic of weakly bound ligaments of water were either adsorbed on the surface or trapped in large cavities.

2.5.5 Differential Thermal and Thermogravimetric Analysis

Differential thermal analysis (DTG) and thermogravimetric analysis (TGA) of WCP were carried out to examine the change of weight as a function of temperature. Figures 2.13 and 2.14 display the TGA and DTG curves of various minerals,

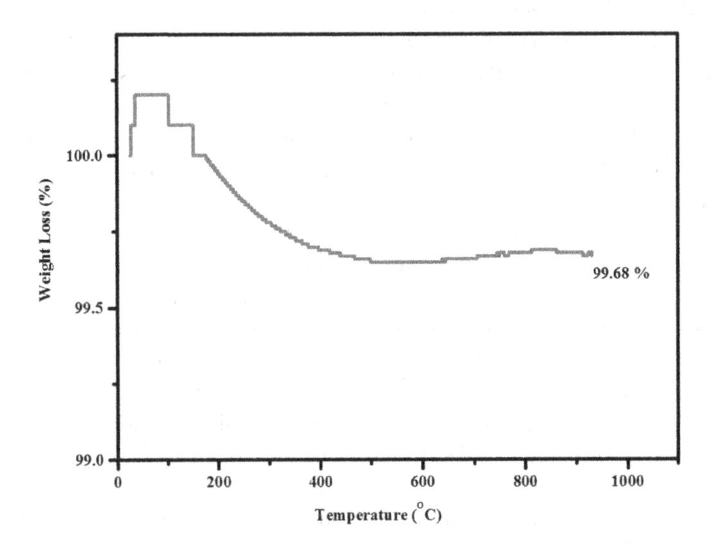

FIGURE 2.13 TGA curves of ceramic powder.

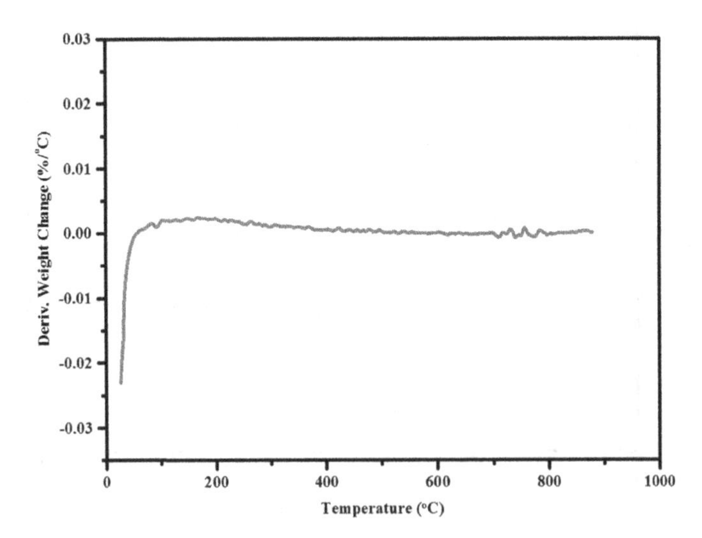

FIGURE 2.14 DTG curves for ceramic powder.

respectively. The weight changes occurred at different temperatures. The results of TGA tests showed that the WCP was stable under high temperatures. Furthermore, WCP revealed a desorption/drying (rerun) stage only and weight loss (0.32%). DTG measurement was performed to assess mass loss during the pre-set heating of the samples in order to measure physical changes in materials. Figure 2.13 depicts the DTG curves for WCP. Results revealed that WCP was more stable because it displayed high strength against the decomposition reaction.

2.6 CONCLUSIONS

Based on the experimental results on the preparation and characterization of ceramic wastes, the following conclusions can be drawn:

i. The physical properties of prepared ceramic aggregates, and micro- or nano-size powder highly depend on the preparation method.
ii. The particle size and distribution of ceramic wastes satisfied the standard of ASTM C33 (for aggregates) and ASTM C618 (for powder).
iii. The WCP presented crystalline structures. However, depending on XRD results, WCP is considered as a semi-amorphous material.
iv. The results of chemical composition showed that quartz is the main element in WCP chemical composition.
v. From SEM results, WCP presented irregular and angular particles with rough surface.
vi. According to TGA and DTG results, WCP shows high stability under elevated temperatures (up to 900°C) and weight loss less than 0.5%.

REFERENCES

1. Huseien, G. F. and K. W. Shah, Durability and life cycle evaluation of self-compacting concrete containing fly ash as GBFS replacement with alkali activation. *Construction and Building Materials*, 2020. **235**: pp. 117458.
2. Mohammadhosseini, H., et al., Enhanced performance of green mortar comprising high volume of ceramic waste in aggressive environments. *Construction and Building Materials*, 2019. **212**: pp. 607–617.
3. Hussein, A. A., et al., Performance of nanoceramic powder on the chemical and physical properties of bitumen. *Construction and Building Materials*, 2017. **156**: pp. 496–505.
4. Hussein, A. A., et al., Physical, chemical and morphology characterisation of nano ceramic powder as bitumen modification. *International Journal of Pavement Engineering*, 2019: pp. 1–14.
5. Huseien, G. F., et al., Evaluation of alkali-activated mortars containing high volume waste ceramic powder and fly ash replacing GBFS. *Construction and Building Materials*, 2019. **210**: pp. 78–92.
6. Senthamarai, R. and P. D. Manoharan, Concrete with ceramic waste aggregate. *Cement and Concrete Composites*, 2005. **27**(9–10): pp. 910–913.
7. Pacheco-Torgal, F. and S. Jalali, Reusing ceramic wastes in concrete. *Construction and Building Materials*, 2010. **24**(5): pp. 832–838.

8. Senthamarai, R., P. D. Manoharan, and D. Gobinath, Concrete made from ceramic industry waste: Durability properties. *Construction and Building Materials*, 2011. **25**(5): pp. 2413–2419.
9. Daniyal, M. and S. Ahmad, Application of Waste Ceramic Tile Aggregates in Concrete. *International Journal of Innovative Research in Science, Engineering and Technology*, 2015. **4**(12): pp. 12808–12815.
10. Ay, N. and M. Ünal, The use of waste ceramic tile in cement production. *Cement and Concrete Research*, 2000. **30**(3): pp. 497–499.
11. Koyuncu, H., et al. Utilization of ceramic wastes in the construction sector. In *Key Engineering Materials*. 2004: Trans Tech Publ.
12. Sánchez de Rojas, M. I., et al., Properties and performances of concrete tiles containing waste fired clay materials. *Journal of the American Ceramic Society*, 2007. **90**(11): pp. 3559–3565.
13. Puertas, F., et al., Ceramic wastes as alternative raw materials for Portland cement clinker production. *Cement and Concrete Composites*, 2008. **30**(9): pp. 798–805.
14. Lavat, A. E., M. A. Trezza, and M. Poggi, Characterization of ceramic roof tile wastes as pozzolanic admixture. *Waste Management*, 2009. **29**(5): pp. 1666–1674.
15. Naceri, A. and M. C. Hamina, Use of waste brick as a partial replacement of cement in mortar. *Waste Management*, 2009. **29**(8): pp. 2378–2384.
16. Vejmelková, E., et al., Properties of high performance concrete containing fine-ground ceramics as supplementary cementitious material. *Cement and Concrete Composites*, 2012. **34**(1): pp. 55–61.
17. Cheng, Y., et al., Test research on effects of ceramic polishing powder on carbonation and sulphate-corrosion resistance of concrete. *Construction and Building Materials*, 2014. **55**: pp. 440–446.
18. Lim, N. H. A. S., et al., Microstructure and strength properties of mortar containing waste ceramic nanoparticles. *Arabian Journal for Science and Engineering*, 2018. **43**(10): pp. 5305–5313.
19. Huseien, G. F., et al., Effects of ceramic tile powder waste on properties of self-compacted alkali-activated concrete. *Construction and Building Materials*, 2020. **236**: p. 117574.
20. Bentz, D. P., Activation energies of high-volume fly ash ternary blends: hydration and setting. *Cement and Concrete Composites*, 2014. **53**: pp. 214–223.
21. Huseien, G. F., et al., Properties of ceramic tile waste based alkali-activated mortars incorporating GBFS and fly ash. *Construction and Building Materials*, 2019. **214**: pp. 355–368.
22. Khatib, J. M., Properties of concrete incorporating fine recycled aggregate. *Cement and Concrete Research*, 2005. **35**(4): pp. 763–769.
23. de Brito, J., A. Pereira, and J. Correia, Mechanical behaviour of non-structural concrete made with recycled ceramic aggregates. *Cement and Concrete Composites*, 2005. **27**(4): pp. 429–433.
24. Debieb, F. and S. Kenai, The use of coarse and fine crushed bricks as aggregate in concrete. *Construction and Building Materials*, 2008. **22**(5): pp. 886–893.
25. Gomes, M. and J. de Brito, Structural concrete with incorporation of coarse recycled concrete and ceramic aggregates: Durability performance. *Materials and Structures*, 2009. **42**(5): pp. 663–675.
26. Torkittikul, P. and A. Chaipanich, Utilization of ceramic waste as fine aggregate within Portland cement and fly ash concretes. *Cement and Concrete Composites*, 2010. **32**(6): pp. 440–449.
27. Higashiyama, H., et al., Compressive strength and resistance to chloride penetration of mortars using ceramic waste as fine aggregate. *Construction and Building Materials*, 2012. **26**(1): pp. 96–101.

28. Medina, C., M. Frías, and M. S. De Rojas, Microstructure and properties of recycled concretes using ceramic sanitary ware industry waste as coarse aggregate. *Construction and Building Materials*, 2012. **31**: pp. 112–118.
29. Tavakoli, D., A. Heidari, and M. Karimian, Properties of concretes produced with waste ceramic tile aggregate. *Asian Journal of Civil Engineering*, 2013. **14**(3): pp. 369–382.
30. Jiménez, J., et al., Use of fine recycled aggregates from ceramic waste in masonry mortar manufacturing. *Construction and Building Materials*, 2013. **40**: pp. 679–690.
31. Alves, A., et al., Mechanical properties of structural concrete with fine recycled ceramic aggregates. *Construction and Building Materials*, 2014. **64**: pp. 103–113.
32. Samadi, M., et al., Properties of mortar containing ceramic powder waste as cement replacement. *Jurnal Teknologi*, 2015. **77**(12): pp. 93–97.
33. Binici, H., Effect of crushed ceramic and basaltic pumice as fine aggregates on concrete mortars properties. *Construction and Building Materials*, 2007. **21**(6): pp. 1191–1197.
34. Neville, A. M., *Properties of Concrete*. Vol. 4. 1995: Longman.
35. Heidari, A. and D. Tavakoli, A study of the mechanical properties of ground ceramic powder concrete incorporating nano-SiO2 particles. *Construction and Building Materials*, 2013. **38**: pp. 255–264.
36. Huang, B., Q. Dong, and E. G. Burdette, Laboratory evaluation of incorporating waste ceramic materials into Portland cement and asphaltic concrete. *Construction and Building Materials*, 2009. **23**(12): pp. 3451–3456.
37. Prajapati, L. and I. Patel, *Analysis of the Strength and Durability of the Concrete with Partially Replaced by the Ceramic Slurry Waste Powder*. 2014.
38. ASTM, C., 618 (1993) Standard specification for coal fly ash and raw or calcined natural pozzolan for use as a mineral admixture in concrete. In *Annual Book of ASTM Standards*. 1999: ASTM.
39. Sanchez, F. and K. Sobolev, Nanotechnology in concrete: A review. *Construction and Building Materials*, 2010. **24**(11): pp. 2060–2071.
40. Raki, L., J. Beaudoin, and R. Alizadeh, Nanotechnology applications for sustainable cement-based products. In *Nanotechnology in Construction 3*. pp. 119–124. 2009: Springer.
41. Scrivener, K. L. and R. J. Kirkpatrick, Innovation in use and research on cementitious material. *Cement and Concrete Research*, 2008. **38**(2): pp. 128–136.
42. Aydın, A. C., V. J. Nasl, and T. Kotan, The synergic influence of nano-silica and carbon nano tube on self-compacting concrete. *Journal of Building Engineering*, 2018. **20**: pp. 467–475.
43. Fu, J., et al., Comparison of mechanical properties of CSH and portlandite between nano-indentation experiments and a modelling approach using various simulation techniques. *Composites Part B: Engineering*, 2018. 151: pp. 127–138.
44. Cheah, C. B. and M. Ramli, Mechanical strength, durability and drying shrinkage of structural mortar containing HCWA as partial replacement of cement. *Construction and Building Materials*, 2012. **30**: pp. 320–329.
45. Jackiewicz-Rek, W., et al., Properties of cement mortars modified with ceramic waste fillers. *Procedia Engineering*, 2015. **108**: pp. 681–687.
46. Karim, M., et al., On the utilization of pozzolanic wastes as an alternative resource of cement. *Materials*, 2014. **7**(12): pp. 7809–7827.
47. Tangpagasit, J., et al., Packing effect and pozzolanic reaction of fly ash in mortar. *Cement and Concrete Research*, 2005. **35**(6): pp. 1145–1151.
48. Mymrin, V., et al., Red ceramics enhancement by hazardous laundry water cleaning sludge. *Journal of Cleaner Production*, 2016. **120**: pp. 157–163.
49. Samyn, F., et al., Characterisation of the dispersion in polymer flame retarded nanocomposites. *European Polymer Journal*, 2008. **44**(6): pp. 1631–1641.

50. Zhou, W., et al., A comparative study of high-and low-Al 2 O 3 fly ash based-geopolymers: The role of mix proportion factors and curing temperature. *Materials & Design*, 2016. **95**: pp. 63–74.
51. Sitarz, M., W. Mozgawa, and M. Handke, Vibrational spectra of complex ring silicate anions: Method of recognition. *Journal of Molecular Structure*, 1997. **404**(1–2): pp. 193–197.
52. Phair, J. W., J. Van Deventer, and J. Smith, Mechanism of polysialation in the incorporation of zirconia into fly ash-based geopolymers. *Industrial & Engineering Chemistry Research*, 2000. **39**(8): pp. 2925–2934.
53. Guo, X., H. Shi, and W. A. Dick, Compressive strength and microstructural characteristics of class C fly ash geopolymer. *Cement and Concrete Composites*, 2010. **32**(2) pp. 142–147.

3 Utilizing Ceramic Wastes in the Concrete Industry

3.1 INTRODUCTION

To manufacture one ton of ordinary Portland cement (OPC), about one ton of carbon dioxide (CO_2) is emitted in the atmosphere [1, 2]. The world produces more than 3.5 billion tons of cement; this means about the same quantity of CO_2 is emitted from cement industry alone [3, 4]. Reusing the waste materials of the concrete industry is one of the latest technologies attempting to decrease the usage of OPC. Geopolymer and alkali-activated mortars are emerging, new, cement-less binders to make concrete [5–7]. They have environmentally sustainable characteristics [8]. Using industrial waste products such as ground blast furnace slag (GBFS), palm oil fuel ash (POFA), fly ash (FA), and ceramic wastes to produce modified cement and alkali-activated mortar/concrete show excellent results to be alternatives to (OPC) mortar/concrete [9–12].

Ceramics are manufactured by firing clay, feldspar, and quartz at high temperatures. According to Nepomuceno et al. [13] and Mohammadhosseini et al. [14], the overall worldwide manufacture of various types of ceramic tiles was approximately 12 billion m^2 in 2012. During the same period, the total production in Malaysia was approximately 92 million m^2 [15], and the production rate is growing by 2.3% yearly. Generally, in ceramic industries, about 30% of the production goes to waste [16]. The waste products from ceramic industries are durable and highly resistant to both chemical and physical degradation forces. However, only a part of these ceramic wastes is used (in very low volumes) in different applications, such as tartan floor or for gardening, while a huge amount of ceramic wastes is sent to landfills. Consequently, there is a need for ceramic industries to find an alternative way for ceramic waste disposal [17]. In this regard, Matias et al. [18] and Torgal and Jalali [19] pointed out that the use of ceramic waste in sustainable construction materials, such as mortar and concrete, aids by saving energy and cost in construction, decreasing ecological impacts and reducing the consumption of raw materials.

Many factors influence the final fresh and hardened properties of modified cement and alkali-activated binder in preparation of mortar/concrete, including chemical composition of recycled waste materials, amorphous and semi-crystalline patterns, particle size and water absorption, ratio of binder to aggregate, and water content [20–22]. The durability of modified cement and alkali-activated binders tend to improve by including the waste materials as partial or full replacement of cement [23, 24]. Waste materials high in calcium oxide such as GBFS can be used to produce alkali-activated mortar/concrete cured at ambient temperatures [25]. Kumar at el. [26] reported the alkali activation of GBFS results in precipitation of calcium silicate hydrate (C-S-H) and calcium-alumina silicate hydrate (C-A-S-H) gels for alkali-activated concrete and mortar cured at ambient conditions.

A substantial number of researchers have developed a variety of raw materials, such as metakaolin, FA, POFA, and ceramic wastes incorporated with GBFS, to synthesize alkali-activated mortar/concrete with excellent performance characteristics [8, 27–29]. All these materials could supply the hydration and polymerization reaction with sufficient silica and aluminium [30, 31]. On the other hand, ceramic waste is a kind of silicon-rich material [32]; using it as source material to manufacture alkali-activated concrete could be an efficient and environment friendly use.

The aim of this chapter is to evaluate fresh and strength properties of modified cement and alkali-activated binder using ceramic wastes as partial replacement of OPC, fine and coarse aggregates, and combined with other wastes such as GBFS and FA. Tests conducted on the modified cement and alkali-activated mortar/concrete were workability, setting time, density, and ultrasonic pulse velocity (UPV). Results obtained were expected to enrich the knowledge of alkali-activated concrete and provide a promising alternative to reutilize ceramic wastes economically and sustainably.

3.2 WORKABILITY PERFORMANCE OF TRADITIONAL CONCRETE

Fresh mix needs to be tested to measure the workability and consistency of mortar/concrete. The workability of mortar was measured by a flow test. Workability of mortar/concrete generally implies the ease that the mix can be handled with from the mixer until it's finally compacted in the formwork or mould without segregation. The measurements of the workability of fresh mortar/concrete are of importance in assisting the practicality of compacting the mix and also in maintaining consistency throughout the job. In addition, a workability test is often used as an indirect check on the water content of mortar/concrete. The workability and consistency of mortar/concrete were determined using flow or slump and setting-time tests, respectively.

3.2.1 Effects of Ceramic Powder on Cement Flowability

Several studies [14, 15] reported that the inclusion of waste ceramic powder (WCP) as OPC replacement led to enhanced workability of prepared mortar/concrete. However, replacing OPC with 40% WCP and up negatively affected on flowability. The comparison of flow between OPC and 40% WCP is shown in Figure 3.1. The WCP mortar was prepared with 0.48 (water-to-cement ratio), 1:3 (binder to fine aggregate) and 100% ceramic wastes as fine aggregate. Unlike, the OPC mortar prepared with river sand as fine aggregate, the measured flow diameter for OPC and ceramic mortars was 14.5 cm and 14.0 cm, respectively. The ceramic mortar shows slightly lower workability compared to the OPC mortar. The workability of ceramic mortar was lower by 5 mm compared with the OPC mortar. This was due to the increase in angular shape of ceramic fine aggregate and finer particles of ceramic powder as cement replacement, which absorb more water. Several researchers reported that a higher amount of water was required to obtain the desired consistency and lower

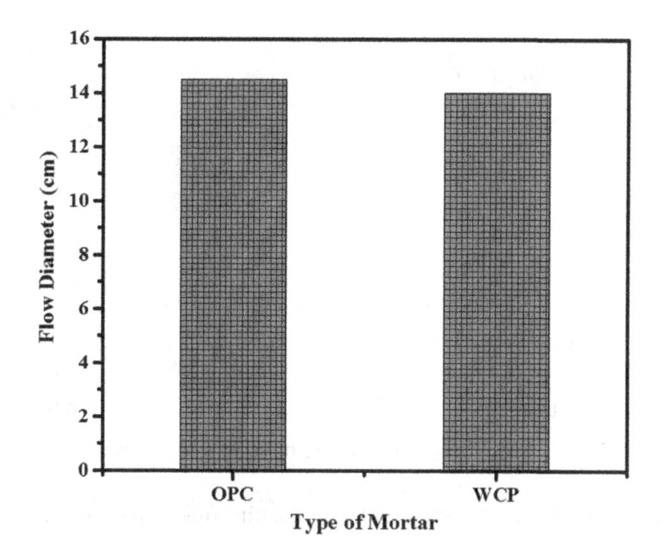

FIGURE 3.1 Comparison of flow diameter between OPC and ceramic mortars.

FIGURE 3.2 Flow test of prepared mortar. (a) OPC, (b) WCP.

workability is common in pozzolanic materials [33–35]. Furthermore, Figure 3.2 shows the flow diameter of OPC and ceramic mortar during the flow test. It can be seen that the flow diameter for ceramic mortar was less than OPC mortar due to the presence of the angular shape of ceramic fine aggregate.

3.2.2 Effects of WCP on Cement Setting Time

It's well-known in the concrete industry that OPC performance is highly influenced by the chemical composition of waste materials used as partial replacement of OPC. Cement has a high content of calcium and any waste materials added to the cement matrix effect the total calcium content, which directly affects the setting time of the prepared concrete. Figure 3.3 shows the effect of percentage of WCP replacement

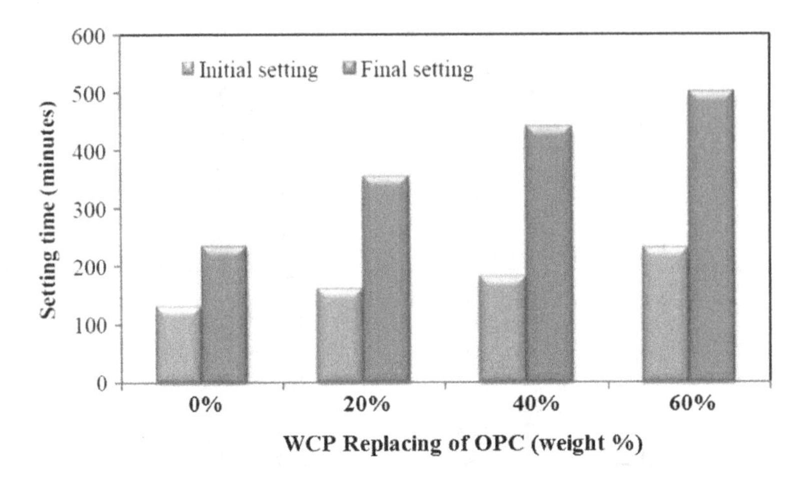

FIGURE 3.3 Effect of WCP replacement level on setting time of prepared mortar.

on the setting time of OPC mortars. As can be seen from the figure, the initial and final setting times of the OPC paste were 130 and 235 minutes, respectively. The OPC paste needed 105 minutes to reach final setting time after the initial setting time. The initial and setting time of ceramic mortar increases with the increase in percentage of ceramic powder replacement. This was due to the fact that the C_3S in the mix reduces as ceramic powder replacement increases, thus reducing the cement hydration process of mortar. The initial and final setting time of the 20% ceramic paste were 160 and 355 minutes, respectively. In contrast to the 20% ceramic paste, the initial and final setting time of 60% ceramic paste were 230 and 500 minutes, respectively. The 20% ceramic paste needed 195 minutes to reach its final setting time after the initial setting time, while the 60% ceramic paste required longer setting time, which was 270 minutes to reach the final setting time. However, the setting time of ceramic paste is longer than the OPC paste. This is due to the pozzolanic reaction between SiO_2 from ceramic and $Ca(OH)_2$ evolved from cement hydration, which only occurs after the cement started to react with water.

3.2.3 Effect of Ceramic Aggregates on Concrete Workability

Several studies [9, 36, 37] reported that the workability of concrete or mortar tends to decrease with increasing amount of ceramic aggregates. This is probably due to the particle shape that has an effect on the workability of concrete, where the angular particles of ceramic waste aggregates result in the low workability [38]. The angularity of the ceramic aggregates cannot be controlled to achieve an angularity similar to that of the natural aggregates, as the aggregates were crushed from ceramic tile wastes. A study by Vishvakarma et al. [39] also reported the use of ceramic waste as filler (10–20%) in mortar led to a decrease in the workability and consistency of the mortar, which causes a rise in water demand. This may be due to the higher water absorption of sanitary ceramic filler compared to that of sand. Furthermore,

Alves et al. [40] reported that the workability of concrete with 20, 50, and 100% ceramic aggregate replacement was affected negatively compared to concrete using natural aggregates. They suggest changing the water-to-cement ratio as an effective solution to overcome the lower workability problem.

In addition, the reduced workability can be effectively compensated with the use of superplasticizer [41, 42]. Similarly, Tavakoli et al. [43] stated that by using ceramic waste aggregates, the amount of slump decreases as the amount of aggregates increases. This is due to the higher water absorption and the angular shape of the ceramic aggregates. Also, as the replacement of recycled aggregates level increased, the workability of the mixes decreased probably because of the grading, shape, and water absorption of ceramic aggregates [44]. However, Senthamarai and Manoharan [45] reported a higher slump value for 100% ceramic aggregate replacement compared to normal concrete, which is due to the lower water absorption and smooth surface texture of the ceramic aggregate. However, so far there are few studies found in the literature on the use of ceramic waste as aggregate and cement replacement simultaneously in mortar and concrete production.

3.3 WORKABILITY OF CEMENT-FREE CONCRETE

3.3.1 Effect of WCP on Alkali-Activated Flowability

To evaluate the effect of WCP on workability performance of GBFS-based alkali-activated mortar, GBFS was replaced with 0, 25, 50, 75, and 100% of WCP. Table 3.1 displays of various binary binders incorporating WCP and GBFS to prepare the alkali-activated mixes. The ratio of binder to fine aggregate (B:A), solution to binder (S:B), and sodium silicate (NS) to sodium hydroxide (NH) was fixed at 0.5, 0.35, and 1.5, respectively. A total of five alkali-activated mixes were considered along with standard OPC mortar. WCP- and GBFS-based alkali-activated mortar mixes were manufactured separately. Additionally, the OPC control sample (Grade 25) mortar mix was designed as per ASTM C109. WCP-GBFS–based alkali-activated mortar was mixed first for 2 minutes, then added to fine aggregate and mixed for another 3 minutes. Binary alkali-activated binders were activated using alkaline activator solutions. The alkaline activator solutions were prepared by combining sodium

TABLE 3.1
Alkali-Activated Mortars Mix Design by Weight %

	Binder (%)		Alkali Solution		Chemical Composition (%)				
Mix	WCP	GBFS	NS:NH	Molarity	SiO_2	Al_2O_3	CaO	Si:Al	Ca:Si
1	0	100	1.5	4	30.0	10.90	51.80	2.83	1.68
2	25	75	1.5	4	41.30	11.22	38.85	3.68	0.94
3	50	50	1.5	4	51.80	11.55	25.90	4.48	0.50
4	75	25	1.5	4	62.30	11.87	12.95	5.25	0.21
5	100	0	1.5	4	72.80	12.20	0	5.97	0

hydroxide and sodium silicate. Cubic 50 × 50 × 50 mm, cylinder 75 × 150 mm, and prism 40 × 40 × 160 mm moulds were used to prepare the specimens of each alkali-activated mortar mix.

The workability of fresh alkali-activated mortar mixtures was tested by flow test. The flow of fresh alkali-activated mortars was measured in accordance with ASTM C1437-07. The flow test was conducted immediately after mixing. The effect of ceramic on workability of alkali-activated mortars was investigated. Results showed that the workability of mortar increased with more ceramic added to the GBFS ratio. The results of the flow tests recorded 13, 15.25, 17, 20, and 24 cm with 0, 25, 50, 75, and 100% WC:GBFS ratio, respectively. Figure 3.4 shows the effect ceramic percentage has on workability of alkali-activated mortars. The increase in content of ceramic led to an increased content of silicate and reduced the content of calcium and positively affected the workability of mortar.

3.3.2 Effect of WCP on Prepared Mortar Setting Time

The setting time of alkali-activated mortar was tested in accordance with ASTM C191. The mortar was prepared by mixing the binders and the alkaline solutions manually in a bowl and the setting time test using the Vicat apparatus. Figure 3.5 depicts the influence of WCP content on setting time. The setting time tests were carried out at a controlled temperature of 27 ± 1.5°C. In such a situation, alkali-activated mortar containing WCP only as the binder generally takes significantly longer time to set due to the slow rate of chemical reaction at low ambient temperature. In this experiment, the alkali-activated mortar mixture without WCP as control specimen, designed with GBFS only as the binder, showed a very fast setting time. When WCP was incorporated in the mixture, alkali-activated mortar's setting time improved significantly. Both initial and final setting time increased with the increasing WCP content. Mixture prepared with 25% WCP as the GBFS replacement achieved the initial

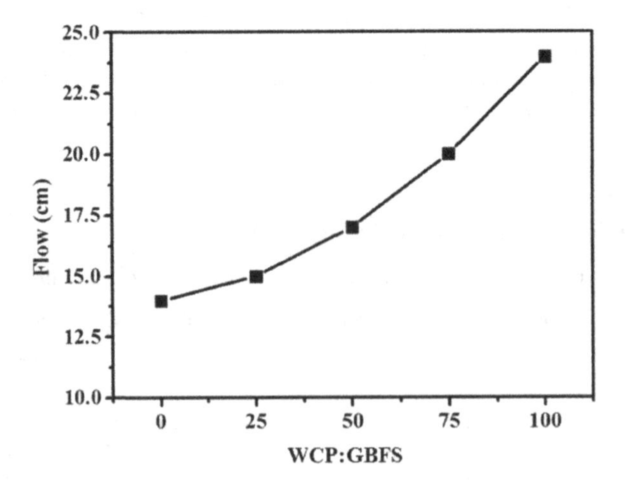

FIGURE 3.4 Effect of WCP on workability of GBFS-based alkali-activated mortar.

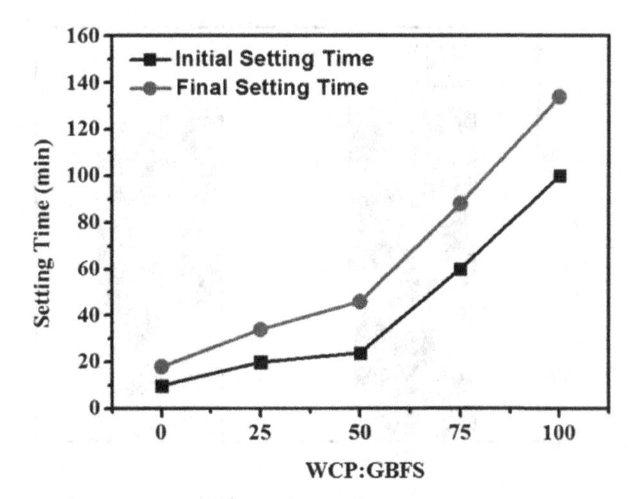

FIGURE 3.5 Setting time of alkali-activated mortar prepared with various WCP content.

setting time of 20 minutes. However, it was observed with the rising level of replacement to 50 and 75%, the setting time increased to 24 and 64 minutes, respectively. The rate of setting increased significantly as indicated by the substantial difference in the initial setting time. The difference between initial and final setting time also increased with increasing WCP content in the alkali-activated matrix. It also supports the fact that the higher content of silicate content led to a slower rate of setting. The results established that GBFS as part of the binary blended binder is effective to decelerate setting time of alkali-activated mortar in ambient conditions [46, 47].

3.4 DENSITY OF TRADITIONAL CONCRETE

3.4.1 Effect of WCP

In study by Mohammadhosseini et al. [14], it was found that the inclusion of ceramic waste as OPC and natural aggregate replacement reduced the weight of prepared specimens. The addition of finer particles, such as ceramic powder, into the mortar and concrete mixtures would generally densify the matrix by filling the cavities and reducing the porosity. In addition, owing to the lower loose bulk density (LBD) of the ceramic fine aggregates (0.94 g/cm^3) than that of sand (1.43 g/cm^3), the replaced volume of WCP would be higher, and consequently, it made the matrix stiffer, and this resulted in lower flowability of the ceramic mortar. Figure 3.6 shows that the fresh-state density of the mortar mix reduced the replacement of ceramic powder and fine aggregates. This was expected because of the lower density of the ceramic particles, compared to that of OPC and natural river sand. It was also found that the ceramic mortar obtained lower dry density. Results indicated that the fresh and dry density of the ceramic mortar reduced by approximately 4% and 5%, respectively, compared to that of the OPC mix.

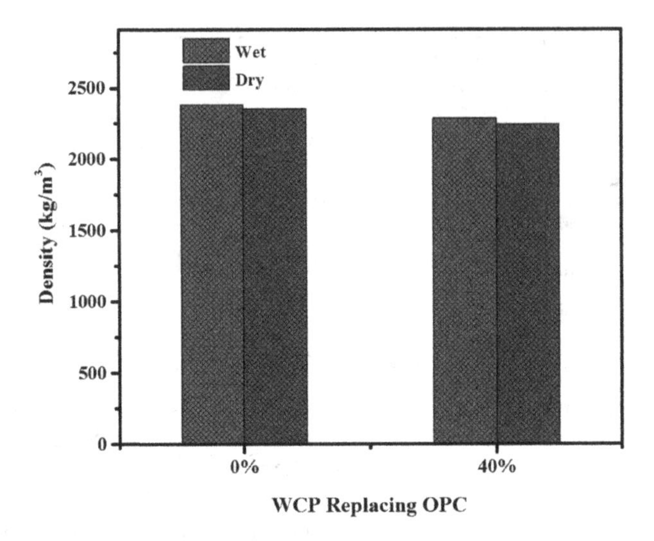

FIGURE 3.6 Effect of WCP on modified mortar density.

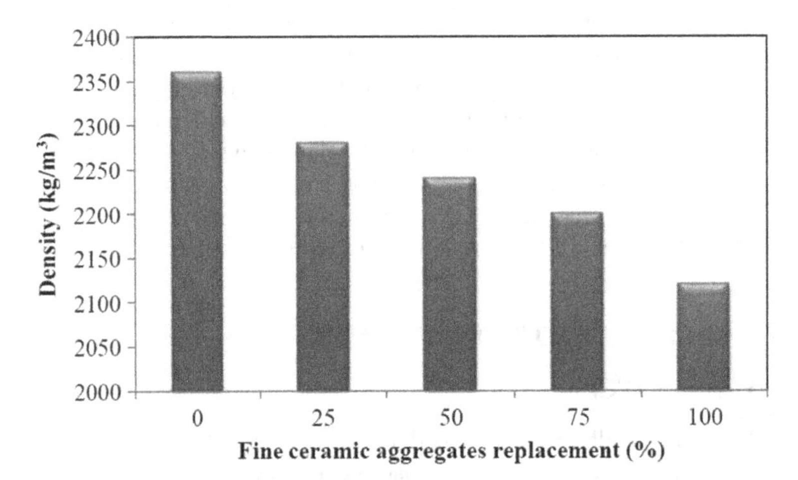

FIGURE 3.7 Effect of ceramic fine aggregate content on density of mortar.

3.4.2 Effect of Ceramic Aggregates

The density of mortar with ceramic fine aggregates was measured as the ratio of the average weight value of three specimens after being surfaced dry to the volume of the specimen. The dry density values of the specimen are shown in Figure 3.7. The density for OPC mortar is 2360 kg/m^3 and the density of ceramic fine aggregates mortar with 25, 50, 75, and 100% replacement is 2280, 2240, 2200, and 2120 kg/m^3, respectively. As can be seen in the figure, there is a decrease in dry density of mortar with the increase in ceramic fine aggregates content as compared to OPC mortar. The density of 100% ceramic fine aggregates mortar decreased by 10% from the

OPC mortar. This is due to the lower density of ceramic fine aggregate itself, thus when using it in large quantities reduces the density of mortar. As a result, the mortar produced is lighter than the one formed using a lower amount of ceramic fine aggregate as the sand replacement.

3.5 DENSITY OF CEMENT-FREE MORTAR

The density of waste ceramic powder replaced GBFS-based alkali-activated mortar was also investigated. Figure 3.8 shows the effect of an increased ceramic ratio on binder alkali-activated mortar; the effect of curing temperature was also studied at the same time. Samples were cured at three different temperatures: 27, 60, and 90°C. The results indicated the density of alkali-activated mortar reduced with increasing WCP content as the specific gravity of ceramic powder was lower than GBFS. The results of the curing regime showed that alkali-activated mortar cured at elevated temperatures were more lightweight than samples cured at ambient temperatures.

3.6 TRADITIONAL CONCRETE UPV READINGS

3.6.1 Effect of WCP

The effect of ceramic powder replacement on the ultrasonic pulse velocity (UPV) value of mortar is shown in Figure 3.9. The mortar with ceramic powder replacement produced lower UPV value at early age and increased as the hydration process increased. The UPV values at 90 days for OPC, 20, 40, and 60% ceramic mortars were 4896, 5474, 5719, and 4210 m/s, respectively. At 90 days curing, 40% ceramic powder replacement showed higher UPV values compared to OPC mortar by 14%. In this study, these excellent UPV values could be attributed to the improved pore structure of mortar as a result of optimum particle packing and the enhanced filling

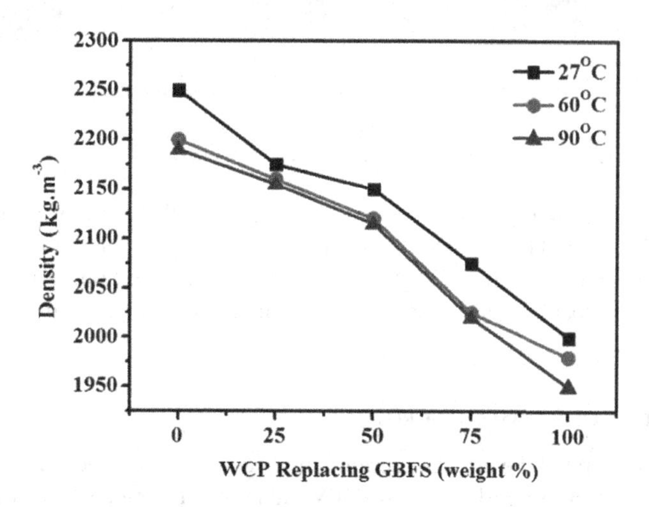

FIGURE 3.8 Effect of WCP replacing GBFS on density of alkali-activated mortar.

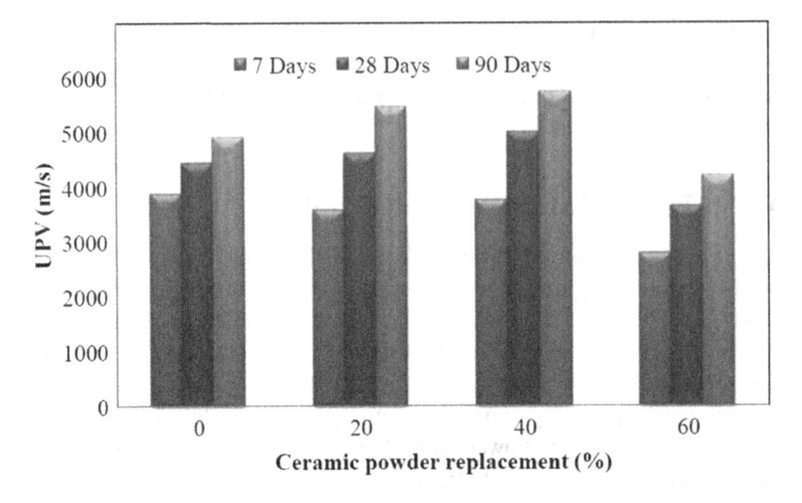

FIGURE 3.9 Effect of ceramic powder on UPV values.

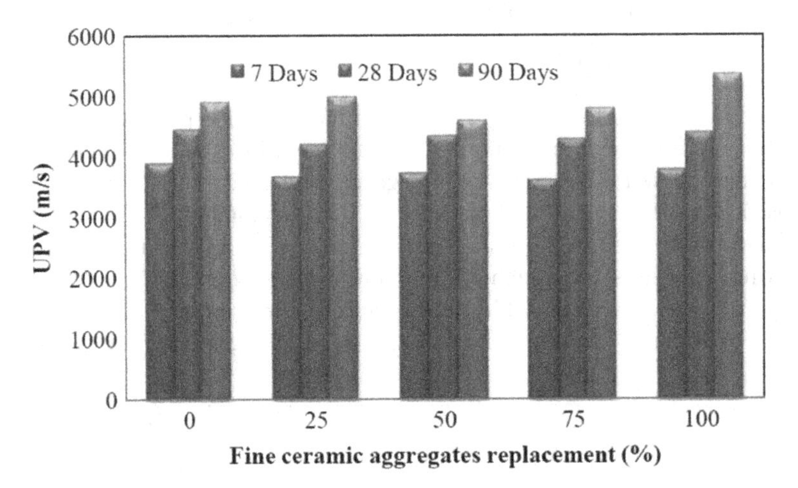

FIGURE 3.10 Effect of ceramic fine aggregate replacement on UPV values.

ability of finer ceramic powder. Furthermore, higher UPV values were obtained at later age, which could be attributed to the increased densification of the paste resulting from secondary hydration of the reactive ceramic powder. Similar findings were reported by Tangpagasit et al. [48] using fly ash as the pozzolanic material on the packing effect.

3.6.2 Effect of Ceramic Aggregates

Figure 3.10 illustrated the effect of ceramic fine aggregates replacement on the UPV value of mortar. At an early age, the UPV value of OPC mortar was higher than others. However, as the age of curing increases, the UPV value also increases. As recorded from the experimental work, at 90 days curing, the 100% ceramic fine aggregates replacement showed an increase of 8% in the UPV value compared with OPC mortar. This is due to the physical properties of ceramic fine aggregates with angular and crushed particles, which increases the interlocking friction with the binder and reduced the void. Since the void content in the mortar reduces, the UPV value increases. Neville [49] reported that a UPV value more than 4500 m/s generally indicates mortar with excellent quality. In this study, these excellent UPV values could be attributed to the improved pore structure of mortar as a result of optimum particle packing and enhanced filling ability of the angular shape of ceramic fine aggregates.

3.7 UPV READINGS OF CEMENT-FREE SPECIMENS

In a study by Huseien et al. [50], the inclusion WCP in alkali-activated matrix effect on UPV readings of prepared specimens was observed. Figure 3.11 shows the effect of high volume WCP on the UPV of alkali-activated specimens. In Figure 3.11a, an increasing content of WCP from 50 to 60 to 70% led to a reduced UPV reading from 3732 to 3703 to 3144 m/s, respectively. This observation was mostly attributed to the high particle size of WCP (35 μm) compared to GBFS (12.4 μm). Moreover, the higher solid phase of GBFS affected the microstructure of the prepared specimens containing a high level of WCP. The alkali-activated structure with higher WCP content became more porous compared to those prepared with a high content of GBFS.

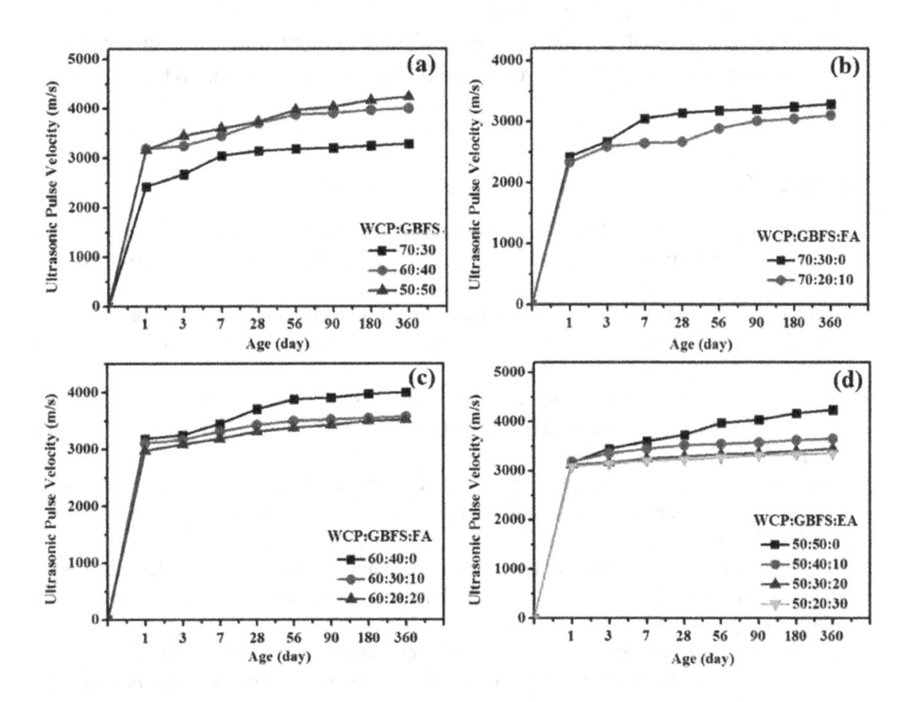

FIGURE 3.11 Effect of high-volume WCP content on the UPV of alkali-activated mortars.

Meanwhile, the structure of specimens revealed the development with the increase of age because more C-S-H, C-A-S-H, and N-A-S-H (sodium aluminium silicate hydrate) are formed with time. This in turn enhanced the velocity reading after 360 days compared to other lower ages. The effect of GBFS replaced by FA on each level of high volume WCP is shown in Figure 3.11b–d. Reduction in the GBFS contents and an increase in the FA amount (increased porosity of the specimens) diminished the UPV of specimens.

3.8 CONCLUSIONS

Based on the aforementioned characterization of the morphological and mechanical properties of ceramic waste–based modified cement and alkali-activated mortar/concrete the following conclusions can be drawn:

i. Replacing OPC with 40% or more of WCP lowered the workability of synthesized mortar/concrete.
ii. The ceramic mortar and concrete prepared with ceramic wastes as aggregate replacement display lower workability compared to OPC mortar/concrete due to the angular shape of the ceramic fine aggregates.
iii. Setting time testing revealed that at all levels of ceramic powder replacement in modified cement and alkali-activated mortar, the setting time of the paste containing ceramic powder was much longer than that of the OPC paste due to the pozzolanic reaction.
iv. The density and UPV readings of modified cement and alkali-activated mortar/concrete tend to decrease with increasing level of WCP in the matrix.

REFERENCES

1. Ogunbode, E. B., et al., Microstructure and mechanical properties of green concrete composites containing coir fibre. *Chemical Engineering Transactions*, 2017. **61**: pp. 1879–1884.
2. Kubba, Z., et al., Impact of curing temperatures and alkaline activators on compressive strength and porosity of ternary blended geopolymer mortars. *Case Studies in Construction Materials*, 2018. **9**: p. e00205.
3. Huseien, G. F., K. W. Shah, and A. R. M. Sam, Sustainability of nanomaterials based self-healing concrete: An all-inclusive insight. *Journal of Building Engineering*, 2019. **23**: pp. 155–171.
4. Mohammadhosseini, H., et al., Durability performance of green concrete composites containing waste carpet fibers and palm oil fuel ash. *Journal of Cleaner Production*, 2017. **144**: pp. 448–458.
5. Huseien, G.F., et al., Alkali-activated mortars blended with glass bottle waste nano powder: Environmental benefit and sustainability. *Journal of Cleaner Production*, 2019. **243**: p. 118636.
6. Kubba, Z., et al., Effect of sodium silicate content on setting time and mechanical properties of multi blend geopolymer mortars. *Journal of Engineering and Applied Sciences*, 2019. **14**(7): pp. 2262–2267.

7. Samadi, M., et al., Influence of glass silica waste nano powder on the mechanical and microstructure properties of alkali-activated mortars. *Nanomaterials*, 2020. **10**(2): p. 324.
8. Huseien, G. F., et al., Effects of POFA replaced with FA on durability properties of GBFS included alkali activated mortars. *Construction and Building Materials*, 2018. **175**: pp. 174–186.
9. Mohammadhosseini, H., et al., Enhanced performance of green mortar comprising high volume of ceramic waste in aggressive environments. *Construction and Building Materials*, 2019. **212**: pp. 607–617.
10. Mohammadhosseini, H., et al., Enhancement of strength and transport properties of a novel preplaced aggregate fiber reinforced concrete by adding waste polypropylene carpet fibers. *Journal of Building Engineering*, 2020. **27**: p. 101003.
11. Asaad, M. A., et al., Enhanced corrosion resistance of reinforced concrete: Role of emerging eco-friendly *Elaeis guineensis*/silver nanoparticles inhibitor. *Construction and Building Materials*, 2018. **188**: pp. 555–568.
12. Huseien, G. F., et al., Compressive strength and microstructure of assorted wastes incorporated geopolymer mortars: Effect of solution molarity. *Alexandria Engineering Journal*, 2018. **57**(4): pp. 3375–3386.
13. Nepomuceno, M. C., R. A. Isidoro, and J. P. Catarino, Mechanical performance evaluation of concrete made with recycled ceramic coarse aggregates from industrial brick waste. *Construction and Building Materials*, 2018. **165**: pp. 284–294.
14. Mohammadhosseini, H., et al., Effects of waste ceramic as cement and fine aggregate on durability performance of sustainable mortar. *Arabian Journal for Science and Engineering*, 2019: pp. 1–12.
15. Lim, N. H. A. S., et al., Microstructure and strength properties of mortar containing waste ceramic nanoparticles. *Arabian Journal for Science and Engineering*, 2018. **43**(10): pp. 5305–5313.
16. Torkittikul, P. and A. Chaipanich, Utilization of ceramic waste as fine aggregate within Portland cement and fly ash concretes. *Cement and Concrete Composites*, 2010. **32**(6): pp. 440–449.
17. Suzuki, M., M. S. Meddah, and R. Sato, Use of porous ceramic waste aggregates for internal curing of high-performance concrete. *Cement and Concrete Research*, 2009. **39**(5): pp. 373–381.
18. Matias, G., P. Faria, and I. Torres, Lime mortars with heat treated clays and ceramic waste: A review. *Construction and Building Materials*, 2014. **73**: pp. 125–136.
19. Pacheco-Torgal, F. and S. Jalali, Reusing ceramic wastes in concrete. *Construction and Building Materials*, 2010. **24**(5): pp. 832–838.
20. Kabir, S., et al., Influence of molarity and chemical composition on the development of compressive strength in POFA based geopolymer mortar. *Advances in Materials Science and Engineering*, 2015. 2015(2): pp. 1–5.
21. Huseien, G. F., et al., Utilizing spend garnets as sand replacement in alkali-activated mortars containing fly ash and GBFS. *Construction and Building Materials*, 2019. **225**: pp. 132–145.
22. Huseien, G. F., et al., Geopolymer mortars as sustainable repair material: A comprehensive review. *Renewable and Sustainable Energy Reviews*, 2017. **80**: pp. 54–74.
23. Huseien, G. F. and K. W. Shah, Durability and life cycle evaluation of self-compacting concrete containing fly ash as GBFS replacement with alkali activation. *Construction and Building Materials*, 2020. **235**: p. 117458.
24. Ismail, I., et al., Modification of phase evolution in alkali-activated blast furnace slag by the incorporation of fly ash. *Cement and Concrete Composites*, 2014. **45**: pp. 125–135.

25. Phoo-ngernkham, T., et al., High calcium fly ash geopolymer mortar containing Portland cement for use as repair material. *Construction and Building Materials*, 2015. **98**: pp. 482–488.
26. Kumar, S., R. Kumar, and S. Mehrotra, Influence of granulated blast furnace slag on the reaction, structure and properties of fly ash based geopolymer. *Journal of Materials Science*, 2010. **45**(3): pp. 607–615.
27. Huseien, G. F., et al., Effect of metakaolin replaced granulated blast furnace slag on fresh and early strength properties of geopolymer mortar. *Ain Shams Engineering Journal*, 2016. **9**(4): pp. 1557–1566.
28. Li, C., H. Sun, and L. Li, A review: The comparison between alkali-activated slag (Si+ Ca) and metakaolin (Si+ Al) cements. *Cement and Concrete Research*, 2010. **40**(9): pp. 1341–1349.
29. Vejmelková, E., et al., Properties of high performance concrete containing fine-ground ceramics as supplementary cementitious material. *Cement and Concrete Composites*, 2012. **34**(1): pp. 55–61.
30. Al-Majidi, M. H., et al., Development of geopolymer mortar under ambient temperature for in situ applications. *Construction and Building Materials*, 2016. **120**: pp. 198–211.
31. Al-mashhadani, M. M., et al., Mechanical and microstructural characterization of fiber reinforced fly ash based geopolymer composites. *Construction and Building Materials*, 2018. **167**: pp. 505–513.
32. Ariffin, M. A., et al., Effect of ceramic aggregate on high strength multi blended ash geopolymer mortar. *Jurnal Teknologi*, 2015. **77**(16): pp. 33–36.
33. Karim, M., et al., On the utilization of pozzolanic wastes as an alternative resource of cement. *Materials*, 2014. **7**(12): pp. 7809–7827.
34. Hassan, I. O., et al., Flow characteristics of ternary blended self-consolidating cement mortars incorporating palm oil fuel ash and pulverised burnt clay. *Construction and Building Materials*, 2014. **64**: pp. 253–260.
35. Chandara, C., et al., The effect of unburned carbon in palm oil fuel ash on fluidity of cement pastes containing superplasticizer. *Construction and Building Materials*, 2010. **24**(9): pp. 1590–1593.
36. de Brito, J., A. Pereira, and J. Correia, Mechanical behaviour of non-structural concrete made with recycled ceramic aggregates. *Cement and Concrete Composites*, 2005. **27**(4): pp. 429–433.
37. Medina, C., et al., Durability of recycled concrete made with recycled ceramic sanitary ware aggregate. Inter-indicator relationships. *Construction and Building Materials*, 2016. **105**: pp. 480–486.
38. Anderson, D. J., S. T. Smith, and F. T. Au, Mechanical properties of concrete utilising waste ceramic as coarse aggregate. *Construction and Building Materials*, 2016. **117**: pp. 20–28.
39. Prajapati, L. and I. Patel, Analysis of the strength and durability of the concrete with partially replaced by the ceramic slurry waste powder. 2014.
40. Alves, A., et al., Mechanical properties of structural concrete with fine recycled ceramic aggregates. *Construction and Building Materials*, 2014. **64**: pp. 103–113.
41. Rana, A., et al., Recycling of dimensional stone waste in concrete: A review. *Journal of Cleaner Production*, 2016. **135**: pp. 312–331.
42. Shafigh, P., et al., Lightweight concrete made from crushed oil palm shell: Tensile strength and effect of initial curing on compressive strength. *Construction and Building Materials*, 2012. **27**(1): pp. 252–258.
43. Tavakoli, D., A. Heidari, and M. Karimian, Properties of concretes produced with waste ceramic tile aggregate. *Asian Journal of Civil Engineering*, 2013. **14**(3): pp. 369–382.

44. Hebhoub, H., et al., Use of waste marble aggregates in concrete. *Construction and Building Materials*, 2011. **25**(3): pp. 1167–1171.
45. Senthamarai, R. and P. D. Manoharan, Concrete with ceramic waste aggregate. *Cement and Concrete Composites*, 2005. **27**(9–10): pp. 910–913.
46. Huseien, G. F., et al., Influence of different curing temperatures and alkali activators on properties of GBFS geopolymer mortars containing fly ash and palm-oil fuel ash. *Construction and Building Materials*, 2016. **125**: pp. 1229–1240.
47. Huseien, G. F., et al., The effect of sodium hydroxide molarity and other parameters on water absorption of geopolymer mortars. *Indian Journal of Science and Technology*, 2016. **9**(48): pp. 1–7.
48. Tangpagasit, J., et al., Packing effect and pozzolanic reaction of fly ash in mortar. *Cement and Concrete Research*, 2005. **35**(6): pp. 1145–1151.
49. Neville, A. M., *Properties of Concrete*. Vol. 4. 1995: Longman.
50. Huseien, G. F., et al., Properties of ceramic tile waste based alkali-activated mortars incorporating GBFS and fly ash. *Construction and Building Materials*, 2019. **214**: pp. 355–368.

4 Effects of Ceramic Waste on Durability Performance of Traditional Concrete

4.1 INTRODUCTION

The manufacturing of Portland cement results in the release of an enormous amount of greenhouse gases to the atmosphere, which, in turn, pollute the environment and are a significant contributing factor to climate change. To mitigate such negative environmental effects, alternative green construction materials have continuously been researched to improve sustainable development [1, 2]. Worldwide, the majority of cement produced is consumed in the production of concrete and mortar. For each tonne of cement manufactured, approximately one tonne of greenhouse gases is released, which equates to a total of up to 8% of all greenhouse gases released worldwide [3]. The main part of these released greenhouse gases is the result of the high-temperature processing of cement clinker. Thus, continuing the production of cement at the current rate may cause irreversible damage to ecological systems worldwide unless alternative materials to produce cement are introduced [4, 5].

Recently, the use of alternatives to cement, including solid wastes from the construction industry, has rapidly increased. These waste materials are preferred because of advantages such as environment-friendliness, abundance, low energy consumption, zero cost, and excellent binding properties, which are able to fulfil the demands of a suitable construction material. Reusing these waste materials saves natural resources, contributes to sustainability, and prevents the dumping of these wastes into landfills [6]. Despite the use of various industrial wastes as a replacement for cement, some essential technical features, including the economy, stability, durability, environmental responsiveness, and sustainability of the wastes, need to be addressed. Rana et al. [7] and Mohammadhosseini et al. [8] stated that the concepts of sustainability and green manufacturing promote the development of the construction industry through the utilization of different wastes as substitutes for natural raw materials, such as extra cementitious materials and aggregates. This has led to the production of environmentally friendly and low-cost materials that are useful in various construction sectors. Ceramic tiles, for instance, are produced at high temperatures from fire clay, feldspar and quartz.

The Association of Italian Manufacturers of Machinery and Equipment for the Ceramic Industry (ACIMAC) acknowledged that the worldwide production of different kinds of ceramic tiles was around 13.7 billion m^2 in 2018 [9]. During the same time period, Malaysia produced a total of nearly 92 million m^2, and the annual manufacturing rate has been growing at a constant rate of 2.3%. Typically, approximately 30% of the materials in the ceramic industry go to waste. These ceramic wastes are highly durable and resistant both chemically and physically to harsh environmental conditions. So far, the use of such ceramic wastes has not been widely explored (in very high volumes) in various applications, including tartan floors and gardening, but vast quantities of these wastes are being dumped in landfills. Thus, the ceramics industry worldwide must find an alternative to the discarding of ceramic waste. According to Pacheco-torgal and Jalali [10], the reuse of ceramic waste in concrete and mortar as a sustainable construction material can lead to energy-saving and cost-effective construction with much lower ecological impacts due to the remarkable decrease in the consumption of raw materials. Higashiyama et al. [11] explored the feasibility of employing ceramic wastes in the synthesis of concrete and mortar. They showed that partial substitution of river sand (RS) by such wastes significantly improved the workability of fresh mortars due to the excess amount of water absorption that occurred during synthesis. Higashiyama et al. [12] showed that fine ceramic aggregates (up to 50%) incorporated into concrete improved the compressive strength (CS) and durability performance. Suzuki et al. [13] used up to 40% ceramic waste as a coarse aggregate to produce high-performance concrete and showed a considerable reduction in autogenous shrinkage. Zegardło et al. [14] studied ultra-high strength concrete that incorporated recycled ceramic aggregates. Anderson et al. [15] studied ultra-high strength concrete and reported an enhancement in the CS of concrete that incorporated recycled ceramic aggregates.

Anderson et al. [15] examined the CS performance of a concrete that utilized up to 100% ceramic waste as a fine aggregate substitution and found a decrease in the CS of only 10%. Senthamarai et al. [16] evaluated the influence of the replacement of cement with ceramic powder (WCP) and showed the potential for the enhancement of the CS and durability performance, especially over longer curing durations owing to the ceramic waste's pozzolanic characteristics. Matias et al. [17] and Higashiyama et al. [18] demonstrated the CS enhancement of mortar made from the ceramic wastes of pottery and premium red roof tiles. Typically, when concrete and mortar are exposed to harsh environmental conditions (i.e. acidic conditions), they are vulnerable to expansion and cracking. Most of the problems with concretes and mortars are related to the development of cracks and an increase in porosity. The attack of sulphate ions negatively impacts concrete and makes it less durable. Generally, groundwater contains a substantial quantity of sulphates due to the presence of excessive amounts of clay in the soil. In addition, both sulphates and chlorides are primary elements of seawater.

The characteristics of sulphate solutions are important, as the reaction mechanisms can be affected by the existence of other chemicals. Zivica and Bajza [19] analyzed the influence of sulphate attacks on the components of concrete and showed how those attacks in concrete or mortar can lead to the formation of excess gypsum and ettringite. Accordingly, the hydration products of cement, including calcium aluminate, deteriorate. Supplementary reactions that occur amongst the generated gypsum and cement particles, such as calcium aluminate, can also produce an extra quantity of ettringite. This, in turn, enables the formation of additional cracks, causing a decrease in the CS and durability. Numerous strategies [20, 21] have been adopted to improve the long-standing durability performance of mortar and concrete against various chemical attacks. One strategy is to create dense microstructures by incorporating graded particles with a narrow size distribution so that the ingress of chemicals into the mortar and concrete can be substantially minimized [22]. It was found that the durability and CS of specimens exposed to aggressive environments can be appreciably improved using such a strategy. Higashiyama et al. [12] used fine particle ceramic waste as an aggregate to improve the CS and durability of concrete and mortar. The results revealed remarkable improvement in the long-term durability performance of the designed mortar exposed to a sulphate and chloride attack. However, the studies conducted thus far on the use of ceramic waste in mortar as a substitute for cement and the impact of aggressive environments (i.e. acid attacks) have been deficient.

In addition to the mechanical properties, the aspect of durability is important in the assessment of the behaviour and potential use of any new waste material in construction applications. Thus far, and to the best of the authors' knowledge, there is limited literature on the durability and environment benefits of mortar containing ceramic waste as both a cement and fine aggregate substitute. Considering the immense importance of the reuse of inexpensive and abundant industrial ceramic waste in mortar and concrete as a partial replacement for cement, this study prepared certain new types of mortars by incorporating ceramic waste. Locally available Malaysian ceramic waste was used to evaluate the feasibility of developing sustainable and durable mortars for civil engineering and construction applications. The as-prepared mortar specimens were characterized via different analytical tools to evaluate their sustainability and durability performance. The role played by the sustainable and environmentally friendly ceramic waste in improving the mechanical traits and durability of the mortar was emphasized. The proposed ceramic waste-blended mortar demonstrated enhanced performance in harsh environments. The use of recycled ceramic waste as a cement substitute in the mortar showed several advantages, and the proposed mortar made using ceramic waste as a fine aggregate could appreciably lower greenhouse gas emissions, thereby contributing to environmental sustainability.

4.2 MATERIALS AND MIX DESIGN

4.2.1 MATERIALS

This study used type I ordinary Portland cement (OPC) with a specific gravity of 3.15 and strength class of 42.5 in accordance with ASTM C150–07, and waste ceramic tiles, which were collected from White Horse Ceramic Industries Sdn. Bhd., Johor, Malaysia. First, the tiles were crushed to make fine particles of varying size and shape. Next, the fine particles were sieved in accordance with ASTM C33 to eliminate large particles, and a volume of fine aggregate was obtained. Additionally, the produced fine aggregate (4 kg each time) was ground for 4 hours using a modified Los Angeles abrasion test machine to produce the WCP. In accordance with ASTM C618, grinding was performed until approximately 95% of the WCP passed through the 45 μm sieve (Figure 4.1). Table 4.1 depicts the physical and chemical properties of the OPC and WCP. According to ASTM C618, the WCP was categorized as a class F pozzolan because the overall content of $SiO_2+Al_2O_3+Fe_2O_3$ was above 70%. In addition, Table 4.2 shows the physical properties of the aggregates.

4.2.2 Mix Proportions

The mix proportions for the mortar were prepared based on the weight of the materials and in accordance with ASTM C1329. The ratio of water to cement (w/c) was chosen to be 0.48 in order to achieve acceptable flowability following the requirements of ASTM C320, as well as the desired strength. A total of 12 batch samples (i.e. mortar mixes) were cast. The first sample, without any replacement of the binder or ceramic fine aggregate (CFA), acted as the control specimen (0GC0CFA). Table 4.3 presents the mixes prepared with WCP and CFA. All the designed mixes were unmoulded after 24 h and then placed inside a water container to cure until the tests were conducted at a mean ambient temperature (AT) of 27±2°C and relative humidity (RH) of 85±5%. To examine the impacts of the harsh environmental conditions on the performance of the mortar, the samples were subjected to a sulphate solution consisting of 5% Na_2SO_4. For the sulphate attack test, the specifications of ASTM C267 were followed. First, the mortar samples were cast in the form of cubes that were 100 × 100 × 100 mm. Then, the samples were immersed in the sulphate solution for 18 months. Three cubes were cast and tested in each batch,

FIGURE 4.1 The process to produce ceramic fine aggregate and ground ceramic from ceramic.

TABLE 4.1
Physical and Chemical Characteristics of WCP and OPC

	Material	
Physical Property	**WCP**	**OPC**
Specific gravity	2.35	3.15
% Passing through 45 μm wet sieve	99	90.0
Medium particle size (μm)	35	40
Chemical contents (% by mass)		
CaO	1.13	68.30
SiO_2	74.10	16.40
Al_2O_3	17.80	4.24
K_2O	0.44	0.22
Fe_2O_3	3.58	3.53
SO_3	0.023	4.39
MgO	1.24	2.39
LOI	0.10	2.40

TABLE 4.2
Physical Properties of the Fine Aggregates

Property	RS	CFA	Relevant Standard and Reference
Mass passing 75 μm sieve (%)	0.0	0.0	ASTM C33
Oven dry basis, bulk density (kg/m^3)	1624	1450	ASTM D1895
Specific gravity	2.62	2.38	
Water absorption at 24 h (%)	1.8	1.3	

and the average of the three values was considered to be the result. Following a certain duration of exposure, the sulphate resistance of the mortar samples was evaluated through changes in the visual appearance, dimensions, mass, and compressive strength.

4.3 SPECIMEN PREPARATION AND TEST METHODS

The microstructures and surface morphologies of the prepared samples were characterized using field emission scanning electron microscopy (FESEM), X-ray diffraction (XRD), and Fourier-transform infrared spectroscopy (FTIR) measurements. Oven-dried particles with a mean size less than 10 mm were made and coated with

TABLE 4.3
Mix Design for All Mortar Batches (kg/m³)

	Binder		Fine Aggregates	
Mix Label	**OPC**	**WCP**	**RS**	**CFA**
0GC0CFA	550	—	1460	—
10GC0CFA	495	55	1460	—
20GC0CFA	440	110	1460	—
30GC0CFA	395	165	1460	—
40GC0CFA	330	220	1460	—
50GC0CFA	275	275	1460	—
60GC0CFA	220	330	1460	—
0GC25CFA	550	—	1095	365
0GC50CFA	550	—	730	730
0GC75CFA	550	—	365	1095
0GC100CFA	550	—	—	1460
40GC100FCA	330	220	—	1460

double platinum prior to FESEM imaging. Samples with a mean particle size below 45 mm were subjected to XRD measurement to determine the existence of the mineral phases. The typical XRD patterns were recorded for 2 h in the angular range (2theta) of 5–75° at a scan of 0.02°/step with a scan speed of 0.5 s/step.

4.4 ENGINEERING PROPERTIES

4.4.1 Compressive Strength

The CSs of the mortar samples with different percentages of WCP as a cement substitute are presented in Figure 4.2a. At 7 days of curing, the CS of samples 20GC0CFA, 40GC0CFA, and 60GC0CFA were recorded as 37.8 MPa, 40.1 MPa, and 29.4 MPa, respectively. However, after 28 days of curing, samples 20GC0CFA and 40GC0CFA showed higher CS than sample 0GC0CFA. At 28 days of curing, the CS of samples 0GC0CFA, 20GC0CFA, 40GC0CFA, and 60GC0CFA were found to be 52.01 MPa, 53.32 MPa, 55.72 MPa, and 38.84 MPa, respectively. This was due to the pozzolanic reactions that occurred between the silica (SiO_2) and calcium hydroxide ($Ca(OH)_2$) generated from the hydration of the cement. The CS of the ground ceramic (GC) showed an increase as time passed. Comparable outcomes have been observed in other studies that used pozzolanic materials [20]. However, the strength enhancement was marginal compared to the sample 0GC0CFA, which was likely due to the lower amount of active silicate in the GC. Additionally, the amount of aluminium oxide (Al_2O_3) in the GC was high, which caused a gain in strength at an early age.

When the OPC was mixed with water, it formed a hydrated binding cement paste (HCP) of calcium silicate hydrate (C-S-H), and liberated calcium hydroxide

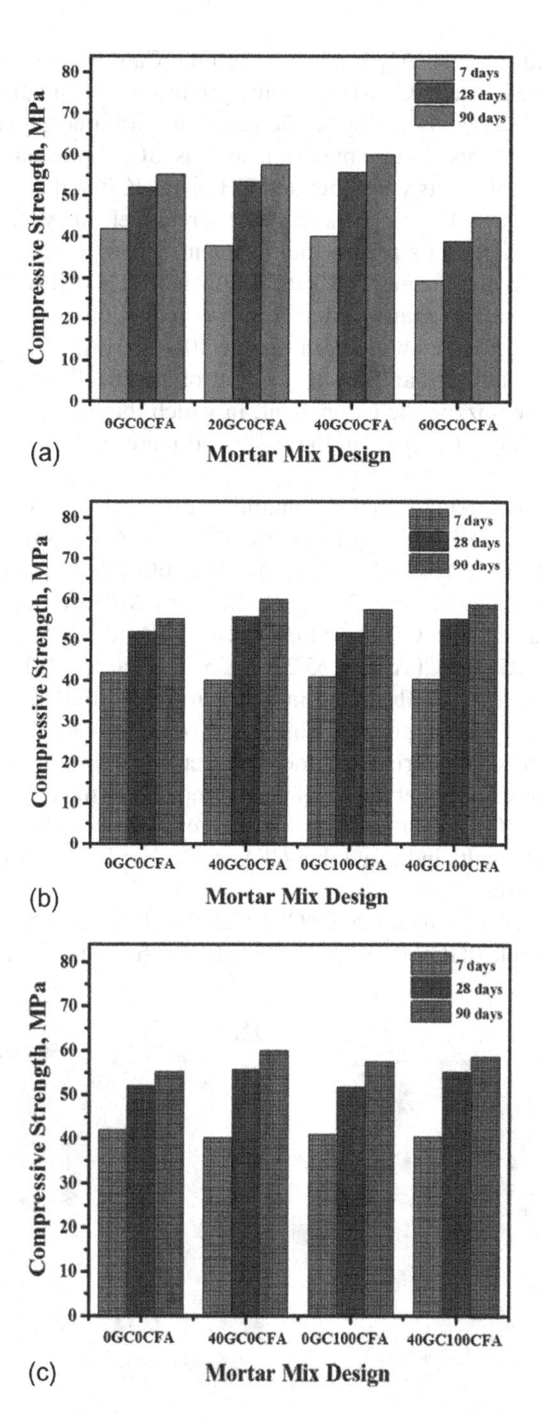

FIGURE 4.2 The compressive strength of the samples (a) with GC, (b) with CFA, and (c) 40GC100CFA compared to others.

$(Ca(OH)_2)$. In addition, the Al_2O_3 reacted with the $Ca(OH)_2$ and H_2O to generate calcium aluminate hydrate (C-A-H) gels and produced calcium aluminate silicate hydrate (C-A-S-H) crystals owing to the reaction with amorphous silicate [23]. Nevertheless, when a pozzolanic material, such as GC, is present, its reactive silica (SiO_2) constituent reacts with liberated CH in an HCP in the presence of water to form secondary C-S-H gels. This reaction is relatively slow, resulting in a low rate of heat liberation and strength improvement. However, sample 60GC0CFA showed a lower CS than samples 20GC0CFA and 40GC0CFA. This was likely due to the dilution effect that resulted from the reduction in the cement content [24]. The best results were obtained in sample 40GC0CFA, which had a CS at 90 days that was 7% higher than that of the control sample. This was attributed to the capacity of the pozzolanic component, in which the incorporation of the GC reduced the amount of $Ca(OH)_2$ in the paste and increased the amount of C-S-H gel generated.

The CSs of the mortar samples containing different percentages of CFA as an RS replacement are illustrated in Figure 4.2b. The CS of samples 0GC0CFA, 0GC25CFA, 0GC50CFA, 0GC75CFA, and 0GC100CFA at 7 days of age were 40.6 MPa, 39.4 MPa, 40.1 MPa, 38.8 MPa, and 41.1 MPa, respectively. However, at 90 days of curing, the CS of samples 0GC0CFA, 0GC25CFA, 0GC50CFA, 0GC75CFA, and 0GC100CFA were 55.2 MPa, 55.6 MPa, 53.7 MPa, 54.9 MPa, and 57.4 MPa, respectively. As illustrated in the figure, the difference in the CS of all the samples was not very significant. This may be due to the fact that the size distribution and physical characteristics of the CFA were very similar to the RS. Sample 0GC100CFA showed a higher CS at all ages compared to the other samples. At 90 days, sample 0GC100CFA showed an increase of 28% from its 7 days' strength. The experimental results indicated that CFA can be used to replace natural aggregates in mortar mixes.

The failure pattern of samples 0GC0CFA and 0GC100CFA is shown in Figure 4.3. The failure of sample 0GC100CFA was attributed to the collapse of the binders and

FIGURE 4.3 Failure pattern of the mortars after compression testing for (a) 0GC0CFA and (b) 40GC100CFA.

fine aggregates. Conversely, the failure of the other samples occurred because of breakages in the fine aggregates. It can be seen that the addition of ceramic particles enhanced the strength of the mortar samples by preventing sudden failure of the specimens. This could be due to the considerable decrease in the size of the ceramic waste aggregate due to the crushing of the larger sized particles, which resulted in the enhanced performance of the mortar samples [25].

Figure 4.2c displays the measured CS values for samples 0GC0CFA and 40GC100CFA, which were 41.8 MPa and 40.3 MPa, respectively, at 7 days of curing. As the curing time increased, the CS of sample 40GC100CFA became higher than that of sample 0GC0CFA. For instance, at 180 days of curing, the CS of sample 40GC100CFA (60.5 MPa) was approximately 8% more than that of sample 0GC0CFA (56.2 MPa). This enhancement in the CS of sample 40GC100CFA was attributed to the pozzolanic reaction that occurred owing to the large amount of SiO_2 in the GC and the hydration products of the OPC, including $Ca(OH)_2$. The reactions between the SiO_2 and $Ca(OH)_2$ generated an excess of C-S-H gel in the 40GC100CFA sample. The development of such additional C-S-H gels led to a reduction in the porosity, thereby causing a higher CS at longer curing times. Matias et al. [17] made a similar observation, wherein the partial substitution of cement by recycled ceramic particles and fine aggregates was shown to considerably enhance the CS of the studied mortar and concrete.

The addition of fine particles, such as GC, into mortar and concrete mixtures generally densifies the matrix by filling the cavities and reducing the porosity. In addition, owing to the lower loose bulk density (LBD) of the ceramic fine aggregates (0.94 g/cm^3) compared to sand (1.43 g/cm^3), the replaced volume of ceramic powder would be higher. Consequently, the matrix is made denser and results in higher strength values [25].

4.4.2 XRD Patterns

Figure 4.4 shows the XRD results for samples 0GC0CFA, 20GC0CFA, 40GC0CFA, and 60GC0CFA after 28 days. The presence of a $Ca(OH)_2$ crystalline phase revealed peaks at approximately 18.05°, 34.11°, and 47.13°. Moreover, the intensity of these peaks was reduced with an increase in the WCP content. The peak intensity for the SiO_2 (quartz) mineral was enhanced because of the WCP replacement. This observation was ascribed to the occurrence of the pozzolanic reaction of SiO_2 with $Ca(OH)_2$ from the cement hydration. The occurrence of a weak peak corresponding to the $Ca(OH)_2$ crystals in sample 60GC0CFA was likely due to the availability of a lower quantity of $Ca(OH)_2$ for the reaction with the excess SiO_2. It can also be seen that the intensity of the CS was reduced in the mixes with a higher percentage of GC [25]. Furthermore, the intensity of the C_3S/C_2S peak decreased with an increase in the GC level of the mortars, which was ascribed to the lower content of OPC within the mixes. The characteristic peak at 21.50° was due to the existence of gypsum minerals in samples 0GC0CFA, 20GC0CFA, 40GC0CFA, and 60GC0CFA. This suggested that the decomposition of the C-S-H phase in the binder had supplied the calcium required for the formation of gypsum [26].

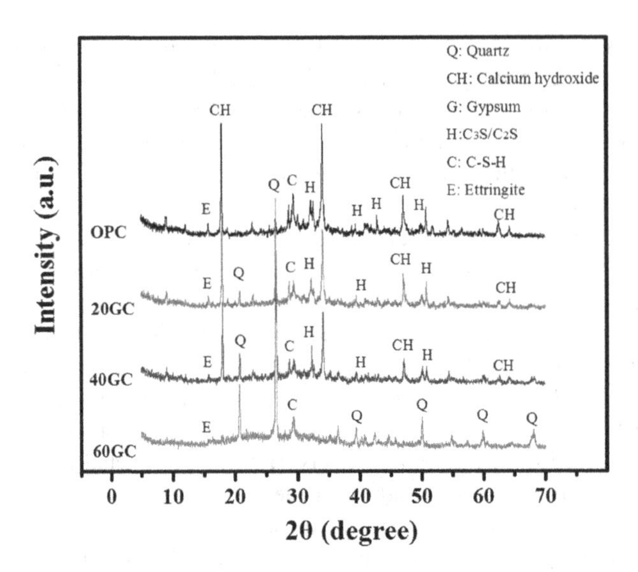

FIGURE 4.4 XRD patterns for samples 0GC0CFA, 20GC0CFA, 40GC0CFA, and 60GC0CFA.

4.4.3 SEM Images

Figure 4.5 displays the SEM micrographs for all of the mortar samples after a curing time of 28 days. The SEM image analyses provided a better understanding of the effects of WCP inclusion on the mortar hydration process. Sample 20GC0CFA revealed a larger amount of crystalline $Ca(OH)_2$ in hexagonal plate-like structures. The hydration of the cement produced an elevated amount of $Ca(OH)_2$ crystals because of the presence of an excess amount of CaO in the OPC, which acted as the source of the higher CS of the samples. However, the low number of $Ca(OH)_2$ crystals in samples 40GC0CFA and 60GC0CFA was caused by the higher percentage of cement substitution. This led to a decrease in the level of C_3S and C_2S in the mixture. The observed low CS for sample 60GC0CFA was attributed to an insufficient amount of $Ca(OH)_2$ to complete the reaction. Additionally, the formation of a larger amount of C-S-H crystals in sample 40GC0CFA made it denser, which was ascribed to the pozzolanic reaction of SiO_2 with $Ca(OH)_2$ during hydration of the cement. Consequently, these dense microstructures were responsible for enhancing the CS at the final curing age. The generated C-S-H gels contained crystals of varying shapes, such as fine granules, fibrous, massive, decorated and plumose, as seen in the SEM images and detected in the energy dispersive spectroscopy (EDS) spectral peaks (Figure 4.6). The EDS spectral analyses verified the existence of the C-S-H crystalline phases in the mortar samples with very high contents of Ca and Si. In fact, such crystals were responsible for the higher CS in the proposed cement paste [27].

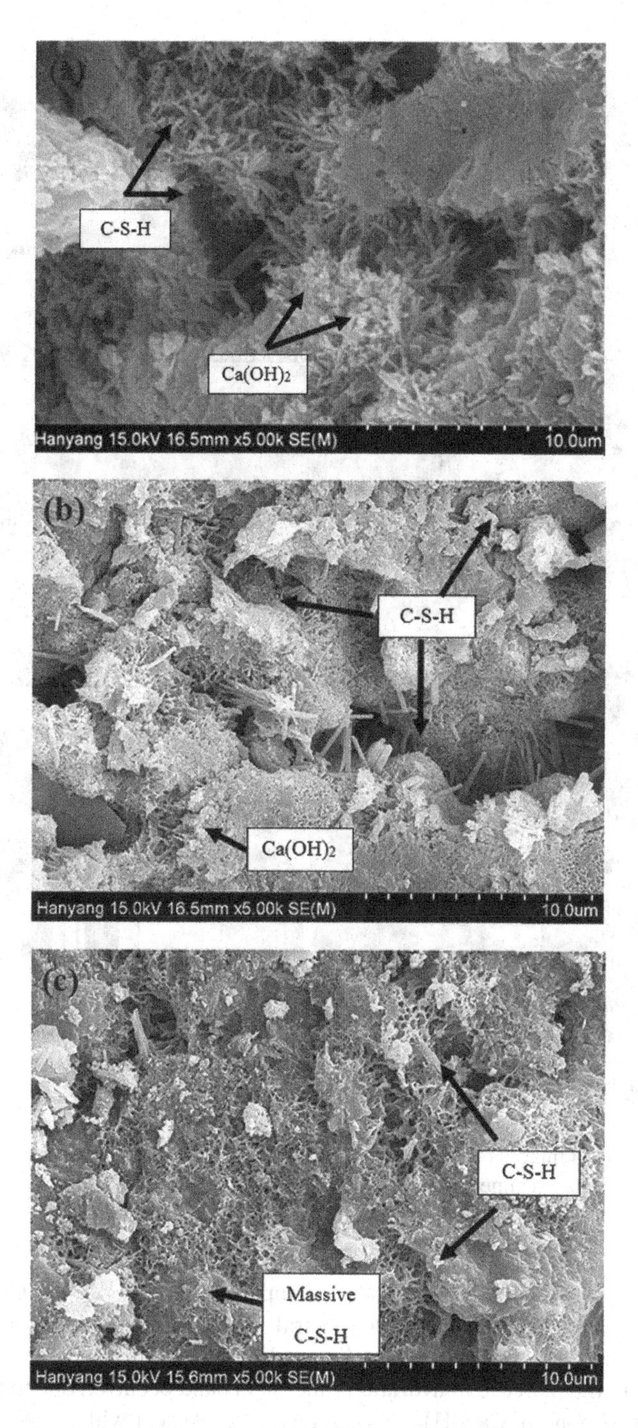

FIGURE 4.5 SEM images for samples (a) 20GC0CFA, (b) 40GC0CFA, and (c) 60GC0CFA.

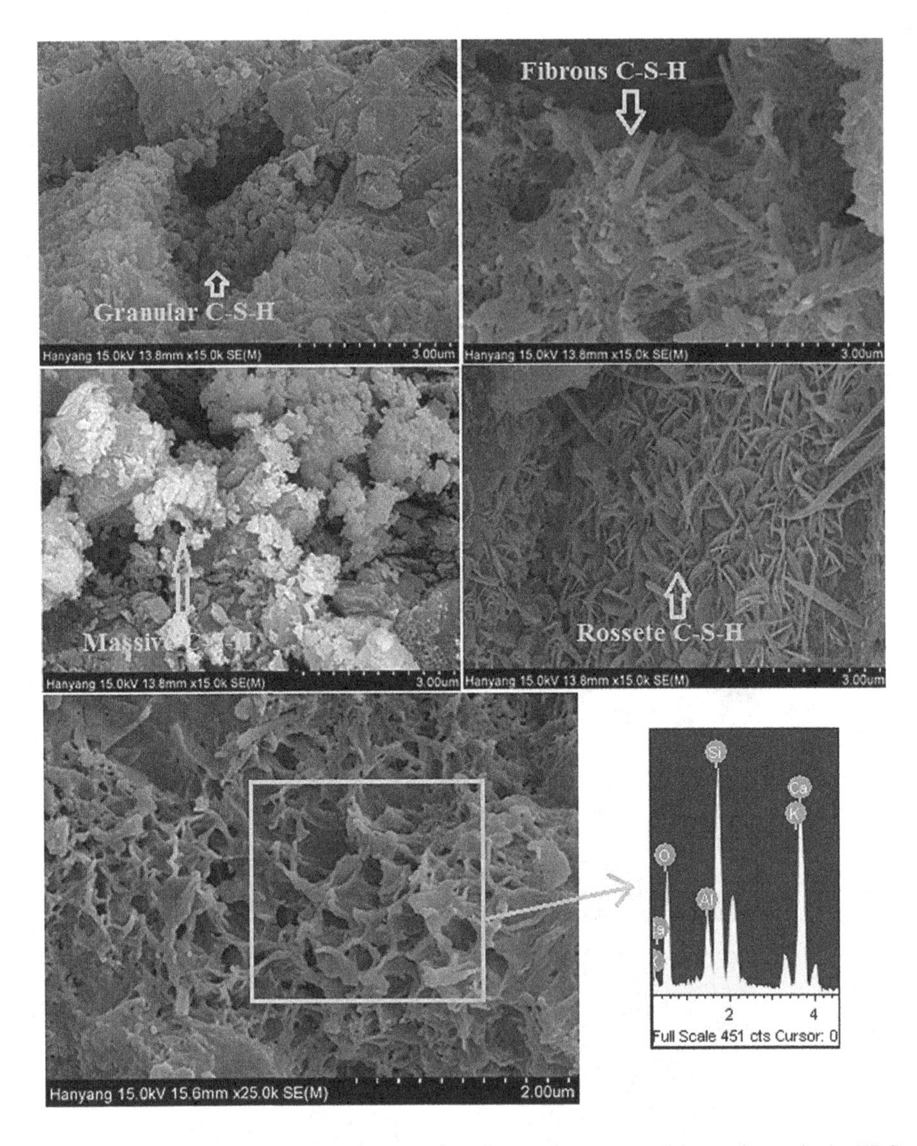

FIGURE 4.6 SEM images of the various C-S-H crystal shapes. The inset picture is the EDS spectra of the selected scanned area for the elemental analysis.

4.4.4 FTIR Spectral Analysis

Figure 4.7 shows the FTIR spectra of samples 20GC0CFA, 40GC0CFA, and 60GC0CFA, which had very similar spectral profiles. In the higher wavenumber region, a prominent vibration band occurred at 3645 cm^{-1}, where the intensity increased with the increase in hydration time. This band occurred due to the stretching of the O–H bond in $Ca(OH)_2$. This result was consistent with the XRD analysis for sample 20GC0CFA, which had a maximum $Ca(OH)_2$ content. The broad vibration

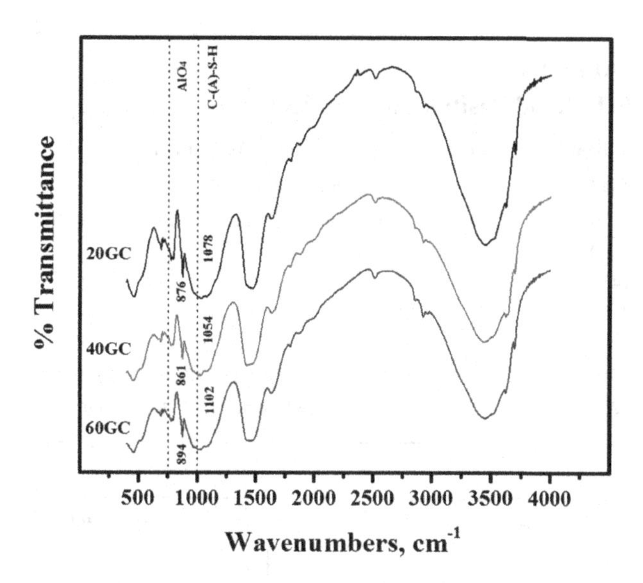

FIGURE 4.7 FTIR spectra for samples 20GC0CFA, 40GC0CFA, and 60GC0CFA.

band in the range of 3100–3400 cm^{-1} was due to the symmetric and asymmetric stretching of the O-H bond present in water. The appearance of a band at around 1650 cm^{-1} was due to the H-O-H distortion mode in the water molecules absorbed by the samples. The appearance of bands at approximately 1480 cm^{-1}, 875 cm^{-1}, and 712 cm^{-1} was due to the vibration of CO_3^{-2}. The existence of $CaCO_3$ in the samples was due to the absorption of atmospheric CO_2 during the curing process by the high content of OPC. The dislocation of the Si-O stretching mode from 925 cm^{-1} (present in the non-hydrated cement) to 1040 cm^{-1} (existed in the hydrated cement) was an interesting observation that was ascribed to the polymerization of the SiO_4^{4-} units that occurred in the C_3S and C_2S. At 1040 cm^{-1}, sample 40GC0CFA displayed higher absorbance compared with the others. This was due to the addition of the appropriate percentage of GC that, in part, produced higher strength compared with the others, which were similar to the CS results. Therefore, adding more GC reduced the CS due to the presence of excessive unreacted SiO_2 [28]. Table 4.4 summarizes the FTIR band positions and assignments identified in the proposed mortar samples.

4.4.5 TGA Thermograms

Figure 4.8 illustrates the thermogravimetric analysis (TGA) curves of samples 0GC0CFA, 20GC0CFA, 40GC0CFA, and 60GC0CFA, and shows the loss of mass experienced by each sample in the range of 30 to 1000°C. The mass loss was classified into four main categories, which were the evaporation of water moisture around 100 ± 5°C, C-S-H crystal dehydration in the range of 100 to 300°C, dehydration of Ca(OH)2 between 450 and 550°C and the development of CaCO3 decarbonization in the range of 600 to 850°C [29]. A rapid loss of mass was observed in all samples

TABLE 4.4
FTIR Band Positions and Assignments

Wavenumber (cm^{-1})	Band Assignment
3645	Stretching O-H of $Ca(OH)_2$
3100–3400	Symmetric and asymmetric stretching of O-H
1650	Deformation H-O-H
1480	CO_3^{2-}
875	CO_3^{2-}
712	CO_3^{2-}
1040	Stretching Si-O (in a polymeric unit of SiO_4^{4-})
925	Stretching Si-O (in non-hydrated cement)

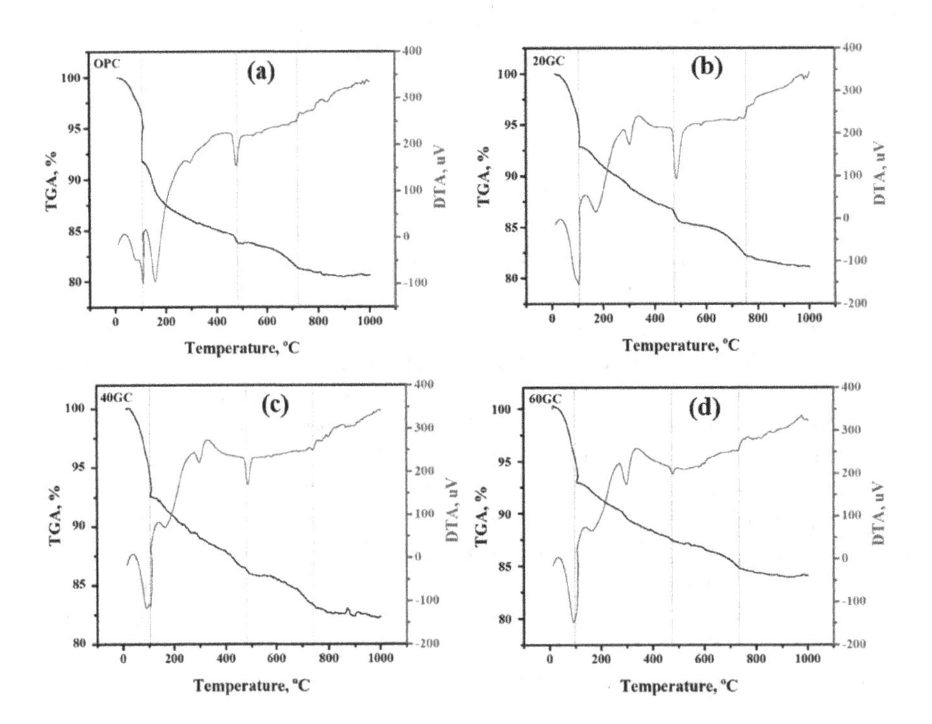

FIGURE 4.8 TGA curves for samples (a) 0GC0CFA, (b) 20GC0CFA, (c) 40GC0CFA, and (d) 60GC0CFA.

around 100 ± 5°C, which was related to the evaporation of moisture (i.e. free water molecules). The water that was not used during the cement hydration and mixing process contributed to the workability of the mixture. Weakly bonded water particles approximately 5 nm in diameter and greater existed inside the sample's pores. The mass loss due to moisture evaporation at 100 ± 5°C was enhanced with an increase in the GC substitution percentage. The mean weight (%) for samples 0GC0CFA,

20GC0CFA, 40GC0CFA, and 60GC0CFA that remained after exposure to 850°C was 80%, 81%, 82%, and 84%, respectively. A comparison of the differential thermal analysis (DTA) results with the TGA data revealed they were consistent (Figure 4.8).

The dehydration process of the $Ca(OH)_2$ and the evaporation of the H_2O in this process followed the reaction pathway [29]

$$Ca(OH)_2\ (s) \rightarrow CaO\ (s) + H_2O(g)$$

The mass loss of the samples due to the $CaCO_3$ decarbonization at its melting temperature of 825°C followed the reaction path [30]

$$CaCO_3\ (s) \rightarrow CaO\ (s) + CO_2(g)$$

4.5 RESISTANCE TO SULPHATE ATTACK

The effects of varying amounts of WCP and fine ceramic aggregates on the working performance and durability of the mortar samples were tested by exposing them to sulphate solutions. The resistance of the mortar samples to sulphate was obtained by visual inspections of the specimens (i.e. colours and physical conditions), loss of mass, residual CS and a microstructural analysis. Table 4.5 outlines the general properties of the mortar samples that were subjected to a 5% Na_2SO_4 solution. Alterations in the sample's dimensions and/or conditions were observed and recorded. No alterations were observed on the surface of sample 40GC100CFA. However, sample 0GC0CFA displayed considerable surface deformation where the formation of gypsum was found because of the sulphate attack. This was attributed to the chemical reaction between the sulphate solution and the $Ca(OH)_2$ present in the cement. In contrast, the amount of gypsum formed in sample 40GC100CFA due to sulphate attack was somewhat lower, which indicated the existence of a lower amount of $Ca(OH)_2$. As the main products of sulphuric acid and cement particles are gypsums, expansion of the mortar samples was also observed. This, in turn, generated cracks and caused spalling in the samples. In sample 40GC100CFA, the $Ca(OH)_2$ was used in the hydration process to form excess C-S-H gels that were relatively stable against an acid attack. Additional reactions of the gypsum with other particles in the cement, including calcium aluminate, might have formed ettringite, thereby increasing the volume of the samples and generating excess cracks [31].

TABLE 4.5
Description of the Physical Appearance of the Specimens after 18 Months of Exposure to a 5% Na_2SO_4 Solution

Mix	Surface Texture	Size	Colour	Edge	Shape	Mass Change (%)	Strength Loss (%)
0GC0CFA	Deteriorated	Increased	Whitish	Cracks	Distorted	+2.9	41.1
40GC100CFA	Smooth	Unchanged	Greyish	Perfect	Perfect	–5.1	16.8

Figure 4.9a shows the exposure time (18 months) dependent variation of the mass and residual CS of samples 0GC0CFA and 40GC100CFA exposed to a 5% Na_2SO_4 solution. After three months of exposure, the mass of samples 0GC0CFA and 40GC100CFA compared to their initial masses had decreased by approximately 0.7% and increased by 1.7%, respectively. Early in the immersion period, both samples (0GC0CFA and 40GC100CFA) showed an increase in their mass. After two months of immersion, the mass of sample 0GC0CFA had reduced by almost 5.1% due to the accelerated sulphate attack, which led to the generation of several products as a result of the chemical reaction between the Na_2SO_4 and the cement hydrates. This reaction caused the expansion of the entire specimen, and was ascribed to the

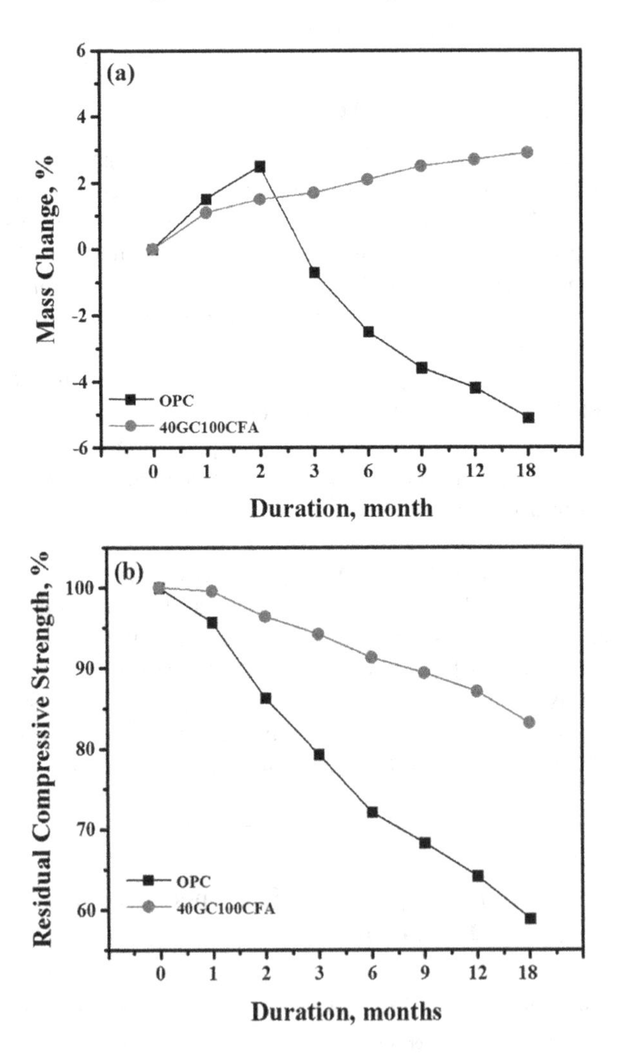

FIGURE 4.9 Exposure time–dependent variation of the (a) mass and (b) residual compressive strength of samples 0GC0CFA and 40GC100CFA exposed to a 5% Na_2SO_4 solution.

generation of various products from the sulphate reactions, including ettringite and gypsum, and led to cracking of the specimen. Meanwhile, the mass gain of sample 40GC100CFA was attributed to the sulphate reaction that generated more gypsum by absorbing the Na_2SO_4 solution. This chemical reaction resulted in an enhancement of the total volume of the specimen via expansion. However, at longer exposure ages, sample 40GC100CFA continued to gain mass, while sample 0GC0CFA lost its shape due to expansion, cracking, and spalling [32].

Figure 4.9b shows the residual CS of samples 0GC0CFA and 40GC100CFA under sulphate solution exposure after 18 months. Both samples 0GC0CFA and 40GC100CFA lost mass due to continuous exposure to the sulphate solution. In addition, the residual CS of the specimens decreased as the immersion time increased. At the end of 18 months of immersion, the residual strength of samples 0GC0CFA and 40GC100CFA had reduced by 41.1% and 16.8%, respectively (Table 4.5). This considerable decrease in the CS was ascribed to the chemical reaction between the Na_2SO_4 solution and $Ca(OH)_2$, which led to the degradation of the hydration products of the cement. Consequently, the CS was remarkably reduced.

The mortar samples designed with WCP and CFA showed a lower rate of CS loss, which was due to the slow pozzolanic reaction of the WCP and $Ca(OH)_2$ that generated the extra C-S-H gels. The formation of extra hydrates enhanced the density of the mortar matrix and reconfigured the pore distributions and structures. Additionally, the use of WCP as replacement for cement not only reduced the quantity of $Ca(OH)_2$ in the mortar matrix, but allowed the residual WCP (i.e. not reacted) to act as a filler. This reduced the porous spaces among the aggregates and binders, leading to denser microstructures. The substitution of WCP and fine ceramic aggregates strengthened the mortar samples and imparted sufficient resistance against a sulphate attack, leading to an enhancement in the performance of the mortar by reducing the CS loss. The generation of gypsum in the mortar samples resulted in their deterioration, whereby the CS was decreased, and the specimens expanded. Indeed, the creation of gypsum was also considered in the early stages of ettringite formation, as it is the primary cause of the deterioration of concrete and mortar subject to attack by a sulphate solution [33].

Figure 4.10 displays the SEM images and EDS spectra of samples 0GC0CFA and 40GC100CFA, respectively. The creation of ettringite was higher in sample 0GC0CFA than 40GC100CFA, which may be due to the production of higher quantities of $Ca(OH)_2$ in the hydration process of the cement. The EDS spectra of sample 0GC0CFA detected the presence of gypsum and ettringite in large amounts by revealing intense Ca, O, Al and S peaks. The EDS spectra of sample 40GC100CFA showed O, Si, and Ca peaks of different intensities, which indicated the formation of excess C-S-H gels. This was due to the presence of excess SiO_2 in the GC that produced enhanced performance of the mortar samples subject to sulphate exposure [24].

Figure 4.11 shows the XRD patterns for samples 0GC0CFA and 40GC100CFA both before and after exposure to the Na_2SO_4 solution. This observation was consistent with the SEM and EDS results. Prior to exposure, the intensity of the $Ca(OH)_2$ peak for sample 0GC0CFA was higher, indicating the formation of a higher level of gypsum and thereby producing ettringite and thaumasite because of the exposure.

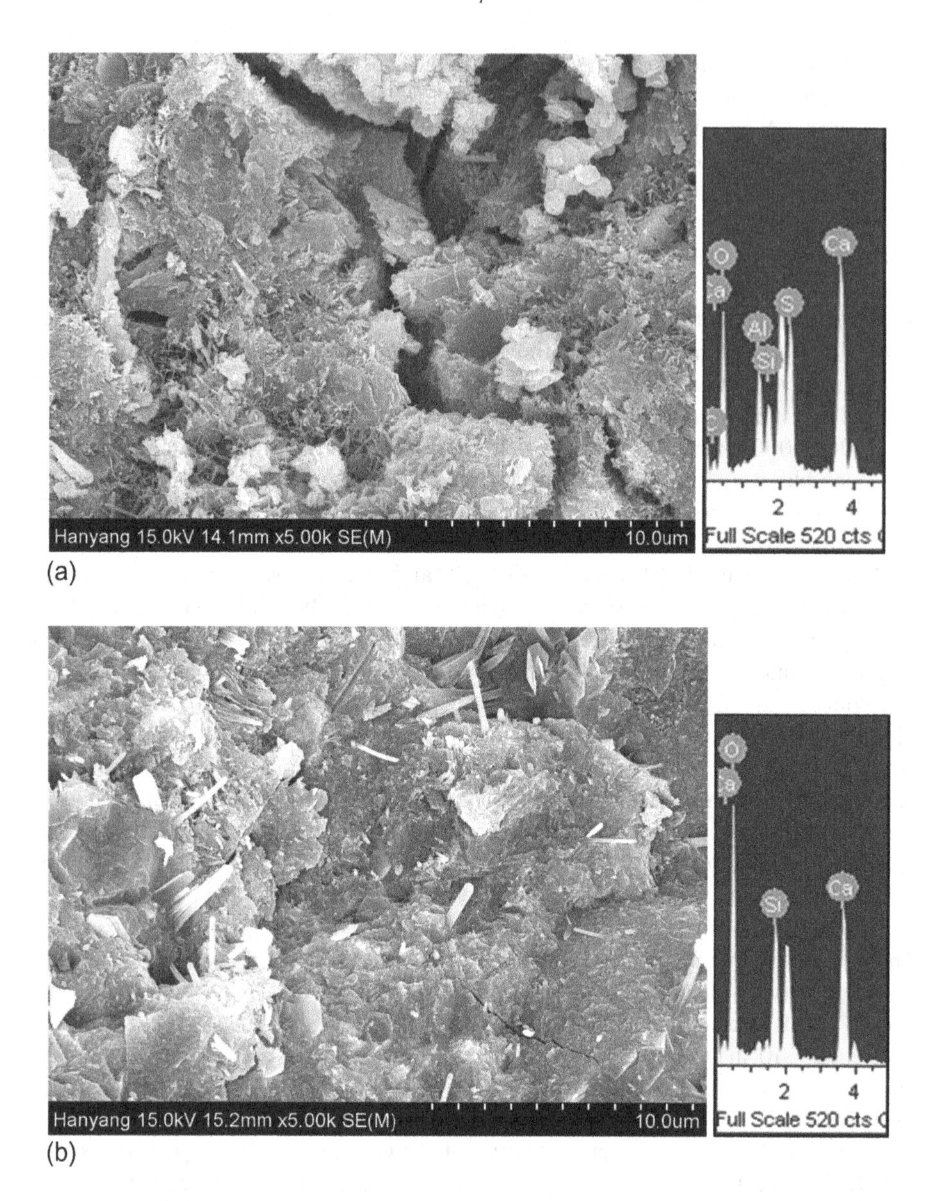

FIGURE 4.10 SEM image and EDS spectra for samples (a) 0GC0CFA and (b) 40GC100CFA under sulphate solution exposure.

Together with the other calcium silicates, thaumasite was also formed during cement hydration, mainly due to sulphate exposure. The peaks relating the separation of thaumasite and ettringite were higher at 2θ (Figure 4.11a). The sulphate attacks on the mortar samples normally occurred due to the formation of gypsum and ettringite crystallites. Thus, the increase in the XRD peak intensities for these minerals represents the weak resistance and deformation capacity of the samples against sulphate attacks. Conversely, the lower level of $Ca(OH)_2$ in sample 40GC100CFA led

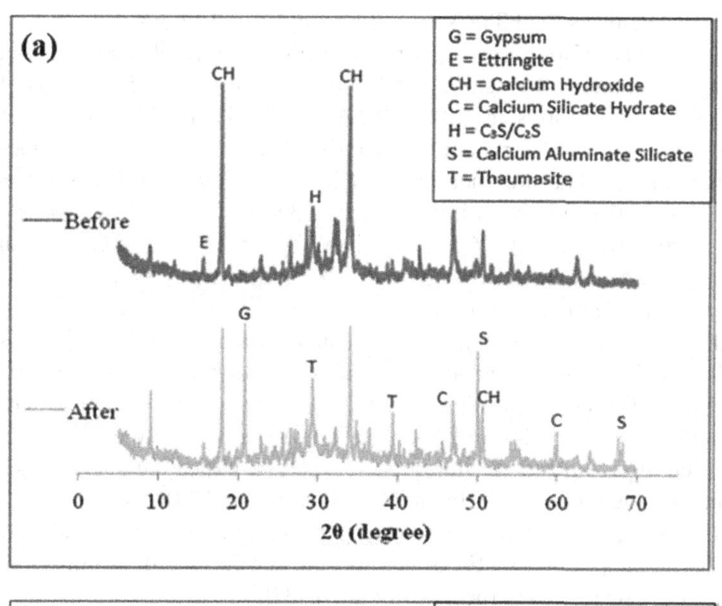

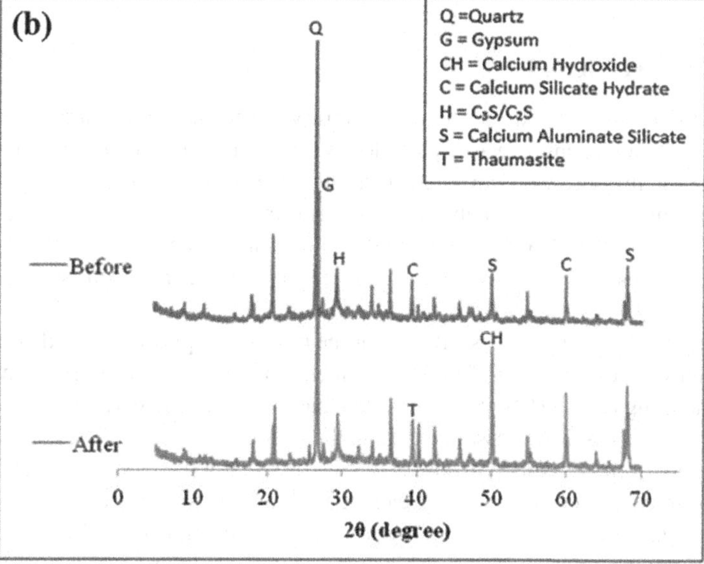

FIGURE 4.11 The XRD patterns for samples (a) 0GC0CFA and (b) 40GC100CFA under sulphate solution exposure.

to a lower production of thaumasite and ettringite compared to sample 0GC0CFA. Additionally, the intense peaks of C-S-H gel observed at 39° and 60° indicated that a strong reaction of the silica in the GC with the excess $Ca(OH)_2$ present in the hydration of the OPC occurred [34]. Figure 4.11b revealed an intense peak for quartz in sample 40GC100CFA even after 18 months of sulphate exposure, which suggested the existence of a large quantity of crystalline silica in the mortar sample.

Figure 4.12 displays the FTIR spectra for samples 0GC0CFA and 40GC100CFA subject to Na_2SO_4 solution exposure, respectively. Several vibration bands can be seen at approximately 3645 cm^{-1}, 3450 cm^{-1}, 1645 cm^{-1}, 1425 cm^{-1}, 1110 cm^{-1}, 985 cm^{-1}, and 470 cm^{-1} for sample 0GC0CFA, and 3450 cm^{-1}, 1645 cm^{-1}, 1425 cm^{-1}, 1015 cm^{-1}, and 463 cm^{-1} for sample 40GC100CFA before they were subjected to the sulphate attack. The presence of water in the mortar matrix was detected by the stretching of the O-H band in the range of 3200–3700 cm^{-1}, which originated from the decomposition of $Ca(OH)_2$. The bending vibration of H-O-H occurred at 1645 cm^{-1} and was attributed to the bound water in the hydration products. In addition, the appearance of bands in the region of 1005–1035 cm^{-1} was due to the presence of the crystalline phases of quartz in the mortar samples [32]. The FTIR spectra of sample 0GC0CFA before and after sulphate exposure were considerably different (Figure 4.12a). The decomposition of the C-S-H crystallites in sample 0GC0CFA was related to the band shifts at 3645 cm^{-1}, 1015 cm^{-1}, 786 cm^{-1}, and 463 cm^{-1} that occurred after exposure to the sulphate solution, indicating a structural alteration in the designed mortar specimen. Nevertheless, the FTIR spectrum for sample 40GC100CFA mortar after exposure to the sulphate solution showed a slight change. The results of sample 0GC0CFA after exposure showed the decomposition of the C-S-H and O-H structures in the mortar matrix [32].

4.6 CONCLUSIONS

In the present study, the strength and durability properties of a sustainable mortar mixture utilizing ceramic waste particles were evaluated, and the environmental benefits of the mixture were revealed. The utilization of waste ceramic particles was shown to contribute to sustainable development and a cleaner environment by producing a green mortar from the recycling of industrial wastes. Based on the detailed characterizations and experimental results, the following conclusions were drawn:

i. Early in the curing process, the CS of the mortar samples was slightly diminished with the addition of GC as a cement replacement. However, for mixes containing GC, the CS was higher than that of the OPC mixes at the end of the curing process. It was asserted that, in the later curing periods, the pozzolanic activity of the GC aided in enhancing the strength of the mortar mixtures.
ii. The inclusion of ceramic particles as both a cement replacement and fine aggregate in the mortar was highly effective in improving the durability performance, with a considerable increase in the sulphate resistance capacity of the mortar samples.
iii. The microstructural analyses of the mortar mixes revealed an enhancement in the matrix microstructure through the formation of additional C-S-H gels. In addition, the number of pores and microcracks were significantly reduced in the ceramic mortar mixes. The voids were filled with hydration products due to the high pozzolanic activity of the GC, particularly in the later stages of the curing process.

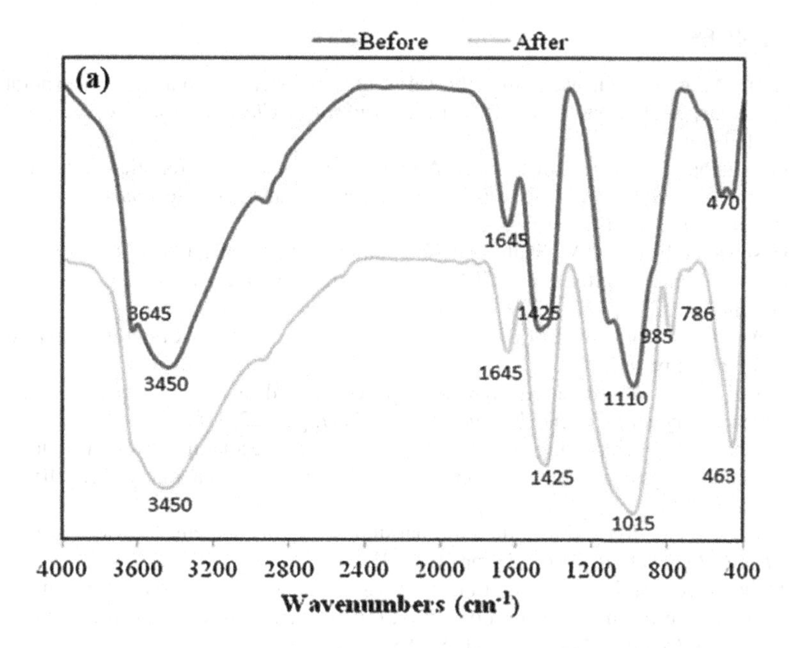

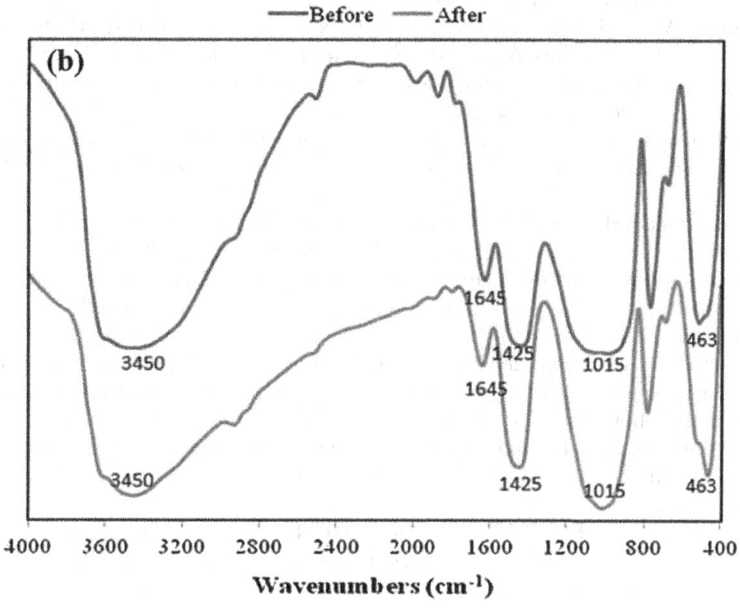

FIGURE 4.12 FTIR spectra of samples (a) 0GC0CFA and (b) 40GC100CFA under sulphate solution exposure.

REFERENCES

1. Awal, A. A. and H. Mohammadhosseini, Green concrete production incorporating waste carpet fiber and palm oil fuel ash. *Journal of Cleaner Production*, 2016. **137**: pp. 157–166.
2. Hemalatha, T. and A. Ramaswamy, A review on fly ash characteristics–Towards promoting high volume utilization in developing sustainable concrete. *Journal of Cleaner Production*, 2017. **147**: pp. 546–559.
3. Huseien, G. F. and K. W. Shah, Durability and life cycle evaluation of self-compacting concrete containing fly ash as GBFS replacement with alkali activation. *Construction and Building Materials*, 2020. **235**: pp. 117458.
4. Meyer, C., The greening of the concrete industry. *Cement and Concrete Composites*, 2009. **31**(8): pp. 601–605.
5. Mo, K. H., et al., Green concrete partially comprised of farming waste residues: A review. *Journal of Cleaner Production*, 2016. **117**: pp. 122–138.
6. Mahpour, A., Prioritizing barriers to adopt circular economy in construction and demolition waste management. *Resources, Conservation and Recycling*, 2018. **134**: pp. 216–227.
7. Rana, A., et al., Recycling of dimensional stone waste in concrete: A review. *Journal of Cleaner Production*, 2016. **135**: pp. 312–331.
8. Mohammadhosseini, H., et al., Enhanced performance for aggressive environments of green concrete composites reinforced with waste carpet fibers and palm oil fuel ash. *Journal of Cleaner Production*, 2018. **185**: pp. 252–265.
9. Samadi, M., et al., Waste ceramic as low cost and eco-friendly materials in the production of sustainable mortars. *Journal of Cleaner Production*, 2020: p. 121825.
10. Pacheco-Torgal, F. and S. Jalali, Reusing ceramic wastes in concrete. *Construction and Building Materials*, 2010. **24**(5): pp. 832–838.
11. Higashiyama, H., et al., Efficiency of ground granulated blast-furnace slag replacement in ceramic waste aggregate mortar. *Cement and Concrete Composites*, 2014. **49**: pp. 43–49.
12. Higashiyama, H., et al., Chloride ion penetration into mortar containing ceramic waste aggregate. *Construction and Building Materials*, 2012. **33**: pp. 48–54.
13. Suzuki, M., M. S. Meddah, and R. Sato, Use of porous ceramic waste aggregates for internal curing of high-performance concrete. *Cement and Concrete Research*, 2009. **39**(5): pp. 373–381.
14. Zegardło, B., M. Szeląg, and P. Ogrodnik, Ultra-high strength concrete made with recycled aggregate from sanitary ceramic wastes—The method of production and the interfacial transition zone. *Construction and Building Materials*, 2016. **122**: pp. 736–742.
15. Anderson, D. J., S. T. Smith, and F. T. Au, Mechanical properties of concrete utilising waste ceramic as coarse aggregate. *Construction and Building Materials*, 2016. **117**: pp. 20–28.
16. Senthamarai, R., P. D. Manoharan, and D. Gobinath, Concrete made from ceramic industry waste: Durability properties. *Construction and Building Materials*, 2011. **25**(5): pp. 2413–2419.
17. Matias, G., P. Faria, and I. Torres, Lime mortars with heat treated clays and ceramic waste: A review. *Construction and Building Materials*, 2014. **73**: pp. 125–136.
18. Higashiyama, H., et al., Compressive strength and resistance to chloride penetration of mortars using ceramic waste as fine aggregate. *Construction and Building Materials*, 2012. **26**(1): pp. 96–101.

19. Zivica, V.R. and A. Bajza, Acidic attack of cement based materials—A review: Part 1. Principle of acidic attack. *Construction and Building Materials*, 2001. **15**(8): pp. 331–340.
20. Frías, M., et al., Properties of calcined clay waste and its influence on blended cement behavior. *Journal of the American Ceramic Society*, 2008. **91**(4): pp. 1226–1230.
21. Kim, H.-J., et al., Durability and purification performance of concrete impregnated with silicate and sprayed with photocatalytic TiO2. *Construction and Building Materials*, 2019. **199**: pp. 106–114.
22. Tangchirapat, W., S. Khamklai, and C. Jaturapitakkul, Use of ground palm oil fuel ash to improve strength, sulfate resistance, and water permeability of concrete containing high amount of recycled concrete aggregates. *Materials & Design*, 2012. **41**: pp. 150–157.
23. Mohammadhosseini, H. and J. M. Yatim, Microstructure and residual properties of green concrete composites incorporating waste carpet fibers and palm oil fuel ash at elevated temperatures. *Journal of Cleaner Production*, 2017. **144**: pp. 8–21.
24. Lim, N. H. A. S., et al., The effects of high volume nano palm oil fuel ash on microstructure properties and hydration temperature of mortar. *Construction and Building Materials*, 2015. **93**: pp. 29–34.
25. Mohammadhosseini, H., et al., Enhanced performance of green mortar comprising high volume of ceramic waste in aggressive environments. *Construction and Building Materials*, 2019. **212**: pp. 607–617.
26. Chen, Y., et al., Resistance of concrete against combined attack of chloride and sulfate under drying–wetting cycles. *Construction and Building Materials*, 2016. **106**: pp. 650–658.
27. Lavat, A. E., M. A. Trezza, and M. Poggi, Characterization of ceramic roof tile wastes as pozzolanic admixture. *Waste Management*, 2009. **29**(5): pp. 1666–1674.
28. Mijarsh, M., M. M. Johari, and Z. Ahmad, Synthesis of geopolymer from large amounts of treated palm oil fuel ash: application of the Taguchi method in investigating the main parameters affecting compressive strength. *Construction and Building Materials*, 2014. **52**: pp. 473–481.
29. Vedalakshmi, R., et al., Quantification of hydrated cement products of blended cements in low and medium strength concrete using TG and DTA technique. *Thermochimica Acta*, 2003. **407**(1–2): pp. 49–60.
30. Brown, R. C. and J. Dykstra, Systematic errors in the use of loss-on-ignition to measure unburned carbon in fly ash. Fuel, 1995. **74**(4): pp. 570–574.
31. Thomas, J. J., et al., Modeling and simulation of cement hydration kinetics and microstructure development. *Cement and Concrete Research*, 2011. **41**(12): pp. 1257–1278.
32. Maes, M. and N. De Belie, Influence of chlorides on magnesium sulphate attack for mortars with Portland cement and slag based binders. *Construction and Building Materials*, 2017. **155**: pp. 630–642.
33. Cheng, Y., et al., Test research on effects of ceramic polishing powder on carbonation and sulphate-corrosion resistance of concrete. *Construction and Building Materials*, 2014. **55**: pp. 440–446.
34. Chindaprasirt, P., C. Jaturapitakkul, and T. Sinsiri, Effect of fly ash fineness on microstructure of blended cement paste. *Construction and Building Materials*, 2007. **21**(7): pp. 1534–1541.

5 Properties of Ceramic Waste-Based Alkali-Activated Mortars

5.1 INTRODUCTION

Over the years, ordinary Portland cement (OPC) has been widely employed as a concrete binder for various building materials worldwide. It is known that manufacturing of OPC causes serious pollution in the environment in terms of considerable amounts of greenhouse gas (GHG) emissions [1–4]. The OPC production alone is accountable for nearly 6 to 7% of total carbon dioxide (CO_2) emissions as estimated by the International Energy Agency (IEA) [5]. In fact, among all the greenhouse gases approximately 65% of the global warming is ascribed to CO_2 emission. It was predicted that the mean temperature of the globe could rise by approximately 1.4–5.8°C over the next 100 years [6]. Globally, in the present backdrop of CO_2 emissions–mediated climate change, the sea level is expected to rise and the frequent occurrence of natural disasters will cause huge economic loss [7–9]. Also, the greenhouse gases emitted from the cement manufacturing industries, including CO_2, SO_3, and NO_X, can cause acid rain and damage the soil fertility [10, 11]. Generally, the industrial consumption of raw materials is around 1.5 tonnes for each tonne of OPC production [12–14]. To surmount such problems, all scientists, engineers and industrial personnel have been continuously dedicating many efforts to develop novel construction materials to achieve alternative binders [15–17].

It is worth noting that millions of tons of natural, industrial and agricultural wastes such as fly ash (FA); coal- and oil-burning by-products; bottom ash; palm oil fuel ash (POFA); bagasse ash (BA); used tyres; dust from cement, marble, and crushed stone; and ceramic waste material are dumped every year in Malaysia. These waste materials cause severe ecological setbacks such as air contamination and leach out of hazardous substances. Several studies [18–21] revealed that many of these wastes may be potentially recycled in the form of innovative concrete materials as an alternative to OPC (often as much as 70%). Also, these newly developed concretes, owing to their green chemical nature, are environmental friendly, durable and inexpensive building materials. Recent research indicated that calcium contents of FA significantly affect the resultant hardening characteristic of the geopolymers where most of the earlier studies revealed promising results [22, 23]. Calcium oxide (CaO) is assumed to generate calcium silicate hydrate (C-S-H) together with sodium aluminium silicate hydrate (N-A-S-H) gel. A major challenge for diverse applications of N-A-S-H–based geopolymers is the need for curing at a higher temperature.

Earlier studies were focused on increasing the reactive nature of these substances through the incorporation of some calcium-based materials [24, 25]. The incorporation of CaO allowed the formation of C-S-H gel together with N-A-S-H networks [22, 26]. The contents of CaO in the precursor substance played a significant role in achieving the final hardening of geopolymers. Meanwhile, an increasing CaO content caused the enhancement in the mechanical characteristics and subsequent reduction in setting time [24, 27]. Palomo et al. [28] developed two models to understand the binding characteristics of geopolymers with alkaline solution activation. The first model concerns the mild alkaline solution activation of silica (Si) plus Ca substances including ground blast furnace slag (GBFS) to produce C-A-S-H gels as main product (called alkali-activated mortars). The second model deals with the alkaline solution activation of Si plus Al substances including FA that needs a robust alkali solution to produce N-A-S-H gels as the major outcome (called geopolymer mortars).

The ceramic industry is known to generate large amounts of wastes each year. So far a huge part is used in landfills. Reusing these wastes in concrete could be a win–win situation. It solves the ceramics industry's waste problem and at the same time leads to a more sustainable concrete [29, 30]. Yearly, the global production of ceramic tiles is more than 10 million square metres [31]. It has been estimated in a survey that about 15–30% of production goes as waste in the ceramics industry [32] and is not reused in any form at present and hence, piling up every year. This percentage of waste is increasing as ceramics are damaged during storage, transportation, and construction as well as housing renovation. Moreover, the deposition processes are becoming expensive due to the ever-growing constraints on landfilling. In that case, industries must look for alternative solutions such as recycling such waste materials as useful products.

Several studies [30, 32–34] have reported that ceramic is highly resistant to biological degradation forces. As ceramic is enriched with silico-aluminium (crystalline materials) and has been a suitable supplementary cementing material to enhance the mechanical and durability properties of concrete [35–37]. Despite its reuse as ceramic waste, the quantities of such waste utilized in the construction sector is still negligible [29, 34]. Thus, its immediate reuse in other industries appears essential. The building sector being the customer of much ceramic waste will continue to play a vital role to overcome some of the environmental issues. The geopolymer industries can use ceramic waste safely without requiring any remarkable change in the production and application processes. Moreover, the cost of deposition of ceramic waste in landfills can be saved together with the replacement of raw materials and natural resources thus saving energy and protecting the environment. Some studies suggested that the construction industry can be more sustainable and beneficial if most of the industrial wastes can be recycled effectively as useful OPC or OPC-free concretes [38–41].

Yet, the development of different alkali-activated mortars (AAMs) as environmental friendly cement-free construction materials is rarely explored by using ceramic waste. The low content of Al_2O_3 and CaO in the chemical composition of waste ceramic powder (WCP) makes it rather impossible to produce AAM-based

WCP at room temperature. Consequently, the present study intends to develop environmental friendly AAMs with broad arrays of applications in the construction industry. This could be achieved by reusing ceramic tile waste as the main binder in AAMs prepared with a low concentration of alkaline solution (for environment benefit). The aim of this study was to assess the performance of AAMs that use a high volume of WCP and partially replaced by GBFS and FA. To achieve this aim, fresh and hardened states as well as microstructure properties of AAMs were investigated.

5.2 MATERIALS

The starting materials used for this study included WCP, GBFS, FA, fine aggregate, and alkaline activator solution. The materials used in this study were obtained from a single source and preserved in airtight plastic containers to avoid any form of contamination. WCP, GBFS and FA were used as binders to produce alkali-activated mortars. The GBFS and FA were used as received; the WCP was crushed and grinded in the laboratory before use as shown in Figure 5.1. In the present work, the cement-free binder is made using pure GBFS (collected from Ipoh, Malaysia) as one of the resource materials, which is further utilized without any laboratory treatment. Low-calcium FA is acquired from a Tanjung bin power station (Malaysia) as a resource aluminosilicate material for making alkali-activated mortars. The ceramic wastes are collected from the White Horse ceramic manufacturer in Pasir Gudang Johor, Malaysia. During the collection stage, only homogeneous ceramic tile wastes were collected (the ceramic tile was of the same thickness with no glassy coating). The ceramic wastes were first crushed using a crushing machine, then sieved through 600 μm to isolate the large particles. The sieved ceramic was again ground for 6 hours using a Los Angeles abrasion

FIGURE 5.1 WCP preparation stages.

machine to achieve the required particle size according to ASTM 618 of 66% passing 45 μm [42]. Next, the powder was collected and used in the mixing process. Table 5.1 demonstrates the physical properties of the binder materials including particle size, specific gravity, and surface area using a laser diffraction particle size analyzer (Mastersizer, Malvern Instruments) and Brunauer–Emmett–Teller (BET) tests. The particle size distributions of the WCP, GBFS and FA obtained using a particle size analyzer (PSA) with the median estimated as 35, 12.8, and 10 μm, respectively, are shown in Figure 5.2.

The chemical composition of the raw materials was determined using a wavelength-dispersive X-ray fluorescence (XRF) instrument. An XRF spectrometry scanning test was estimated by the Quant Express dry method using Spectra Plus software. The fused-bead technique was used to prepare the samples for XRF analysis. Hardening time was also measured by the Vicat apparatus and the moulds were checked in intervals of 5 minutes to evaluate the hardening status. Table 5.1 summarizes the chemical compositions of WCP, GBFS and FA obtained from XRF testing. The main oxide compositions were silica and aluminium totalling 84.8% in WCP, 41.7% in GBFS, and 86% in FA. WCP presented a very high content of silica (72.6%), GBFS presented a very high content of calcium oxide (51.8%) and FA presented a high content of aluminium oxide (28.8%) compared to other materials. It is worth noting that the level of silica, aluminium and calcium oxide played a significant role in the alkali-activated mortars synthesis by forming N-A-S-H and C-(A)-S-H gels in the geopolymerization process. The content of sodium oxide (Na_2O) was observed at a high ratio (13.5%) in WCP chemical composite compared to 0.45% with GBFS

TABLE 5.1
Chemical Composition and Physical Characteristics of WCP, GBFS, and FA as Revealed by XRF Analysis

Material	WCP	GBFS	FA
Chemical Composition (%)			
Silica (SiO_2)	72.6	30.8	57.20
Aluminium oxide (Al_2O_3)	12.2	10.9	28.81
Iron (III) oxide (Fe_2O_3)	0.56	0.64	3.67
Calcium oxide (CaO)	0.02	51.8	5.16
Magnesium oxide (MgO)	0.99	4.57	1.48
Potassium oxide (K_2O)	0.03	0.36	0.94
Sodium oxide (Na_2O)	13.46	0.45	0.08
Loss on ignition (LOI)	0.13	0.22	0.12
Physical Characteristics			
Specific gravity	2.61	2.9	2.20
Surface area-BET (m^2/g)	12.2	13.6	18.1
Average diameter (μm)	35	12.8	10

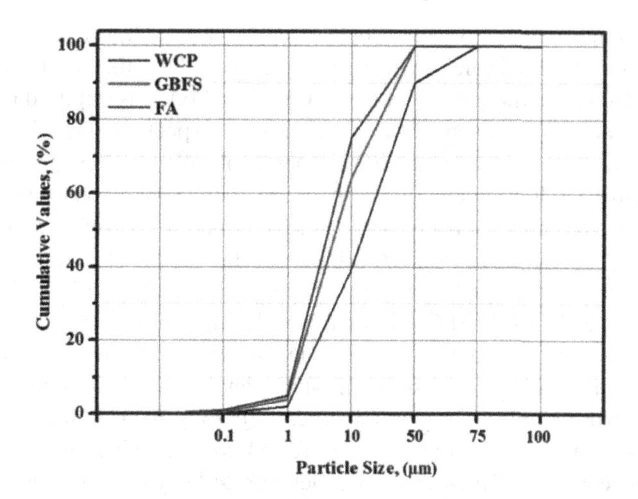

FIGURE 5.2 Particle size distribution of constituent waste materials.

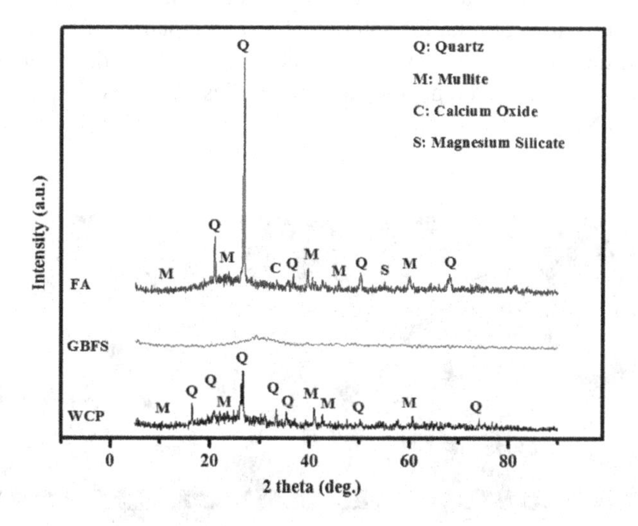

FIGURE 5.3 XRD patterns of WCP, GBFS, and FA.

and 0.08% with FA. The loss on ignition (LOI) contents was observed to be very low in WCP, GBFS and FA.

Figure 5.3 illustrates the XRD patterns of WCP, GBFS, and FA. The XRD pattern of WCP and FA revealed pronounced diffraction peaks around $2\theta = 16–30°$, which were attributed to the crystalline silica and alumina compounds. Nonetheless, the occurrence of other crystalline peaks was ascribed to the presence of crystalline quartz and mullite phases. As reported by several studies [43–45], the amorphous phases of both WCP and FA play a significant role in the hydration process and gel formulation. The XRD pattern of GBFS verified their highly amorphous nature because of the absence of any sharp peak. One of the most important factors in

GBFS formation is the reactive silica and calcium content. This high content of reactive amorphous silica and calcium in GBFS has potential for alkali-activated mortars preparation. However, the incorporation of GBFS and FA was required to overcome the low CaO and Al_2O_3 content (0.02 and 12.2 wt% respectively) in WCP.

Scanning electron microscopy (SEM) images of WCP, GBFS, and FA are presented in Figure 5.4. It is evident that both the WCP and GBFS consisted of irregular and angular particles, whereas FA particles had a smooth spherical surface.

For the proposed application, naturally occurring siliceous river sand was used as fine aggregate to make all mortar specimens. The sand was washed with water following the standard ASTM C117 to reduce silts and impurity contents. Then, the sand was dried in the oven at 60°C for 24 hours to control the moisture content. The estimated specific gravity was 2.6. In this study, analytical grade sodium hydroxide (98% purity, from Quality Reagent Chemical [QREC]–Asia, Malaysia) was used. The pellets were dissolved in water to prepare sodium hydroxide solution of 4 molarity (4 M) concentration. The solution was left for 24 hours to cool, then added to a sodium silicate solution to prepare the final alkaline solution with a 1.02 ratio of SiO_2:Na_2O. The ratio of sodium silicate to sodium hydroxide (NS:NH) was fixed to 0.75 for all mixtures of alkaline solution. The low molarity of sodium hydroxide (4 M) and sodium silicate content were to reduce the cost, hazard, and environmental effects of alkaline solution compared to other studies which used up to 12 M and a 2.5 ratio of NS:NH.

FIGURE 5.4 SEM images of (a) WCP, (b) GBFS, and (c) FA.

5.3 METHOD FOR MIX DESIGN

Alkali-activated mortars were prepared by blending WCP, GBFS, and FA at varying proportions. The WCP content was kept by weight between 50 and 70%, GBFS between 20 and 50%, and FA in the range of 0–30%. This ternary-blend range was considered for the series of tests carried out in line with the aim of studying the impact of a high volume of ceramic waste on alkali-activated mortars. Apart from the blend of WCP, GBFS, and FA, other variables were kept constant in the mix proportions. Details of ternary-blended alkali-activated mortars are presented in Table 5.2.

5.4 SPECIMEN PREPARATION AND TEST PROCEDURE

According to ASTM C579, cube moulds 50×50 × 50 mm were prepared for hardened tests such as compressive strength. Prisms 40×40 × 160 mm were used to prepare the samples for flexural strength tests. For tensile strength and slant shear tests, cylinders were prepared with 75 mm diameter and 150 mm depth. For shrinkage tests, prisms 25×25 × 250 mm were used to prepare the samples. Before casting, the moulds were applied with engine oil to ease the demoulding work. A mixture of NH and NS, measured by weight, was prepared by thoroughly mixing them together. The solution was allowed to cool to room temperature before use due to the production of generated heat during the process. The alkali-activated mortars were prepared by mixing WCP, GBFS, and FA for 2 minutes in dry conditions to achieve a homogeneous mixture, followed by adding the fine aggregates and mixed for another 3 minutes using a mortar mixer machine. Then, the acquired mixture was activated by adding the alkaline solution. The whole matrix is mixed in the machine operating at

TABLE 5.2
Alkali-Activated Mortars Mix Design and Content of Constituents

	Binder (Mass %)			Chemical Composition					
Mix	WCP	GBFS	FA	Si:Al	Ca:Si	Ca:Al	B:A*	S:B*	NS:NH*
$AAMs_1$	0	100	0	2.82	1.68	4.75	1.0	0.40	0.75
$AAMs_2$	50	50	0	4.48	0.50	2.24	1.0	0.40	0.75
$AAMs_3$		40	10	4.08	0.39	1.59			
$AAMs_4$		30	20	3.77	0.29	1.09			
$AAMs_5$		20	30	3.53	0.20	0.70			
$AAMs_6$	60	40	0	4.79	0.37	1.77	1.0	0.40	0.75
$AAMs_7$		30	10	4.35	0.27	1.19			
$AAMs_8$		20	20	4.01	0.18	0.74			
$AAMs_9$	70	30	0	5.09	0.26	1.31	1.0	0.40	0.75
$AAMs_{10}$		20	10	4.62	0.17	0.79			

*Si: SiO_2; Al: Al_2O_3; Ca: CaO; A: fine aggregate; B: binder; S: alkaline solution; NS: sodium silicate; NH: sodium hydroxide.

medium speed for another 4 minutes. Finally, the fresh mortar was cast in moulds in two layers, where each layer was consolidated using a vibration table for 15 seconds for air voids to escape. After casting, the alkali-activated specimens were left for 24 hours for curing at ambient temperature 27 ± 1.5°C under relative humidity of 75%. Then, the specimens were taken out from moulds and left in the same condition till testing time.

The flow ability of various fresh AAMs is measured by using traditional flow tests following ASTM C230 standard. The workability of high-volume WCP-based AAMs incorporating GBFS and FA in different levels were widely assessed, wherein the initial and final setting times were recorded in accordance to ASTM C191 specifications. The compressive strength test was carried out in accordance to ASTM C109. A minimum of three specimens were tested to evaluate the compressive strength at 1, 3, 7, 28, 56, 90, 180, and 360 days. According to standards C496/C496M-11, ASTM C78, ASTM C469/C469M, and ASTM C140-07, the tensile splitting strength, flexural strength, modulus of elasticity, and water absorption, respectively, were tested at 28 days of age.

Fourier-transform infrared (FTIR) spectroscopy is an analytical technique to identify organic and inorganic materials. The vibration modes of the underlying chemical structures of AAMs are determined using the FTIR spectrophotometer. The room temperature FTIR spectra in the wavenumber range of 400–4000 cm^{-1} are collected. Scanning electron microscopy (SEM) with sufficient magnifications is used to examine the surface morphology of the AAMs. SEM testing was conducted using a Hitachi S-3400N to reveal the microstructure and various degrees of reaction at different chemical compositions of mixes. The samples used for SEM/FTIR analysis were collected from specimens tested at 28 days of age then sputtered and coated with gold for 2 minutes using BAL-TEC SCD 005 Sputter Coater machine to increase the electrical conductivity of the fragments.

5.5 FRESH AND HARDENED PROPERTIES

5.5.1 Flow of AAMs

Figure 5.5 displays the results of the flow test of the synthesized AAMs with flow values of 16.5, 18.5, and 23 cm when the ceramic content increased from 50 to 60 to 70%, respectively, compared to the control sample (AAM_1) which presented a 13.2 cm flow diameter. At each level, when WCP content was fixed and GBFS was replaced by FA, the workability of the mortar was reduced. Typically, the flow value decreased with an increase of GBFS and FA content in the mixture. The lower value of the flow diameter of ternary-blended AAMs was recorded with mix AAM_5. The low specific surface area and high particle size distribution of WCP compared to GBFS and FA significantly contributed to the AAMs' flow ability enhancement. Furthermore, it became clear that an increase in FA content decreased the workability of AAMs due to high water adsorption of FA with a porous structure [46–48] as depicted in Figure 5.2. The other way to enhance the workability of mortars is to

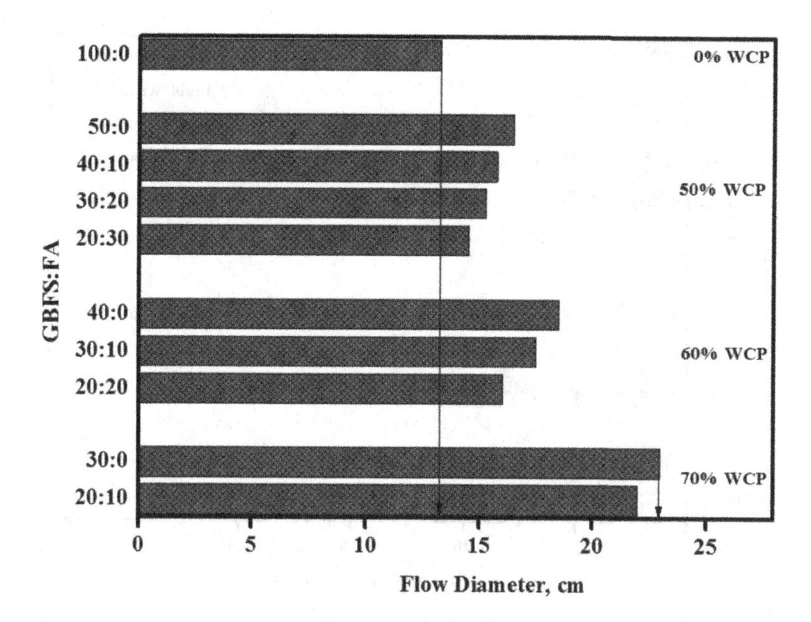

FIGURE 5.5 Flow of AAMs under the effect of high WCP content.

increase the WCP content and reduce the GBFS amount by replacing it with WCP and FA. This affects the rate of chemical reaction [24, 49] and increases the plasticity of the mixture which improves the AAMs' workability.

5.5.2 Setting Time of AAMs

Figure 5.6 shows the high volume WCP content-dependent variation in the setting time of AAMs. Mortars with the highest WCP content took the longest time (92 minutes) to set. A sharp decrease in the setting time was observed when the WCP content was reduced and the GBFS content was increased. The effect of FA on the initial and final setting time was clearly observed at each level of high-volume WCP. An increase in the FA content significantly influenced the setting time. Both the initial and final setting times increased with the increased level of FA replacing GBFS, i.e. a higher setting time (more than 90 minutes) was recorded with AAM_{10} compared to 8 minutes obtained with the control sample (AAM_1). As SiO_2 content increased (64%) and the amount of CaO (10%) decreased in the AAMs mixture containing 70% WCP, 20% GBFS, and 10% FA, a delay in the rate of chemical reaction and an increase in the setting time were observed. The difference between initial and final setting times also increased with the reduction in GBFS content and increased content of both WCP and FA in the mortar. It also supported the fact that the higher the GBFS content in the mortar, the faster the setting rate [46, 50]. These results established that the WCP and FA, as a part of the ternary-blended binder, were effective in reducing the setting time of AAMs at ambient condition.

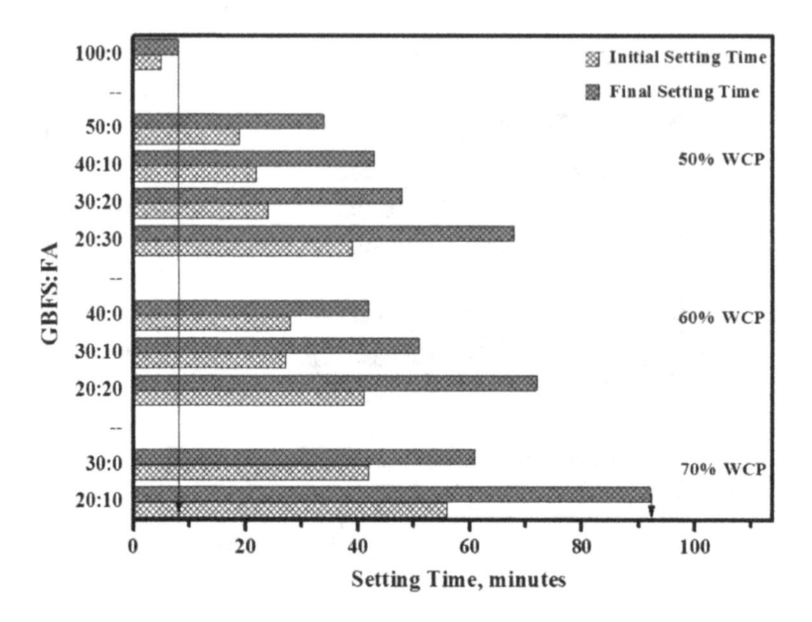

FIGURE 5.6 Effect of GBFS:FA on the setting time of AAMs containing high-volume WCP.

5.5.3 Hardened Density of AAMs

Figure 5.7 summarizes the density of alkali-activated mortars containing high-volume WCP at the age of 28 days. An increase in the WCP content caused a reduction in the alkali-activated mortars' density. This is mostly attributed to the lower density of WCP than GBFS (Table 5.1). An increasing content of WCP (from 50 to 60 to 70%) and reduction of GBFS content (50 to 40 to 30%) have remarkably affected the mortars' density in which an increase in the lightweight ceramic powder particles played a major role. The replacement of GBFS by FA in the blended mixtures reduced the mortars' density. A batch mixture with 50% WCP, 50% GBFS and 0% FA (AAM_2) presented the highest density (2.20 $g.cm^{-3}$) among all the alkali-activated mixtures compared to 2.28 $g.m^{-3}$ observed with the control sample (AAM_1). Moreover, the alkali-activated blend with 70% WCP, 20% GBFS, and 10% FA revealed the lowest density of 1.88 $g.cm^{-3}$. Generally, the WCP presented the highest density compared to FA as depicted in Figure 5.6. This observation was primarily ascribed to the higher density of WCP than FA ash used in the mixture.

5.5.4 Compressive Strength of AAMs (CS)

Figure 5.8 shows the high-volume WCP content-dependent compressive strength development of AAMs. The compressive strength of AAMs was found to vary inversely with the increasing amount of WCP content from 50 to 70%, where the strength dropped from 70.1 to 34.8 MPa at 28 days, respectively. The negative effect

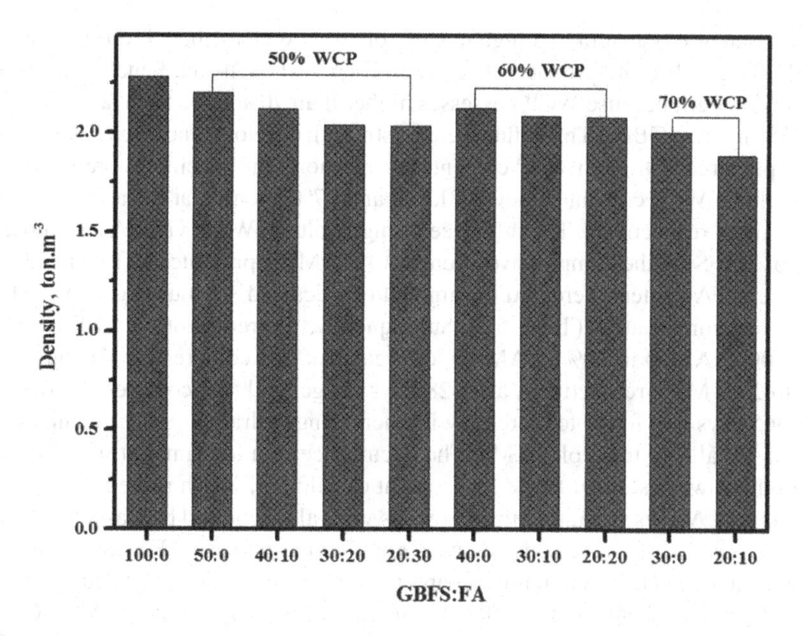

FIGURE 5.7 Density of high-volume WCP containing AAMs at 28 days.

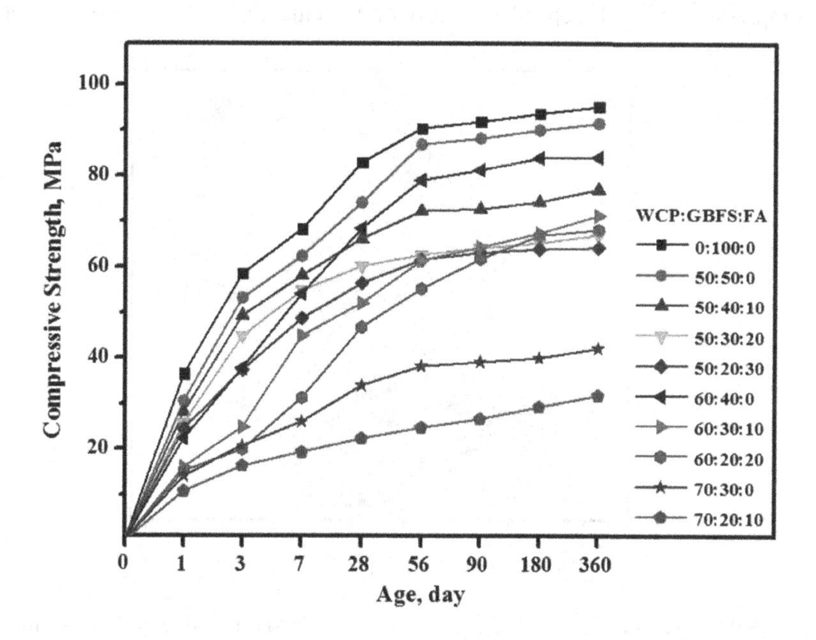

FIGURE 5.8 High-volume WCP content-dependent compressive strength of AAMs.

of low content of calcium and high content of silica was attributed to this drop [41, 47, 51]. This led to the production of fewer C-(A)-S-H gels and hence reduced the AAMs' strength because WCP possesses higher than 70% silica with larger particle size (35 μm) than GBFS. This influenced the strength development of the AAMs specimens prepared with high WCP content. Furthermore, the specimens prepared with high-volume WCP content achieved 81, 94, and 97% strength at the ages of 28, 56, and 90 days, respectively. The influence of high-volume WCP with FA as a replacement of GBFS on the compressive strength of AAMs is presented in Figure 5.9. An increase in FA content increased the amount of silica and Al, thus negatively affecting the calcium content (Table 5.2). Subsequently, the reduction in GBFS content from 100 ($AAMs_1$) to 20% ($AAMs_{10}$) led weakened the compressive strength from 84.6 to 24.8 MPa, respectively, after 28 days. In general, the compressive strength after 365 days was found to increase with increasing hydration time as compared to 28 days for all mixtures of AAMs. The occurrence of a low amount of C-(A)-S-H gel products was ascribed to the low content of calcium, which played a major role to the loss of AAMs strength. Similar results were also reported by Rashad [12] who observed a decrease in the compressive strength with increasing FA content in geopolymer matrix. The reduction in compressive strength could be related to several factors. The first factor is the difference in chemical composition of WCP, GBFS, and FA, which significantly affected the alkali-activation process of the binders. The second factor is the lower reaction rates of WCP and FA than that of GBFS which was partially dissolved [52]. The third factor could be associated with the lowering of compactness and density of the AAM matrix with increasing WCP and FA contents. The fourth factor is linked to the low concentration of sodium hydroxide (4 M) where the compressive strength depended mainly on the calcium oxide content to replace

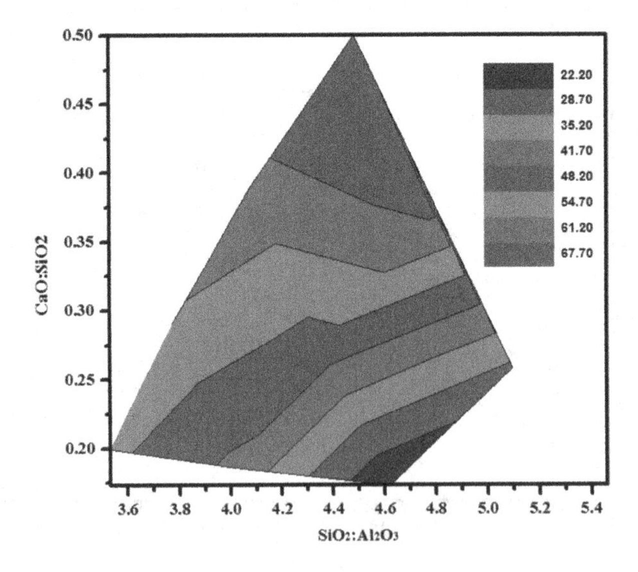

FIGURE 5.9 Effect of varying Ca:Si on the compressive strength of AAMs containing a high volume of WCP.

the low amount of sodium oxide. Thus, the formation of more C-S-H and C-A-S-H gels beside the N-A-S-H gel enhanced the AAMs' strength.

5.5.5 Effect of SiO_2:Al_2O_3 on Strength Development

Figure 5.9 depicts the effect of SiO_2:Al_2O_3 and CaO:SiO_2 on the development of AAMs' compressive strength at 28 days of age. Specimens prepared with SiO_2:Al_2O_3 4.5 or more and a lower ratio of CaO to SiO_2 (less than 0.20) displayed a lower strength of 22.2 MPa. Furthermore, the compressive strength was enhanced with the decrease in SiO_2:Al_2O_3 with a similar ratio of CaO to SiO_2 and recorded values higher than 50 MPa as the silicate-to-aluminium ratio reduced from 4.5 to 3.6. The highest strength (67 MPa) was recorded at 28 days with the alkali-activated matrix prepared with a CaO:SiO_2 ratio higher than 0.40. The influence on strength by percentage of SiO_2, Al_2O_3 and CaO are shown in Figure 5.10. The strength of the AAMs decreased as the percentage of aluminium and silicate increased and calcium content decreased. It was reported by Van et al. [53] that the addition of more calcined source materials led to an increase in compressive strength by improving the microstructure of the AAMs matrix. Thus, the increase in compressive strength of AAM specimens by the inclusion of GBFS is attributed to the formation of a more compact microstructure of the binder [54].

5.5.6 X-Ray Diffraction (XRD)

Figure 5.11 shows the XRD patterns of AAMs containing a high volume of WCP. The intensity of the CSH peak decreased as the WCP content increased, especially the

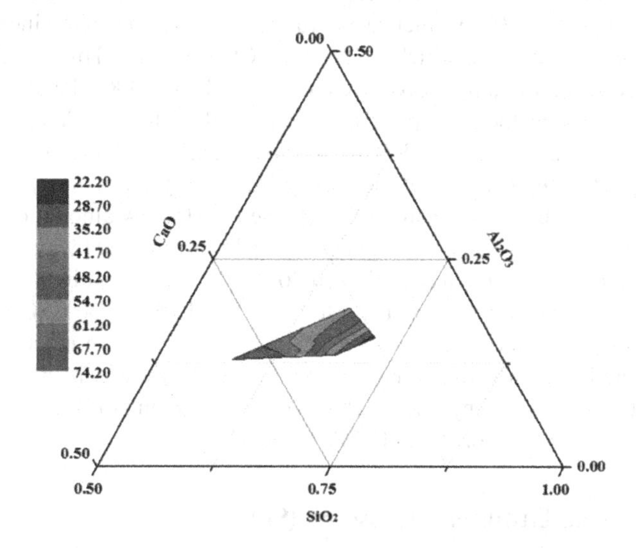

FIGURE 5.10 Effect of varying SiO_2, Al_2O_3, and CaO percentage content on the compressive strength of AAMs containing high-volume WCP at age of 28 days.

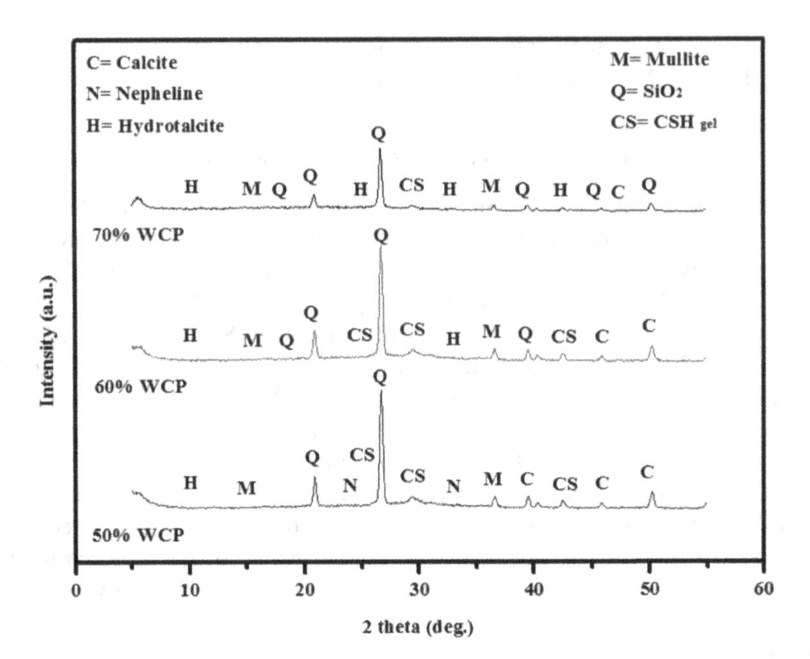

FIGURE 5.11 XRD patterns of AAMs containing high-volume WCP.

peak at 30°. For 70% WCP the observed CSH peak at 43° was found to be replaced by a hydrotalcite peak. Calcite peaks at 38.5 and 51° were observed to be replaced by quartz peaks as the WCP content increased above 50%. Five peaks for quartz, located at 17, 21, 28, 38.5, and 51° were observed at high WCP content, in addition to a weak CSH peak at 30°. An increase in the WCP content led to an increase in the non-reacted silicate amount, which reduced the CSH product. This could be attributed to the reduction of compressive strength from 74.1 to 34.8 MPa after 28 days.

Figure 5.12 presents the XRD patterns displaying the effect of FA replacing GBFS on AAMs' structures containing 50% WCP. The intensity and the number of quartz peaks increased as the FA content increased. The CSH peak intensity was reduced as the FA content increased because of the presence of a low amount of calcium in the alkali-activated matrix. The calcite peak at 46° was replaced by mullite and then by quartz as the FA content increased from 20 to 30%, respectively. The production of CSH gel was reduced with the increase in FA level where the CSH peak at 27° was replaced by a hydrotalcite structure. An increase in the non-reacted silicate or partially reacted content with a reduction in CSH product was evidenced, which led to a drop in the AAMs' strength from 74.1 to 66.2 to 60.2 and to 56.5 MPa as the FA level increased from 0 to 10, 20, and 30%, respectively.

5.5.7 Scanning Electron Microscopy (SEM)

Figure 5.13 displays the SEM images of AAMs containing a high volume of WCP. Three specimens were prepared with 50, 60, and 70% WCP by replacing GBFS

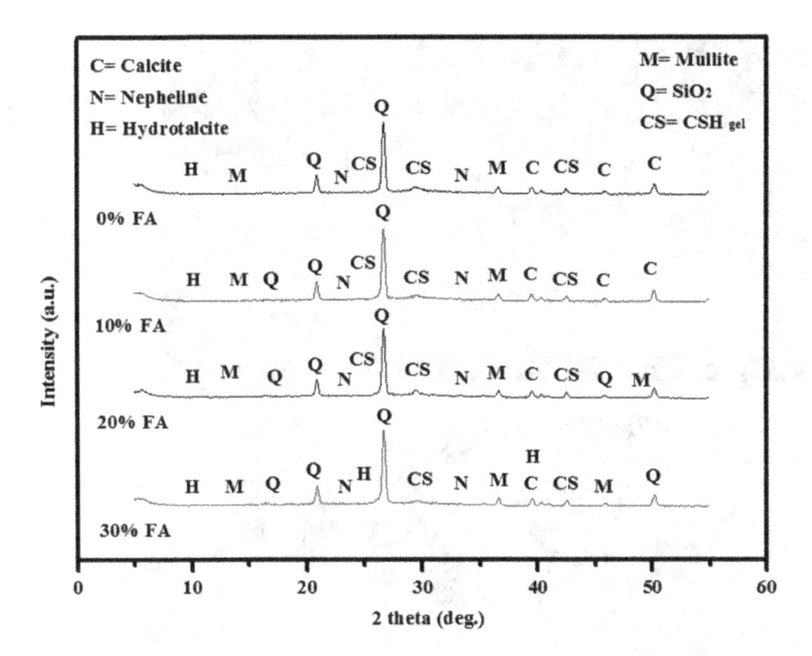

FIGURE 5.12 XRD patterns of FA content replaced with GBFS in AAMs containing 50% WCP.

and tested at 28 days old. A dense surface was observed for AAMs with 50% WCP where a tiny amount of non-reacted and partially reacted particles were present. As the WCP content increased to 60 and 70% the non-reacted content and the partially reacted particles increased (Figure 5.13a,b) compared to the one prepared with 50% WCP (Figure 5.13c). The increasing WCP content led to the creation of more non-reacted silica and poor morphology with highly porous structure. This in turn affected the development of the compressive strength of AAMs where it dropped to 34.8 MPa as compared to 70.1 MPa.

The SEM images in Figure 5.14 show the influence of 60% WCP with FA replaced GBFS on the surface morphology of synthesized AAMs. An increase in the FA content from 0 to 20% generated more non-reacted and partially reacted particles together with poor surface morphology (Figure 5.14b). It is established that a reduction in GBFS content and increasing FA content reduced the C-S-H gel product and generated more partially reacted gel such as mullite and non-reacted particles including quartz [55]. This reduction in the C-S-H gel product finally led to the weakening of the compressive strength from 68.4 to 46.8 MPa.

5.5.8 FTIR Spectral Analysis

The results on compressive strength showed that an increase in WCP content by replacing GBFS has considerably reduced the specimens' strength in the alkali-activated matrix. This progressive decrease in the strength resulted in the reduction

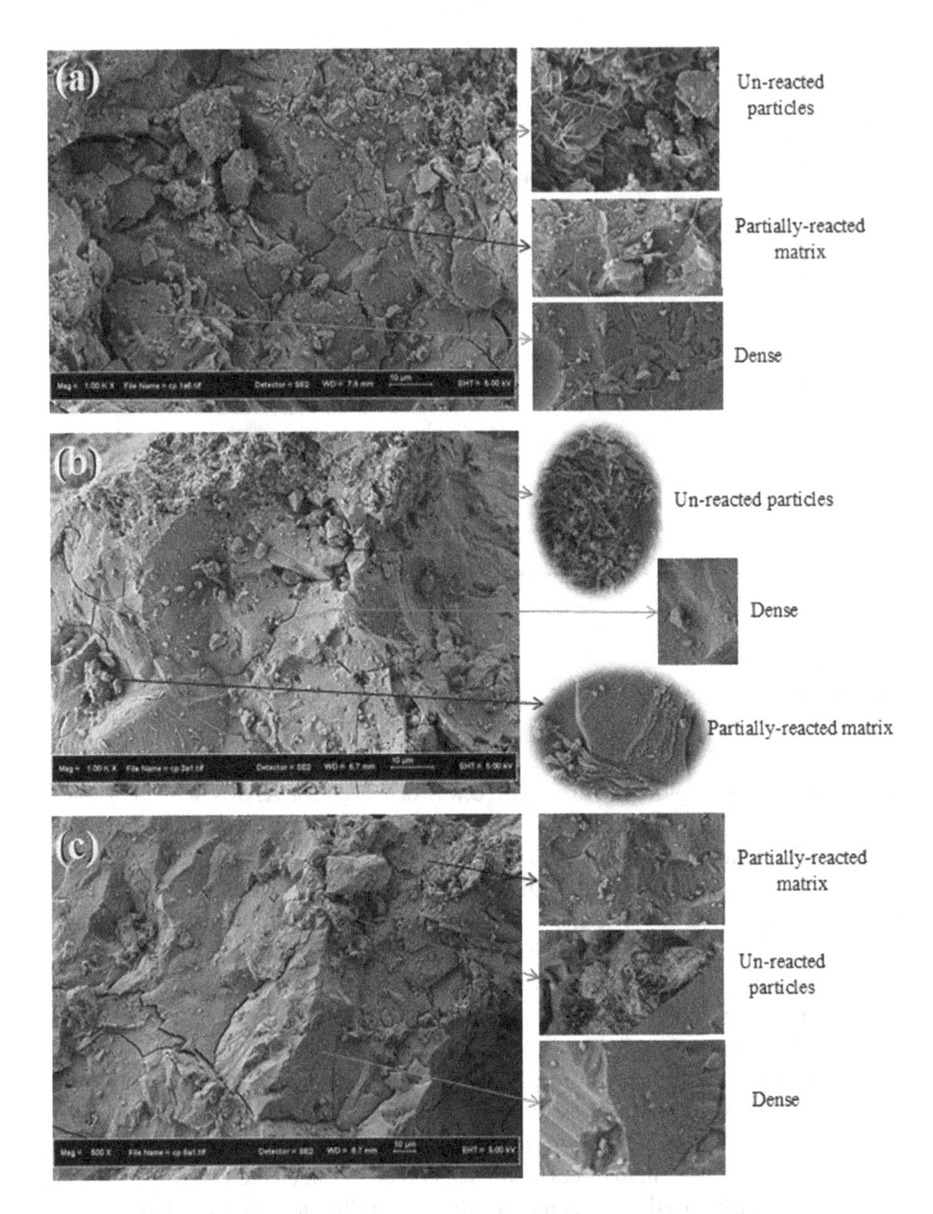

FIGURE 5.13 SEM images of AAMs with WCP:GBFS ratios (a) 70:30, (b) 60:40 and (c) 50:50.

of dissolved Ca and Al as the quantity of GBFS content decreased [56]. A reduction in the dissolved Al has also been reported to reduce the degree of silicate polymerization [56, 57].

Table 5.3 summarizes the FTIR spectral analysis involving the bonding vibrations in AAMs that were responsible for the development of compressive strength. The presence of a broader FTIR spectral band of Si–O–Al in Figure 5.15 supported

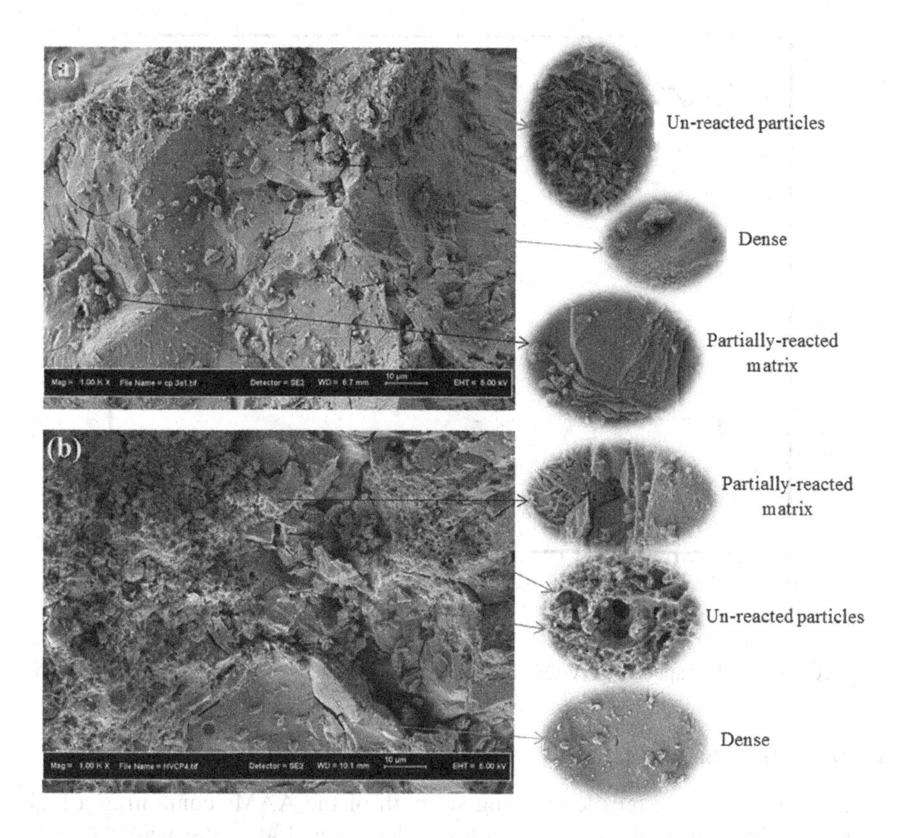

FIGURE 5.14 SEM images depicting the effect of FA replaced GBFS on the surface morphology of AAMs containing 60% WCP and (a) 0% FA and (b) 20% FA.

TABLE 5.3
FTIR Band Positions and Corresponding Band Assignments for AAMs Containing High Volume of WCP

	Ratio of		*fc*	Band Positions (cm^{-1}) and Their Assignments				
Mix	Si:Al	Ca:Si	(MPa)	Al-O	Si-O	AlO_4	CSH	C(N)ASH
$AAMs_9$	5.09	0.26	34.82	669.8	695.9	778.1	875.7	994.2
$AAMs_6$	4.79	0.37	68.44	666.6	693.9	774.6	874.5	966.3
$AAMs_2$	4.48	0.50	74.12	659.1	691.1	773.5	873.9	943.1
$AAMs_5$	3.53	0.20	56.47	660.2	692.7	775.2	875.1	971.5

this observation. The band frequency was found to increase with increasing content of WCP and decreasing content of GBFS. The frequency of bands increased from 943.1 cm^{-1} to 994.2 cm^{-1} with increasing WCP content from 50 to 70%, respectively. This observed increase in the frequency of FTIR vibration could be attributed to the reduction of Al in the AAMs network and thereby weakening of the 3D structure.

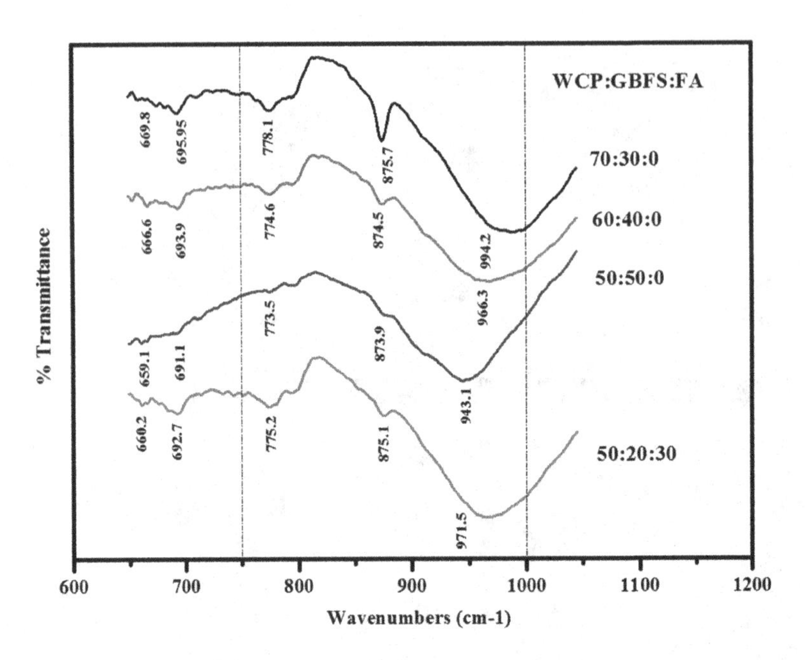

FIGURE 5.15 FTIR spectra of AAMs in the fingerprint zone for high-volume WCP content.

5.5.9 Tensile Splitting Strength of AAMs

Figure 5.16 shows the tensile splitting strength of the AAMs containing a high volume of WCP. The strength values were determined after 28 days. The tensile splitting strength was found to be influenced by the increasing content of WCP and presented a lower strength (2.68 MPa) with high WCP content (70%) as compared to 5.32 MPa achieved for 50% WCP content and 5.84 MPa of the control sample. The tensile splitting strength values at 28 days were 5.32, 5.27, and 2.68 MPa for WCP contents of 50, 60, and 70%, respectively, by replacing the GBFS. An increasing content of WCP led to the reduction of calcium content and slowed the chemical reactions rate to produce C-S-H gel [54, 58]. The influence of a high-volume WCP on the AAMs' tensile splitting strength by the FA replaced by GBFS is shown in Figure 5.17 with 50, 60, and 70% WCP content, respectively. An inverse relationship was found between the tensile splitting strength of AAMs and the FA content. As the FA content increased, the strength dropped. The lower value of tensile splitting strength in this batch was observed for $AAMs_{10}$ which contained 70% WCP, 20% GBFS, and 10% FA and revealed a strength of 1.92 MPa at the age of 28 days. It was reported by Phoo-ngernkham [22] that the strength improved in FA-based AAMs with increased calcium content whereby additional C-S-H and C-A-S-H gels co-existed with N-A-S-H gel. This explained the drop in strength with increase content of WCP and FA, and reduction in GBFS content.

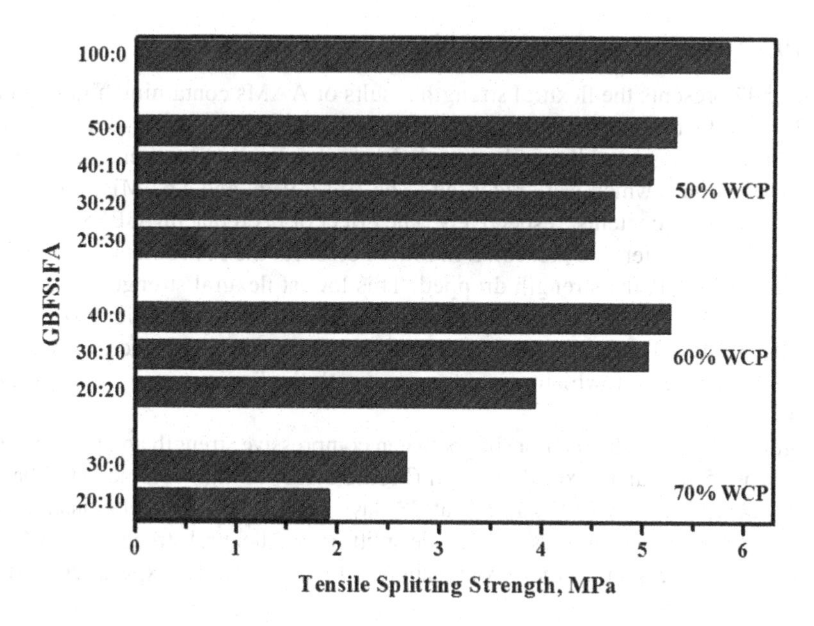

FIGURE 5.16 High-volume WCP content-dependent tensile splitting strength of AAMs at 28 days of age.

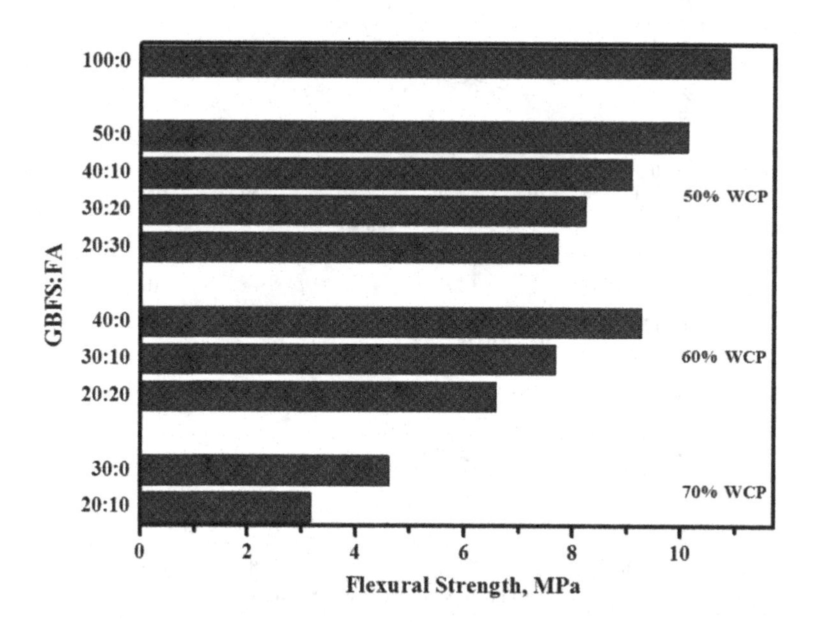

FIGURE 5.17 Flexural strength of AAMs for high volume WCP content at 28 days of age.

5.5.10 Flexural Strength of AAMs

Figure 5.17 presents the flexural strength results of AAMs containing high-volume WCP. The effect of WCP content on the flexural strength was examined at the age of 28 days. The observed flexural strength dropped as the WCP content increased at the age of 28 days which were recorded to be 10.12, 9.26, and 4.62 MPa for 50, 60, and 70% of WCP contents, respectively. The effect of FA replacing GBFS at 50, 60, and 70% WCP contents is presented in Figure 5.18. As the FA content increased in each level of WCP, the strength dropped. This lowest flexural strength (3.17 MPa) observed at the age of 28 days was achieved for $AAMs_{10}$ containing 70% WCP and 20% GBFS. The results indicated that the flexural strength increased as the slag content increased, which is in agreement with the findings of previous studies [24, 47, 59].

Figure 5.18 shows the relationship between compressive strength and tensile splitting (Figure 5.18a) and flexural strength (Figure 5.18b) of high-volume WCP-based AAMs incorporating GBFS and FA at 28 days of age. A direct relationship was observed between the compressive, tensile splitting and flexural strengths of AAMs. Exponential regression methods were applied to correlate the experimental data

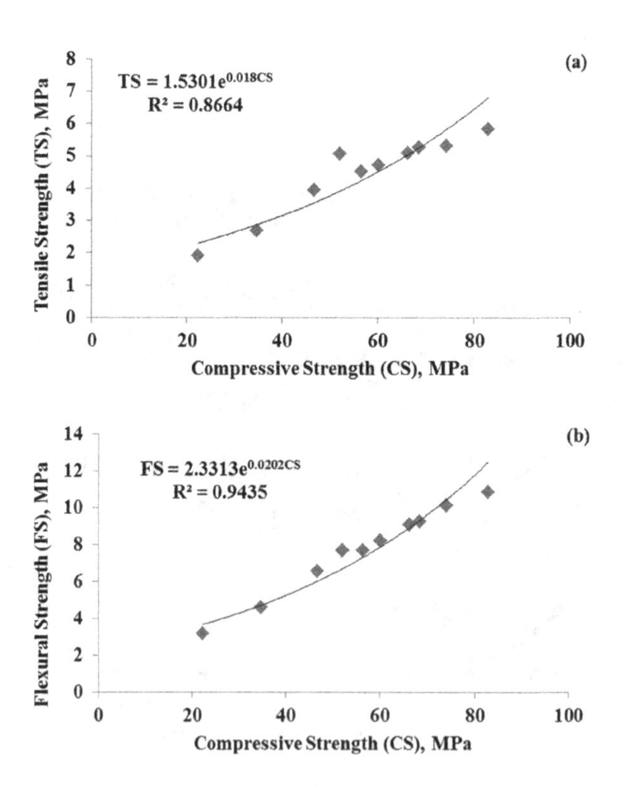

FIGURE 5.18 Relationship between compressive strength and (a) tensile splitting strength and (b) flexural strength of AAMs.

following Equations 5.1 and 5.2, with R^2 values of 0.86 and 0.94, respectively. These values signified high confidence for the relationships.

$$TS = 1.5301e^{0.018\,cs} \tag{5.1}$$

$$FS = 2.3313e^{0.0202cs} \tag{5.2}$$

5.5.11 Modulus of Elasticity (MOE) of AAMs

Figure 5.19 shows the MOE values of AAMs for varying WCP contents (50, 60, and 70%) at 28 days of age. The results showed that as the WCP content replaced GBFS increased, the MOE values decreased. The increase in the WCP percentage replaced GBFS from 50, 60 and 70% led to a drop in the MOE from 16.3, 15.8 to 7.4 GPa as compared to 19.9 GPa recorded with $AAMs_1$ (100% GBFS). In each of the high-level WCP content, the effect of FA replacing GBFS was also evaluated. The results showed that an increasing FA content has increased the silicate content and reduced the amount of calcium. This in turn led to a drop the MOE of prepared AAMs. The lowest value of MOE (4.98 GPa) was found for the $AAMs_{10}$ mixture containing 70% WCP, 20% GBFS, and 10% FA. As the WCP and FA contents were increased, the MOE values of the AAMs remarkably dropped.

5.5.12 Water Absorption

Figure 5.20 shows the effect of high-volume WCP on the water absorption ability of AAMs. The water absorption revealed a direct proportionality with the WCP

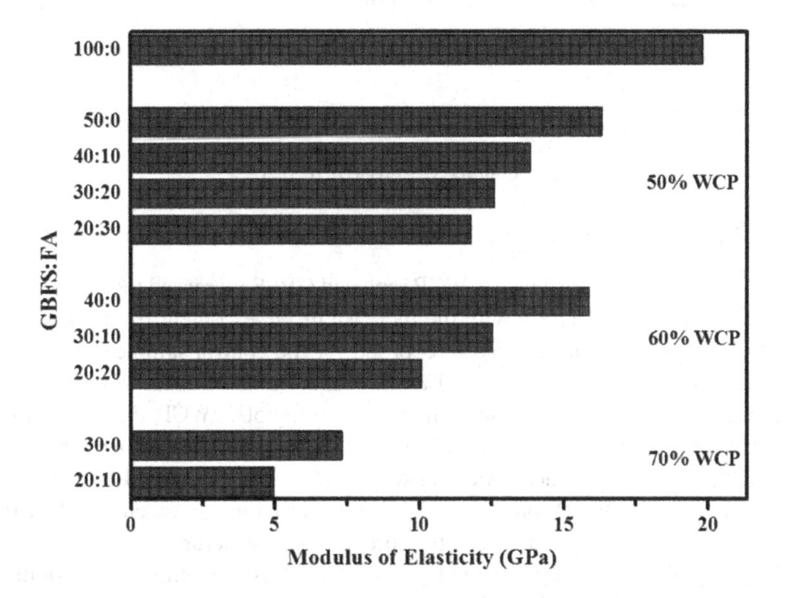

FIGURE 5.19 MOE for AAMs containing high volume of WCP at 28 days of age.

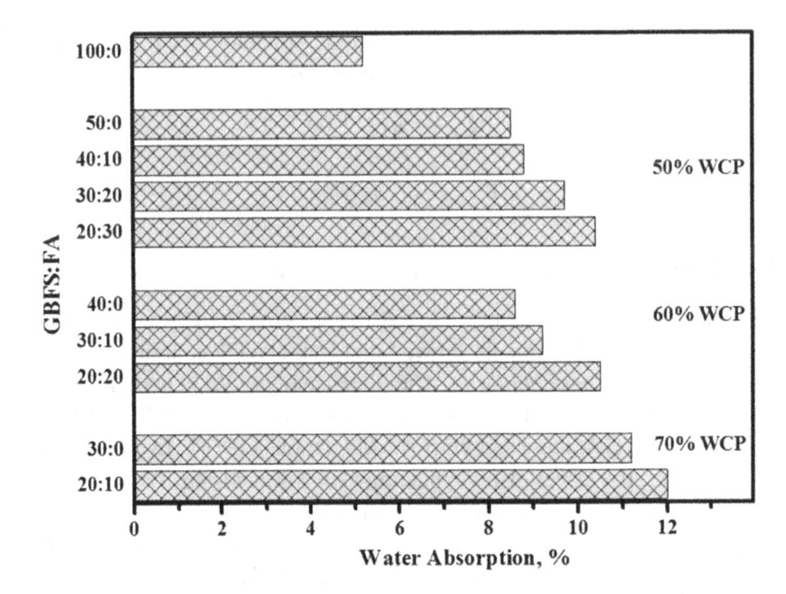

FIGURE 5.20 Water absorption of AAMs containing high volume of WCP at 28 days of age.

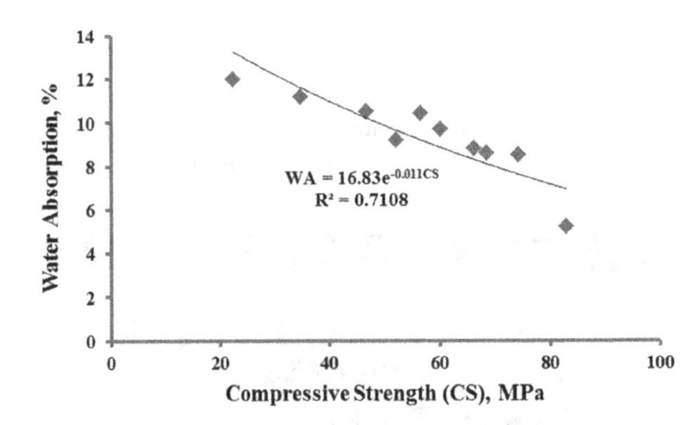

FIGURE 5.21 Relationship between compressive strength and water absorption of AAMs.

content. The increase in the level of WCP replaced GBFS increased the water absorption from 8.5 to 8.6 and 11.2% with the increase in WCP content from 50 to 60 and 70%, respectively compared to 5.6% recorded for the control sample. The effect of FA replaced GBFS on 50, 60, and 70% WCP and on water absorption capacity of AAMs was evaluated. In AAM specimens containing 50% WCP, the water absorption increased with the increase in FA level replacement of GBFS, where an FA level of 0, 10, 20, and 30% produced water absorption of 8.5, 8.8, 9.7, and 10.4%, respectively. An increase in WCP content and FA could cause the non-reacted and partially reacted particles to increase and lead to more a porous structure [60, 61].

At 28 days of age, the relationship between compressive strength and modulus of elasticity and water absorption of AAMs containing a high volume of WCP incorporating GBFS and FA is illustrated in Figure 5.21. The water absorption was found to be inversely proportional to compressive strength, whereby the water absorption dropped from 12 to 8.5% and the strength increased from 24 to 70 MPa, respectively. The exponential regression method was applied to correlate the experimental data as shown in Equation 5.3, with an R^2 value of 0.71. This signified good confidence for the relationships.

$$WA = 16.83e^{-0.011\,cs} \tag{5.3}$$

5.6 CONCLUSIONS

In this study, the ability to reuse ceramic tile waste to produce green mortars was evaluated. This study developed new alkali-activated mortars containing ceramic waste for various construction applications. The following conclusions can be drawn from the current study:

i. Incorporation of WCP with GBFS and FA enhanced the workability of alkali-activated mortars.
ii. The setting time increased with the increase of WCP replacement. The replacement of more than 60% WCP to the mix significantly delayed the setting time.
iii. Using high-volume ceramic tile waste between 50 and 70% showed acceptable compressive strength (36–70 MPa) for construction applications 28 days of age.
iv. Increased WCP content of 70% reduced the calcium oxide content and negatively affected the formation of C-S-H and C-A-S-H gels as shown by microstructure results from XRD, SEM, and FTIR.
v. Compressive strengths of ternary-blended alkali-activated mortars were influenced by the content of FA replaced GBFS at each level of WCP.
vi. A mix prepared with 50% WCP, 40% GBFS, and 10% FA ($AAMs_3$) presented the optimum compressive strength of ternary-blended AAMs at 28 days old (66 MPa) as compared to mixtures prepared with high WCP content and low GBFS content replaced with FA.
vii. All AAM mixtures with high-volume WCP content presented low water absorption compared to the control sample with 0% of WCP and FA. A decrease in GBFS content reduced the reaction process and produced less dense C-S-H and C-A-S-H gels, which further led to a microstructure of mortars with increased voids.
viii. An exponential relationship was observed between compressive, flexural, and splitting tensile strengths and water absorption of AAMs.

REFERENCES

1. Duxson, P., et al., The role of inorganic polymer technology in the development of 'green concrete'. *Cement and Concrete Research*, 2007. **37**(12): pp. 1590–1597.
2. Rashad, A., et al., Hydration and properties of sodium sulfate activated slag. *Cement and Concrete Composites*, 2013. **37**: pp. 20–29.

3. Teng, Y., et al., Reducing building life cycle carbon emissions through prefabrication: Evidence from and gaps in empirical studies. *Building and Environment*, 2018. pp. 132: 125–136.
4. Huseien, G. F., M. Ismail, and J. Mirza, Influence of curing methods and sodium silicate content on compressive strength and microstructure of multi blend geopolymer mortars. *Advanced Science Letters*, 2018. **24**(6): pp. 4218–4222.
5. Palomo, Á., et al., Railway sleepers made of alkali activated fly ash concrete. *Revista Ingeniería de Construcción*, 2011. **22**(2): pp. 75–80.
6. Rehan, R. and M. Nehdi, Carbon dioxide emissions and climate change: Policy implications for the cement industry. *Environmental Science & Policy*, 2005. **8**(2): pp. 105–114.
7. Stern, N. H., *The Economics of Climate Change: The Stern Review*. 2007: Cambridge University Press.
8. Ismailos, C. and M. F. Touchie, Achieving a low carbon housing stock: An analysis of low-rise residential carbon reduction measures for new construction in Ontario. *Building and Environment*, 2017. **126**: pp. 176–183.
9. Xing, R., et al., Estimating energy service demand and CO2 emissions in the Chinese service sector at provincial level up to 2030. *Resources, Conservation and Recycling*, 2018. **134**: pp. 347–360.
10. Zhang, Y., et al., Aspen plus-based simulation of a cement calciner and optimization analysis of air pollutants emission. *Clean Technologies and Environmental Policy*, 2011. **13**(3): pp. 459–468.
11. Asaad, M. A., et al., Improved corrosion resistance of mild steel against acid activation: Impact of novel *Elaeis guineensis* and silver nanoparticles. *Journal of Industrial and Engineering Chemistry*, 2018. **63**: pp. 139–148.
12. Rashad, A. M., Properties of alkali-activated fly ash concrete blended with slag. *Iranian Journal of Materials Science and Engineering*, 2013. **10**(1): pp. 57–64.
13. Onuaguluchi, O. and N. Banthia, Scrap tire steel fiber as a substitute for commercial steel fiber in cement mortar: Engineering properties and cost-benefit analyses. *Resources, Conservation and Recycling*, 2018. **134**: pp. 248–256.
14. Asaad, M. A., et al., Enhanced corrosion resistance of reinforced concrete: Role of emerging eco-friendly *Elaeis guineensis*/silver nanoparticles inhibitor. *Construction and Building Materials*, 2018. **188**: pp. 555–568.
15. Rashad, A. M., A comprehensive overview about the influence of different additives on the properties of alkali-activated slag – a guide for civil engineer. *Construction and Building Materials*, 2013. **47**: pp. 29–55.
16. Verian, K. P., W. Ashraf, and Y. Cao, Properties of recycled concrete aggregate and their influence in new concrete production. *Resources, Conservation and Recycling*, 2018. **133**: pp. 30–49.
17. Huseiena, G. F., et al., Potential use coconut milk as alternative to alkali solution for geopolymer production. *Jurnal Teknologi*, 2016. **78**(11): pp. 133–139.
18. Ismail, M. and B. Muhammad, Electrochemical chloride extraction effect on blended cements. *Advances in Cement Research*, 2011. **23**(5): p. 241.
19. Mehmannavaz, T., et al., Binary effect of fly ash and palm oil fuel ash on heat of hydration aerated concrete. *The Scientific World Journal*, 2014. **2014**: pp. 1–6.
20. Bamaga, S., et al., Evaluation of sulfate resistance of mortar containing palm oil fuel ash from different sources. *Arabian Journal for Science and Engineering*, 2013. **38**(9): pp. 2293–2301.
21. Huseien, G. F., et al., Compressive strength and microstructure of assorted wastes incorporated geopolymer mortars: Effect of solution molarity. *Alexandria Engineering Journal*, 2018. 57: pp. 3375–3386.

22. Phoo-ngernkham, T., et al., High calcium fly ash geopolymer mortar containing Portland cement for use as repair material. *Construction and Building Materials*, 2015. **98**: pp. 482–488.
23. Huseien, G. F., et al., Geopolymer mortars as sustainable repair material: A comprehensive review. *Renewable and Sustainable Energy Reviews*, 2017. **80**: pp. 54–74.
24. Al-Majidi, M. H., et al., Development of geopolymer mortar under ambient temperature for in situ applications. *Construction and Building Materials*, 2016. **120**: pp. 198–211.
25. Huseien, G. F., et al., Influence of different curing temperatures and alkali activators on properties of GBFS geopolymer mortars containing fly ash and palm-oil fuel ash. *Construction and Building Materials*, 2016. **125**: pp. 1229–1240.
26. Phoo-ngernkham, T., et al., Effects of sodium hydroxide and sodium silicate solutions on compressive and shear bond strengths of FA–GBFS geopolymer. *Construction and Building Materials*, 2015. **91**: pp. 1–8.
27. Karakoç, M. B., et al., Mechanical properties and setting time of ferrochrome slag based geopolymer paste and mortar. *Construction and Building Materials*, 2014. **72**: pp. 283–292.
28. Pacheco-Torgal, F., J. Castro-Gomes, and S. Jalali, Alkali-activated binders: A review. Part 2. About materials and binders manufacture. *Construction and Building Materials*, 2008. **22**(7): pp. 1315–1322.
29. Pacheco-Torgal, F. and S. Jalali, Reusing ceramic wastes in concrete. *Construction and Building Materials*, 2010. **24**(5): pp. 832–838.
30. Senthamarai, R., P. D. Manoharan, and D. Gobinath, Concrete made from ceramic industry waste: Durability properties. *Construction and Building Materials*, 2011. **25**(5): pp. 2413–2419.
31. Daniyal, M. and S. Ahmad, Application of waste ceramic tile aggregates in concrete. *International Journal of Innovative Research in Science, Engineering and Technology*, 2015. **4**(12): pp. 12808–12815.
32. Senthamarai, R. and P. D. Manoharan, Concrete with ceramic waste aggregate. *Cement and Concrete Composites*, 2005. **27**(9–10): pp. 910–913.
33. Huang, B., Q. Dong, and E. G. Burdette, Laboratory evaluation of incorporating waste ceramic materials into Portland cement and asphaltic concrete. *Construction and Building Materials*, 2009. **23**(12): pp. 3451–3456.
34. Hussein, A. A., et al., Performance of nanoceramic powder on the chemical and physical properties of bitumen. *Construction and Building Materials*, 2017. **156**: pp. 496–505.
35. Samadi, M., et al., Properties of mortar containing ceramic powder waste as cement replacement. *Jurnal Teknologi*, 2015. **77**(12): pp. 93–97.
36. Huseien, G. F., et al., Waste ceramic powder incorporated alkali activated mortars exposed to elevated Temperatures: Performance evaluation. *Construction and Building Materials*, 2018. **187**: pp. 307–317.
37. Fernandes, M., A. Sousa, and A. Dias, Environmental impacts and emissions trading-ceramic industry: A case study. In Coimbra: Technological Centre of Ceramics and Glass, Portuguese Association of Ceramic Industry (in Portuguese), APICER, 2004: pp. 1–10.
38. Limbachiya, M., M. S. Meddah, and Y. Ouchagour, Use of recycled concrete aggregate in fly-ash concrete. *Construction and Building Materials*, 2012. **27**(1): pp. 439–449.
39. Heidari, A. and D. Tavakoli, A study of the mechanical properties of ground ceramic powder concrete incorporating nano-SiO 2 particles. *Construction and Building Materials*, 2013. **38**: pp. 255–264.

40. Huseien, G. F., et al., The effect of sodium hydroxide molarity and other parameters on water absorption of geopolymer mortars. *Indian Journal of Science and Technology*, 2016. **9**(48): pp. 1–7.
41. Huseien, G. F., et al., Effect of metakaolin replaced granulated blast furnace slag on fresh and early strength properties of geopolymer mortar. *Ain Shams Engineering Journal*, 2016. **9**(4): pp. 1557–1566.
42. ASTM, Standard specification for coal fly ash and raw or calcined natural pozzolan for use in concrete. 2013: ASTM.
43. Yusuf, M. O., et al., Evolution of alkaline activated ground blast furnace slag–ultrafine palm oil fuel ash based concrete. *Materials & Design*, 2014. **55**: pp. 387–393.
44. Temuujin, J., A. van Riessen, and K. MacKenzie, Preparation and characterisation of fly ash based geopolymer mortars. *Construction and Building Materials*, 2010. **24**(10): pp. 1906–1910.
45. Rickard, W. D., et al., Assessing the suitability of three Australian fly ashes as an aluminosilicate source for geopolymers in high temperature applications. *Materials Science and Engineering: A*, 2011. **528**(9): pp. 3390–3397.
46. Nath, P. and P. K. Sarker, Effect of GGBFS on setting, workability and early strength properties of fly ash geopolymer concrete cured in ambient condition. *Construction and Building Materials*, 2014. **66**: pp. 163–171.
47. Huseien, G. F., et al., Synergism between palm oil fuel ash and slag: Production of environmental-friendly alkali activated mortars with enhanced properties. *Construction and Building Materials*, 2018. **170**: pp. 235–244.
48. Huseien, G. F., et al., Effects of POFA replaced with FA on durability properties of GBFS included alkali activated mortars. *Construction and Building Materials*, 2018. **175**: pp. 174–186.
49. Sugama, T., L. Brothers, and T. Van de Putte, Acid-resistant cements for geothermal wells: Sodium silicate activated slag/fly ash blends. *Advances in Cement Research*, 2005. **17**(2): pp. 65–75.
50. Kumar, S., R. Kumar, and S. Mehrotra, Influence of granulated blast furnace slag on the reaction, structure and properties of fly ash based geopolymer. *Journal of Materials Science*, 2010. **45**(3): pp. 607–615.
51. Huseien, G.F., J. Mirza, and M. Ismail, Effects of high volume ceramic binders on flexural strength of self-compacting geopolymer concrete. *Advanced Science Letters*, 2018. **24**(6): pp. 4097–4101.
52. Puertas, F., et al., Alkali-activated fly ash/slag cements: Strength behaviour and hydration products. *Cement and Concrete Research*, 2000. **30**(10): pp. 1625–1632.
53. Van Jaarsveld, J., J. Van Deventer, and G. Lukey, The effect of composition and temperature on the properties of fly ash-and kaolinite-based geopolymers. *Chemical Engineering Journal*, 2002. **89**(1–3): pp. 63–73.
54. Deb, P. S., P. Nath, and P. K. Sarker, The effects of ground granulated blast-furnace slag blending with fly ash and activator content on the workability and strength properties of geopolymer concrete cured at ambient temperature. *Materials & Design (1980-2015)*, 2014. **62**: pp. 32–39.
55. Huseiena, G. F., et al., Effect of binder to fine aggregate content on performance of sustainable alkali activated mortars incorporating solid waste materials. *Chemical Engineering*, 2018. **63**: pp. 667–672.
56. Puligilla, S. and P. Mondal, Role of slag in microstructural development and hardening of fly ash-slag geopolymer. *Cement and Concrete Research*, 2013. **43**: pp. 70–80.
57. García-Lodeiro, I., et al., Effect of calcium additions on N–A–S–H cementitious gels. *Journal of the American Ceramic Society*, 2010. **93**(7): pp. 1934–1940.

58. Huseiena, G. F., et al., Performance of sustainable alkali activated mortars containing solid waste ceramic powder. *Chemical Engineering*, 2018. **63**: pp. 673–678.
59. Song, S. and H. M. Jennings, Pore solution chemistry of alkali-activated ground granulated blast-furnace slag. *Cement and Concrete Research*, 1999. **29**(2): pp. 159–170.
60. Kubba, Z., et al., Impact of curing temperatures and alkaline activators on compressive strength and porosity of ternary blended geopolymer mortars. *Case Studies in Construction Materials*, 2018. **9**: p. e00205.
61. Huseien, G. F. and K. W. Shah, Durability and life cycle evaluation of self-compacting concrete containing fly ash as GBFS replacement with alkali activation. *Construction and Building Materials*, 2020. **235**: p. 117458.

6 Bond Strength Performance of Alkali-Activated Mortars in Aggressive Environments

6.1 INTRODUCTION

Nowadays, the environmental benefits of alkali-activated mortars (AAMs), such as low carbon dioxide (CO_2) emission, energy savings, reduction of landfill problems, saving of natural resources, lowering of the total demand, and cost-effectiveness, make them suitable alternative materials for traditional concretes in civil construction sectors [1–3]. Generally, these AAMs are produced by mixing various waste materials containing high amounts of the aluminosilicates (ASs) with other calcium-based components via the alkaline solution activation [4–6]. Diverse cheap and abundant industrial and agricultural by-products as the waste materials, such as class F and C fly ash (FA), palm oil fuel ash (POFA), waste ceramic powder (WCP), and ground blast furnace slag (GBFS), are introduced as the main source of ASs to produce AAMs. Thus, it is realized that various industrial and agricultural waste by-products as the alkali-activated binder can efficiently be used in AAMs to reduce the CO_2 emission level by up to 75% compared to the ordinary Portland cement (OPC) [7–9].

Several studies [10–13] stated that ceramics are extremely tolerant materials against various forces or factors of the degradation. Being enriched with crystalline ASs these ceramics are the appropriate supplements for the cement materials that can improve the mechanical strength and durability performance of the concretes [14–16]. Despite the recycling and reusing of various kinds of ceramic wastes, the total amount used in the construction sectors in general and concrete industries in particular are yet insignificant [13, 17, 18]. Therefore, the instantaneous recycling of these ceramics for other industrial applications is needed. Presently, the construction sectors need a substantial amount of ceramic waste to surmount various environmental problems. The safe use of the ceramic waste is recommended without any significant changes in the manufacturing and application of the free cement such as the geopolymer and alkali-activated systems. The use of ceramic waste has many environmental implications, where the replacement of the natural resources and raw materials by these wastes can remarkably reduce the total demand and consumption, and save energy and expenditure in terms of the waste dumping in the landfill thereby protecting the environment. Intensive studies showed that the building and concrete industries may get many benefits in terms of the durability and sustainability wherein the industrial and agricultural waste by-products can be reused efficiently as the practical concretes without or with OPC [19–22].

The high content of the amorphous and crystalline form of AS, abundance and low environmental cost make FA the most widespread resource material for the fabrication of the geopolymers and AAMs [23]. In developed countries, some million tonnes of the coals are consumed annually for the generation of the electric power [24]. Therefore, the recycling of these wastes for making geopolymers and alkali-activated systems may directly solve the environmental problems, thereby bring sustainability to the concrete industries [25, 26]. It is important to note that GBFS as waste is usually acquired from molten iron slag quenching (a spin-off of iron and steel fabrication) inside a blast furnace in the presence of water. In this procedure, some granular amorphous products are obtained which are further dried and crushed into a fine powder to get GBFS [27]. Depending on the initial resources utilized for iron manufacturing, the chemical compositions of the slag may vary significantly [28]. Because of the high CaO and SiO_2 contents, GBFS shows both cementing and pozzolanic characteristics [29]. Earlier, GBFS was intensively utilized for construction purposes to enhance the durability and compressive strength (CS) of traditional concrete [30]. The alkali-activated slag (AAS) was found to exhibit very high durability [31, 32], workability, and CS. These factors encouraged the broad usage of GBFS for making durable and strong concretes.

The surfaces of concrete structures, such as sidewalks, pavements, parking decks, bridges, runways, canals, dykes, dams, spillways, deteriorate progressively due to a variety of physical, chemical, thermal, and biological processes. Actually, the permanence of concrete's composition is greatly affected by the improper usage of substances, and their physical as well as chemical conditioning in the environment [33]. The immediate consequence is the anticipated need of maintenance and execution of the repairs [34]. A few million tonnes of solid waste in the form of spin-offs from agricultural and other industries (palm oil FA, bottom ash, ceramic tiles, FA, and GBFS) are discarded annually as landfill in Malaysia [35, 36]. Such waste results in serious ecological problems in terms of air pollution and leaching of harmful products. Several studies have shown the feasibility of recycling these wastes to get novel concretes as a substitute for OPC (above 60%) [37, 38]. Furthermore, such new types of concrete due to their green chemistry have environmental friendliness, durability and cheap characteristics for applying as construction materials. So far, the progress of diverse AAMs (containing the aforementioned wastes) as the repair material especially for deteriorated concrete surfaces have been rarely explored.

Morgan et al. [39] acknowledged that the compatibility between the concrete substrate and repair mortars must meet certain requirements such as compressive, tensile, flexural, and bond strength. However, the bonding strength among the concrete substrate and the repairing system [40–42] determines the binding efficiency of the AAMs as the repair material. Durability of the repair materials is characterized by their tolerance against the declination. A durable mortar almost porosity free reveals strong resistance against a sulphate and chloride attack, abrasion, and environmental conditions [43]. In this study, the effect of an aggressive environment on the slant shear bond strength between the normal cement (NC) and prepared AAMs was assessed. These AAMs were prepared by incorporating three waste materials: WCP, GBFS, and FA (called the ternary blend). These blends were designed at various levels of WCP, GBFS, and FA to determine the viability of reusing solid waste from diverse industries. The idea is to turn these wastes into environmentally responsive and long-lasting repairable mortars/binders for sustainable development. In this procedure, the WCP was kept in high volume (between 50 and 70% by weight) and substituted with different contents of GBFS and FA in the practical operational range with the appropriate physical conditions needed to fabricate such WCP–GBFS–FA ternary mortars with alkali activation. All the proposed mixes were analyzed using different analytical techniques to assess the slant shear bond stability performances (between NC and AAMs) under aggressive environments including sulfuric acid, high heating, and freeze–thaw and wet–dry cycles. The experimental findings were analyzed, interpreted and validated to determine the optimal compositions.

6.2 PHYSICAL AND CHEMICAL PROPERTIES OF MATERIALS

In this work, the starting materials WCP, GBFS, FA, river sand, sodium hydroxide (NH), and sodium silicate (NS) were utilized. The WCP, FA, and GBFS were stored in air-tight plastic storage boxes to avoid any contamination. The ternary components WCP, GBFS, and FA were mixed to get the AAMs as binders. GBFS and FA were utilized as received, but the WCP was treated in the laboratory prior to use (Figure 6.1). The alkali-activated binder was prepared using pure GBFS (obtained from Ipoh, Malaysia) as one of the source materials, which was used without any treatment in the laboratory. The class Fly ash class F (low content of calcium) was collected from the Tanjung bin power station (Malaysia) as the source AS to make

FIGURE 6.1 The WCP preparation stages.

the AAMs. The WCP was obtained from the White Horse ceramic manufacturer (Pasir Gudang Johor, Malaysia). The waste from homogeneous ceramic tiles (with identical thickness and without glass coating) were acquired to make the binders. First, the collected WCP was ground in a crushing machine before being sieved through 600 μm to isolate the large particles. Again, the sieved WCP was ground for 6 hours by a Los Angeles abrasion machine to get the desired particle size in accordance to ASTM 618 of 66% passing 45 μm [44]. Next, the powder was obtained and utilized to make the mortar mixes. Table 6.1 summarizes the physical characteristics of the resource materials in terms of their particle sizes, specific gravities and surface areas obtained via a laser diffraction particle size analyzer (PSA; Mastersizer, Malvern Instruments) and Brunauer–Emmett–Teller (BET) test. The WCP, GBFS and FA particles' size distributions were achieved via the PSA (Figure 6.2) with the median values of 35, 12.8, and 10 μm, respectively.

The X-ray fluorescence (XRF) spectra of the prepared mixes and raw materials (WCP, GBFS, and FA) were recorded to verify their chemical compositions where a wavelength-dispersive XRF (WDXRF) spectrometer was used (the Quant Express dry method interfaced with Spectra Plus software). For the XRF analysis, the samples were prepared using the fused-bead method. Furthermore, the time of hardening was monitored via a Vicat apparatus and the moulds were checked in 5 minute intervals to assess the status of hardening. The chemical composition of the raw materials (WCP, GBFS, and FA) determined from the XRF test are summarized in Table 6.1. The silica and aluminium are found to be the main oxides in the WCP (84.8%) and FA (86%) composition compared to 41.7% in the GBFS. The WCP revealed the high concentration of silicates (72.6%) and GBFS showed a very high

TABLE 6.1
Chemical and Physical Properties of the Raw Materials

Materials	WCP	GBFS	FA
Chemical Composition(%)			
SiO_2	72.64	30.82	57.20
Al_2O_3	12.23	10.91	28.81
CaO	0.02	51.82	5.16
Fe_2O_3	0.56	0.64	3.67
Na_2O	13.46	0.45	0.08
MgO	0.99	4.57	1.48
K_2O	0.03	0.36	0.94
Loss on ignition (LOI)	0.13	0.22	0.12
Physical Properties			
Specific gravity	2.61	2.9	2.20
Surface area-BET (m^2/g)	12.2	13.6	18.1
Mean diameter (μm)	35	12.8	10

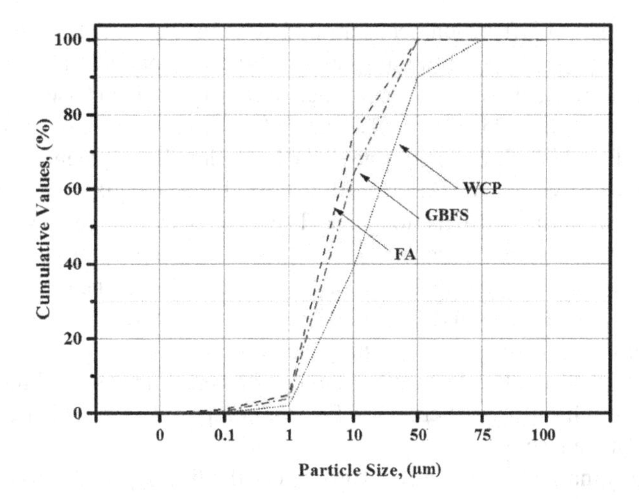

FIGURE 6.2 Particle size distribution of the constituent waste materials.

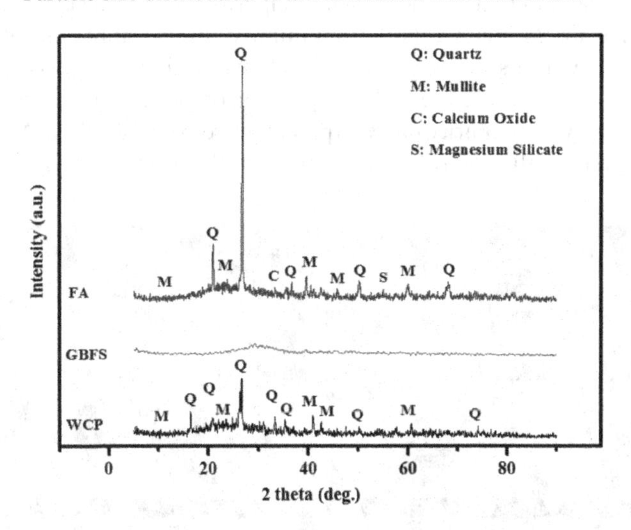

FIGURE 6.3 The XRD patterns of the WCP, GBFS, and FA.

level of CaO (51.8%). Regarding the aluminium oxide content, the FA presented the highest level (28.8%). Also, the amount of silicates, Al and Ca oxides played a vital role in the fabrication of the AAMs wherein the dense gels calcium silicate hydrate (C-S-H), calcium aluminium silicate hydrate (C-A-S-H), and sodium aluminium silicate hydrate (N-A-S-H) were formed via the geopolymerization process. In the WCP chemical composition, the mass ratio of sodium oxide (Na_2O) was much higher (13.5%) than that of in GBFS (0.45%) and FA (0.08%). The loss on ignition (LOI) contents was found to be very small in WCP, GBFS, and FA.

The XRD patterns of WCP, GBFS, and FA are shown in Figure 6.3. The XRD analyses revealed the presence of the low calcium content in WCP and high content in FA with the sharp diffraction peaks around 2θ = 16–30°, which were assigned to the crystalline lattice planes of the silica and alumina compounds. The other characteristic peaks were attributed to the existence of the crystalline phases of quartz and mullite. Several studies [25, 32, 45] also showed that the presence of the amorphous phases of WCP and FA play a major role in the hydration development and gel formation. In contrast to WCP and FA, the GBFS showed a strong glassy character in the absence of any prominent diffraction peak. The existence of the reactive amorphous silica and calcium at high concentration in GBFS was highly prospective for the AAM synthesis. In fact, the inclusion of FA and GBFS was needed to surmount the little Al_2O_3 (12.2 wt%) and CaO (0.02 wt%) concentration that was present in the WCP. Figure 6.4 displays the scanning electron microscope (SEM) micrographs of the WCP, GBFS, and FA. Both WCP and GBFS were comprised of angular particles with the rough surface, whereas the FA was composed of the spherical particles with the smooth surface.

In this study, natural river sand was utilized as the fine aggregates to fabricate all the AAMs. Following the ASTM C117 standard, the collected river sand was first washed using water to remove the silts and impurities. Next, the cleaned river sand was oven-dried for 24 hours at 60°C to minimize the moisture content during the AAM fabrication. The specific gravity of the treated river sand was assessed and estimated with the value of 2.6. For the preparation of the mortar specimens, analytical grade NaOH with the purity of 98% (procured from QREC–Asia, Malaysia) was used to make the alkaline activator solution. First, the NaOH pellets were dissolved

FIGURE 6.4 The SEM micrographs of (a) WCP, (b) GBFS, and (c) FA.

in water to make 4 M of NH solution. Then, the resultant NH solution was cooled for 24 hours before being added with the Na_2SiO_3 solution (obtained from QREC–Asia, Malaysia) to achieve the final alkali mixture with modulus of 1.02 (SiO_2:Na_2O). In the ultimate prepared alkaline solution, the Na_2SiO_3-to-NaOH ratio (NS:NH) was maintained to be 0.75 for all the AAM mixes. The low content of Na_2SiO_3 and molarity of NaOH (4 M) were considered to reduce the cost, energy consumption, carbon dioxide emission, hazard, and environmental effect of the alkaline solution, thereby increasing the sustainability of the AAMs compared to the other studies [46, 47] that used NS:NH of 2.5 and up to 12 M of NaOH.

6.3 MIX DESIGN AND CASTING PROCESS

Various ternary blends were prepared using the raw materials including the WCP, GBFS and FA to produce the AAMs and the details of the mixes are illustrated in Table 6.2. The level of WCP was kept high (between 0.191 to 0.268 m^3 by volume) for all the designed mixes. However, the level of GBFS in the AAM mixes was maintained between 0.068 and 0.171 m^3 (by volume) via the replacement of the FA powder at various levels. For continuity, all the mixtures had equal levels of NH molarity (4 M) with 0.384 m^3 fine aggregates for each batch. The ratio of the binders (such as the WCP, GBFS, and FA) to the fine aggregates (B:A) was kept in the range of 0.94 to 1.03%. For all the AAM mixes, the fixed contents of the alkali activator solution including the sodium silicate (Na_2SiO_3) and sodium hydroxide (NaOH) were 171.43 and 228.57 kg, respectively. The prepared AAMs were characterized via different tests to determine the impact of WCP at elevated volume on their bond strength performance and durability.

6.4 TEST PROCEDURES

The cubic specimens (50 × 50 × 50 mm) of the prepared mortars were subjected to hardening tests (compressive strength) following the ASTM C579 standard. In accordance to the ASTM C109, the CS test was made with the load rate of 2.5 kN/s and a minimum of three specimens were used to assess the average CS at different curing periods (1, 3, 7, 28, 56, 90, 180, and 360 days). The capacity of the slant shear bonding among the NC substrate and different AAMs with the stiffer slant shear angle of 30° was determined in accordance with ASTM C882 [48]. For the sample preparation, the NC was first cast in the cylinder moulds (100 × 200 mm) and cured in water for 3 days. Next, these samples were left at 25 ± 3°C with the relative humidity of 75% for 28 days. Afterwards, these specimens were cut to half following the half slanted dimension (with 30°) before being fixed in the cylindrical moulds (100 × 200 mm) as shown in Figure 6.5. It was cast for the second part (OPC and AAMs) to evaluate their slant shear bond strength (defined as the ratio of the maximum load at the failure to the bond area) performance at different stages of curing (1, 3, 7, and 28 days). The average of the three samples from every AAM mix was considered to report the results of the slant shear bond strength.

TABLE 6.2
Compositions and Proportions of Various Components Used for the AAMs Synthesis

	Binder (Mass), kg			Binder (Volume), m^3				Fine Aggregate		
Mix	WCP	GBFS	FA	WCP	GBFS	FA	Total	Mass (kg)	Volume(m^3)	B:A VolumeRatio
$AAMs_1$	500	500	0	0.191	0.171	0	0.363	1000	0.384	0.94
$AAMs_2$	500	400	100	0.191	0.137	0.045	0.374	1000	0.384	0.97
$AAMs_3$	500	300	200	0.191	0.103	0.091	0.385	1000	0.384	1.00
$AAMs_4$	500	200	300	0.191	0.068	0.136	0.396	1000	0.384	1.03
$AAMs_5$	600	400	0	0.229	0.137	0	0.367	1000	0.384	0.95
$AAMs_6$	600	300	100	0.229	0.103	0.045	0.378	1000	0.384	0.98
$AAMs_7$	600	200	200	0.229	0.068	0.091	0.389	1000	0.384	1.01
$AAMs_8$	700	300	0	0.268	0.103	0	0.371	1000	0.384	0.97
$AAMs_9$	700	200	100	0.268	0.068	0.045	0.382	1000	0.384	0.99

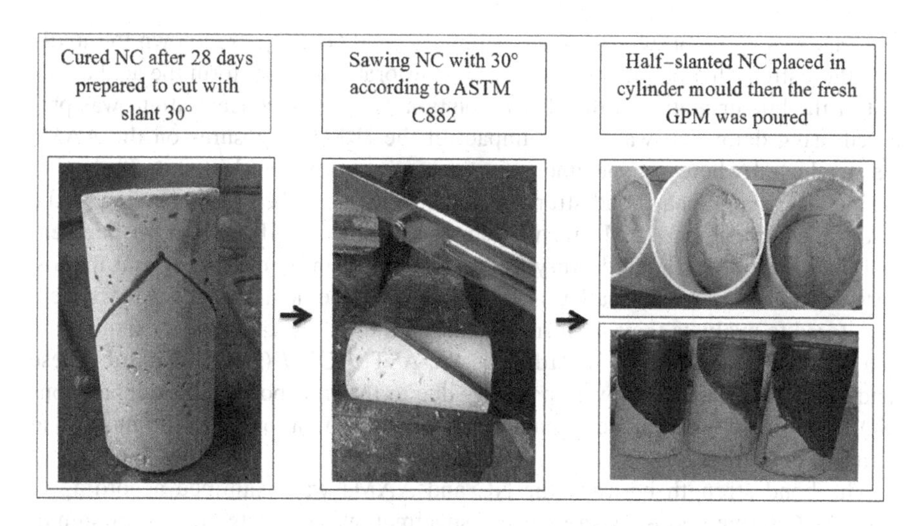

FIGURE 6.5 Procedure for the slant shear bond strength test of the AAMs.

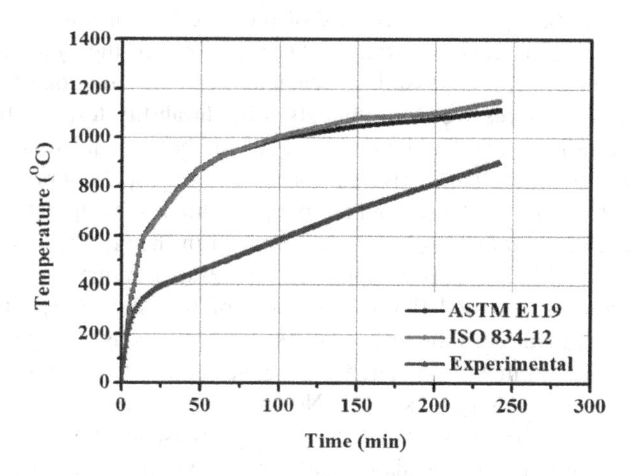

FIGURE 6.6 Time against temperature variation for the ASTM E119 [49] and ISO 834-12 [50].

The evaluation of the bond strength performance of the AAMs under the elevated temperatures is one of the main objectives of this study. The specimens were tested using an automatic electric furnace. The cubical (50 × 50 × 50 mm) specimens of the AAMs were cast according to ASTM C597 and cured for 28 days in ambient temperatures. From each set of AAMs 3 samples were tested at the high temperatures of 400, 700, and 900°C under various time durations (Figure 6.6). The AAMs were chilled via the air cooling method after heating. The slant shears bond strength before and after the high temperature exposure were measured to determine the elevated temperature-mediated bond loss. Finally, the relative quality of the AAMs was assessed after heating. In addition, XRD and SEM analyses of the AAM specimens were performed to determine their microstructures and surface morphologies.

Concretes and mortars exposed to sulphuric acid (H_2SO_4) are known to rapidly lose their strength and undergo structural deterioration. To perform the acid attack test of the binder matrices, the H_2SO_4 solution (with 10% concentration) was prepared using deionized water. The impact of the H_2SO_4 exposures on the AAMs was evaluated where six specimens for each mixture were selected at 28 days of curing. The slant shear bond strength was also assessed before immersing in the acidic medium. Each AAM mix was engrossed in the acid solution for one year and the solution was altered every 90 days to maintain its pH throughout the test period. In each instance (half year and one year), the acid-exposed AAMs were monitored to evaluate their performance depending on the qualitative examination and residual bond strength according to the ASTM C267 (2012) standard. These acid attacks on the AAMs were basically due to the transport of the sulphate ions $(SO_4)^{2-}$ together with calcium, magnesium or sodium cations into the mortars at different concentrations.

The bond strength between the NC and AAMs was examined according to ASTM C666 (method A) wherein every specimen was subjected to a minimum of 300 freeze–thaw cycles to evaluate the slant shear bond strength. In is important to note that by reducing the temperature of the mixes from 5 to –20°C (75% of the cycle time) and increasing it from –20 to 5°C (25% of the cycle time) within 5 hours for every cycle, it is possible to reach at the same situation of the freeze–thaw cycles in the aggressive environments. The durability test can be regarded as the most significant one for the prepared AAMs because no standard method was obtainable for the alkali-activated concrete and geopolymers. Therefore, it was implemented here for the studied temperature domain, temperature ramp and duration of cycles. Two approaches were proposed in the ASTM C666 standard: methods A and B. Method A is comprised of freezing and thawing the specimen in water. In contrast, method B was composed of freezing the specimen in the air and thawing in water. In the present work, method A was used because it was more convenient than method B. The freezing was conducted following approach A where nine mixes (composites of the NC and AAMs) were made in the cylindrical shape (100 × 200 mm). At 28 days of age, these samples were thawed at room temperature to record the slant shear bond strength. Later, these specimens were enclosed in the container to submerge in water. The water temperature was controlled automatically using a timer to monitor every freeze–thaw cycle. After these mixes were subjected to 300 freeze–thaw cycles the durability performance of every AAM was evaluated after every 50 cycles based on the residual bond strength.

No standard method for the wet–dry cycle durability test was available. In a tropical nation like Malaysia, the weather conditions are random and quickly change from hot to dry (for a few days) to rainy. Thus, those few days were very significant during the conduction of such a test. This test was designed to mimic the natural Malaysian environmental condition for accelerating the wet–dry cycles. Figure 6.7 shows the condition for the wet–dry test cycles wherein the data for 50 cycles was recorded to determine any alteration in the loss of the shear bond strength. The total wet–dry cycles were adopted to evaluate the durability of the specimens.

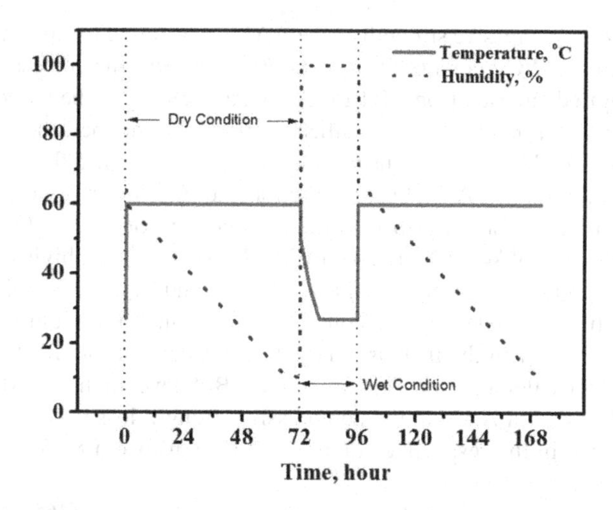

FIGURE 6.7 The AAM wet–dry cyclic process for one cycle.

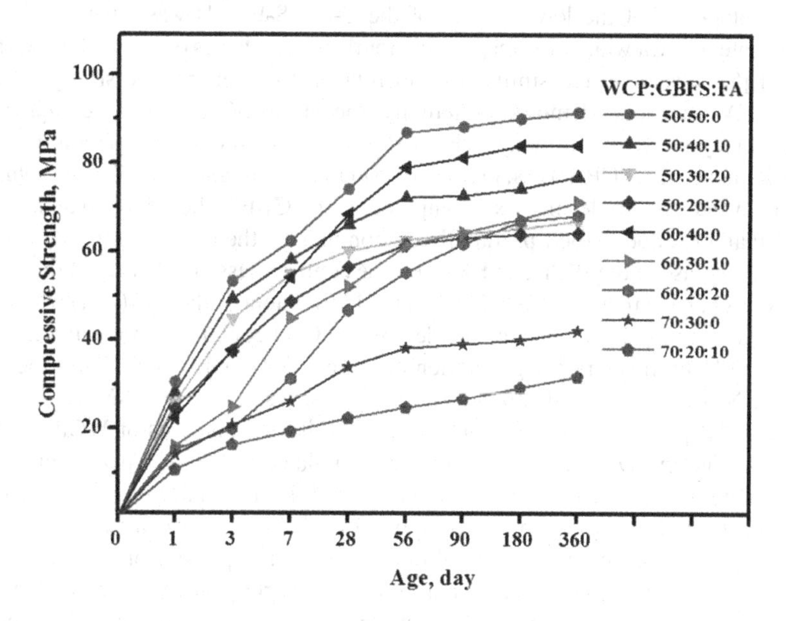

FIGURE 6.8 CS of AAMs tested in different periods.

6.5 RESULTS AND DISCUSSION

6.5.1 Compressive Strength of AAMs

Figure 6.8 displays the influence of the high WCP contents on the early and late compressive strength development of the studied AAMs. The increment in the compressive strength was found to be directly proportional to the curing time of all the AAM specimens. The compressive strength of the AAMs was found to vary inversely with the increase in the WCP content from 50 to 70%, wherein the corresponding compressive strength decreased from 70.1 to 34.8 MPa at 28 days. The reduction in total calcium content and increment in the silica content with the increase in the content of WCP in the AAMs was responsible for this drop [22, 51, 52]. This negatively affected the C-S-H and C-A-S-H gel production and thus lowered the CS of AAMs. This reduction was mostly ascribed to the high content of silica (70%) and larger particle size (35 μm) of WCP. Moreover, the AAMs cast with the high WCP contents produced the compressive strength values of 81, 94, and 97% at 28, 56, and 90 days of age, respectively. Conversely, the content of silica and aluminium was increased in the AAM matrix with the increase in the FA replacement for GBFS which negatively affected the calcium level. Thereafter, the GBFS level in the $AAMs_1$ decreased from 50 to 20% and in $AAMs_{10}$ the compressive strength loss was more than 70%, indicating a drop in the respective compressive strength from 84.6 to 24.8 MPa at 28 days of age.

Comparing the compressive strength at 28 days to a later age (365 days) was generally found to increase with the increase in the hydration time for all the AAMs. The development of the low quantity of the C-(A)-S-H gels was attributed to the small content of Ca which was responsible for the strength loss of the studied AAMs. Rashad [53] also reported similar diminishing in the compressive strength value for the FA-incorporated mortars. Actually, the observed loss in the compressive strength may arise due to many reasons. First, the chemical composition dissimilarities among the WCP, GBFS, and FA considerably influence the alkaline solution activation into the binder matrix. Compared to the GBFS, the poorer reaction rate of WCP and FA due to their partial dissolution may be the other reason [54]. Third, with the increase in the WCP and FA levels the compactness and density of the AAM matrix might have reduced. Fourth, when the low content of the NH (4 M) where the compressive strength was mostly decided by the CaO content for replacing the low amount of NaO. In short, the generation of more C-S-H and C-A-S-H gels together with the N-A-S-H gel could enhance the compressive strength of AAMs.

The XRD patterns of the prepared AAMs with high amounts of WCP are presented in Figure 6.9. The intensity of the formulated C-S-H gel peaks tended to decrease with the increase in the amount of WCP in the binder matrix, especially the peak at 30°. For samples prepared with 70% WCP, the C-S-H peaks at 43° were observed to get substituted by the hydrotalcite ones. The peaks from the calcites at 38.5 and 51° were observed to be replaced by the quartz peaks with the increase in the WCP level above 50%. The AAMs containing elevated amounts of WCP revealed five prominent characteristic peaks of the quartz positioned at 17, 21, 28, 38.5, and 51°. Also, a weak intensity peak at 30° due to C-S-H was appeared. The amount of the non-reacted silicate in the AAMs was increased with the increase in the WCP contents from 50 and 70%, causing a reduction in the corresponding compressive strength value from 74.1 to 34.8 MPa at 28 days of age, which was ascribed to the lowering of the C-S-H generation in the specimens.

Figure 6.10 shows the XRD patterns of the AAMs prepared with 50% of the WCP wherein the consequences of FA substitution for GBFS is clear. The intensity

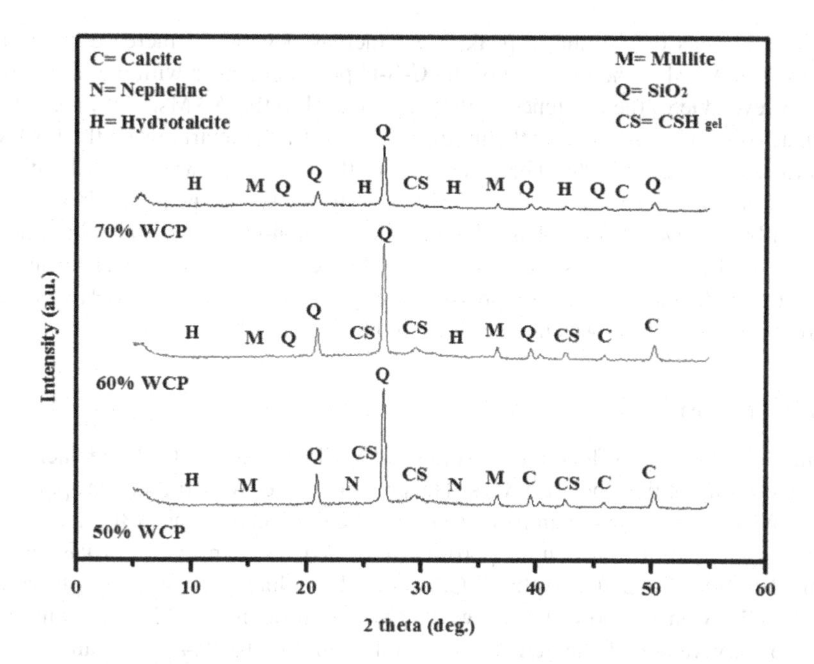

FIGURE 6.9 The XRD patterns of AAMs prepared with high WCP content.

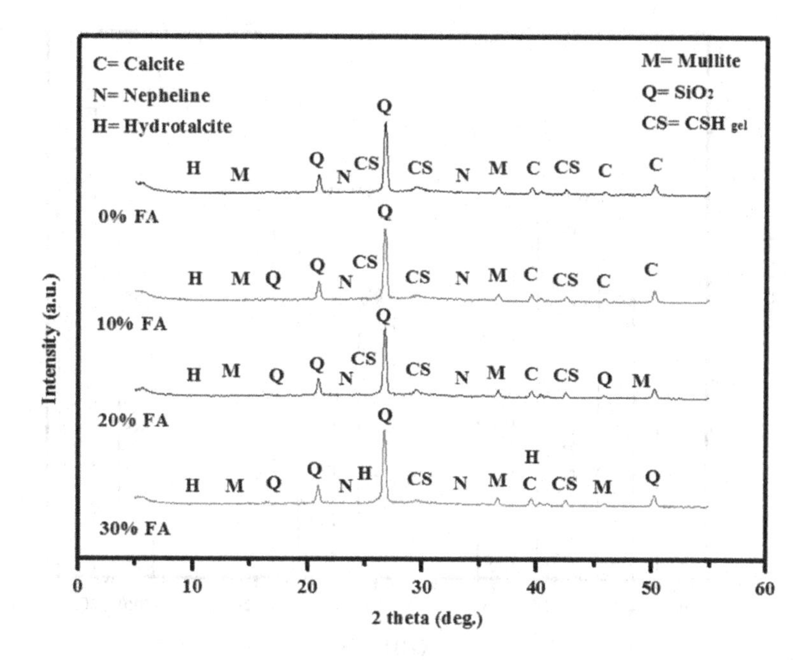

FIGURE 6.10 Effects of FA substitution for GBFS on XRD patterns of AAMs with 50% of WCP.

and occurrences of the quartz peaks were increased with the increase in the FA level in the AAMs. The intensity of the C-S-H peak decreased with the increase in the FA level due to the existence of the low Ca level in the AAMs. First, the calcite peak at 46° was substituted by the mullite and later by the quartz when the FA level was raised from 20 to 30%. The generation of the C-S-H gel was lowered with the increase in the FA content in the AAMs wherein the C-S-H peak at 27° was substituted by the hydrotalcite phase. Meanwhile, the non-reacted or partially reacted silicate level in the AAMs was raised with the decrease in the C-S-H formation, leading to a decrease in the compressive strength from 74.1 to 56.5 MPa when the corresponding FA level was increased from 0 to 30%.

6.5.2 Porosity

Figure 6.11 shows the effect of high volume of WCP content and GBFS replacing FA on the porosity of the studied AAMs. The porosity values were directly proportional to the WCP content, where an increase in the WCP in place of the GBFS from 50 to 70% has led to an increase in the porosity of the AAMs from 13.2 to 21.1%, respectively. For 50, 60 and 70% of the WCP levels, the influence of the FA replacement for the GBFS on the porosity and water absorption of the AAMs was examined. The porosity values of the AAMs increased from 13.3 to 19.4% with an increase in the content of FA replacement for the GBFS from 0 to 30%, respectively, with 50% WCP. A similar trend was observed with the 60 and 70% of WCP-containing

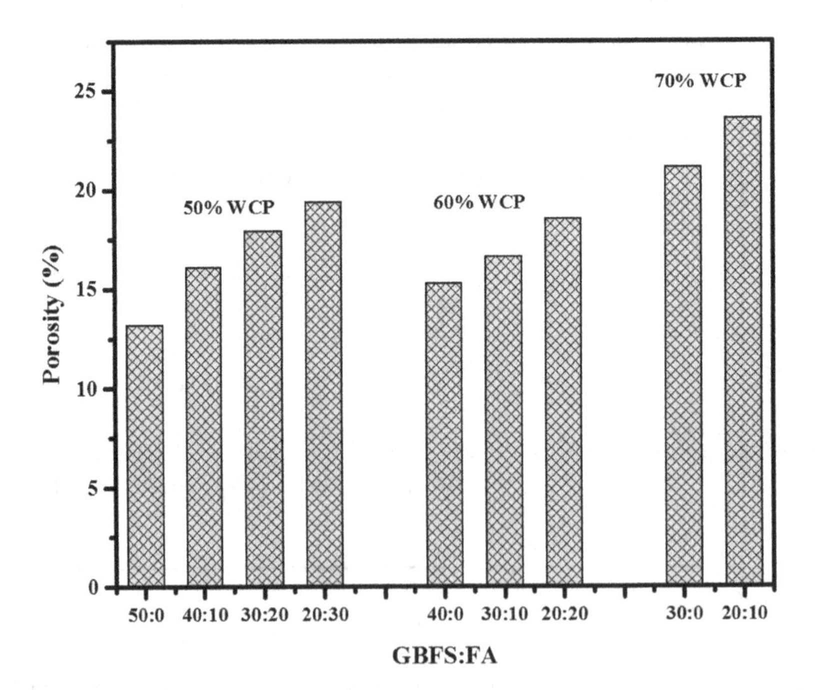

FIGURE 6.11 Effects of the WCP content on the porosity of the AAMs.

AAMs. In general, as the FA content increased and the GBFS content decreased the porosity level of the AAMs increased. The increasing levels of WCP and FA caused an increase in the non-reacted and partially reacted particles, thereby reducing the C-S-H gel products with highly porous structure [55, 56].

6.5.3 Bond Strength

The results of the slant (for 30°) shear bond strength tests of the AAMs containing a high volume of WCP at 1, 3, 7, and 28 days are illustrated in Figure 6.12. The bond strength as increased with the time of aging. However, the bond strength of the AAMs dropped with the increasing content of the WCP replacing GBFS. First, the bond strength of the AAMs containing 50% of WCP at 28 days was 4.2 MPa and then dropped from 3.8 to 2.7 MPa with a rise in the WCP level from 60 to 70%, respectively. Similarly, the bond strength of all the AAMs was higher than that of the normal OPC mortar. It is evident that the AAMs possessed a better bonding character than the cement materials. The effects of the FA replacement for the GBFS in the AAMs matrix were also evaluated and the bond strength was reduced with the rise in the FA level. Furthermore, the bond strength dropped from 4.2 to 2.7 MPa with the increase in the FA level from 0 to 30%, respectively (for specimens prepared with 50% WCP). A similar trend was observed for the mortar specimens prepared with 60 and 70% of the WCP wherein the rise in the FA content resulted in the reduction of the bond strength. This is in line with previous studies [57, 58] on the improved bond

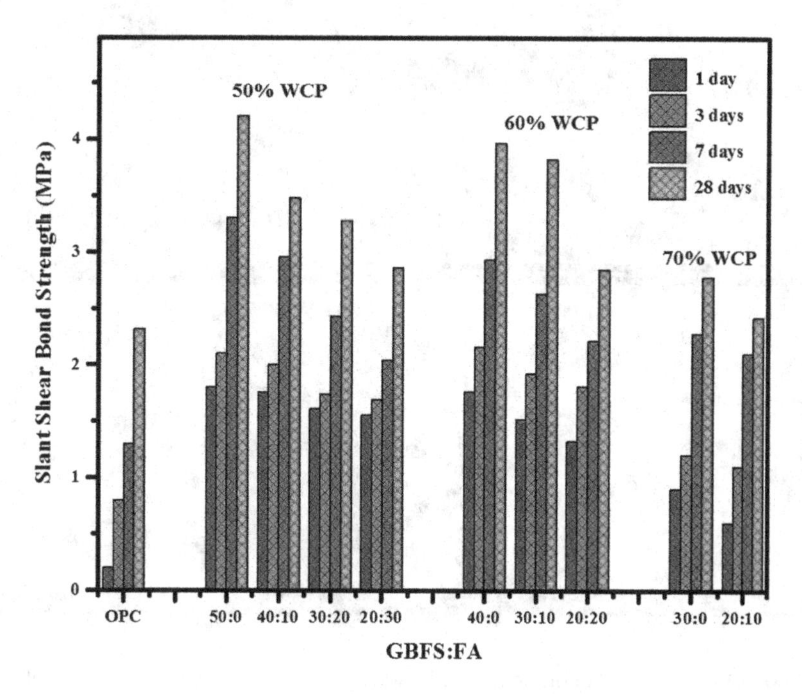

FIGURE 6.12 The 30° slant bond strength of the AAMs containing various levels of WCP.

strength of aluminosilicate-based alkali-activated AAMs, in which with the increase of the calcium content the extra C-(A)-S-H gel co-existed with the N-A-S-H gel. The increase in the reaction products at the interface transition zone between the NC substrate and AAMs was responsible for the enhanced strength at the contact zone [59]. However, the AAMs with a high value of WCP content (70%) exhibited a slight decrease in the shear bond strength.

Figure 6.14 shows the effect of WCP and FA content on the bond zone between the AAMs and concrete substrate. The increase in the content of WCP and FA as GBFS replacement affected the bond zone between the AAMs and concrete substrate. The number and size of the cracks was directly influenced by the WCP and FA content. $AAMs_1$ prepared with 50% of WCP (Figure 6.13a) presented a lower number of cracks with a smaller size compared to $AAMs_5$ (Figure 6.13b) and $AAMs_8$ (Figure 6.13d) prepared with 60 and 70% of WCP, respectively. Likewise, $AAMs_4$ (Figure 6.13c) containing 30% of FA as the GBFS replacement showed a lower performance compared to the $AAMs_1$ mix. The SEM results enabled to understand the sharp fall in the bond strength with the increasing content of the WCP (from 50 to 70%) in the concretes matrix. This observed drop in the slant shear bond strength can be ascribed to the bond defects between the AAMs and the concrete substrate matrix. Meanwhile, all the AAMs presented an excellent bond strength and the failure zone occurred outside the bond zone as depicted in Figure 6.14. It was evident

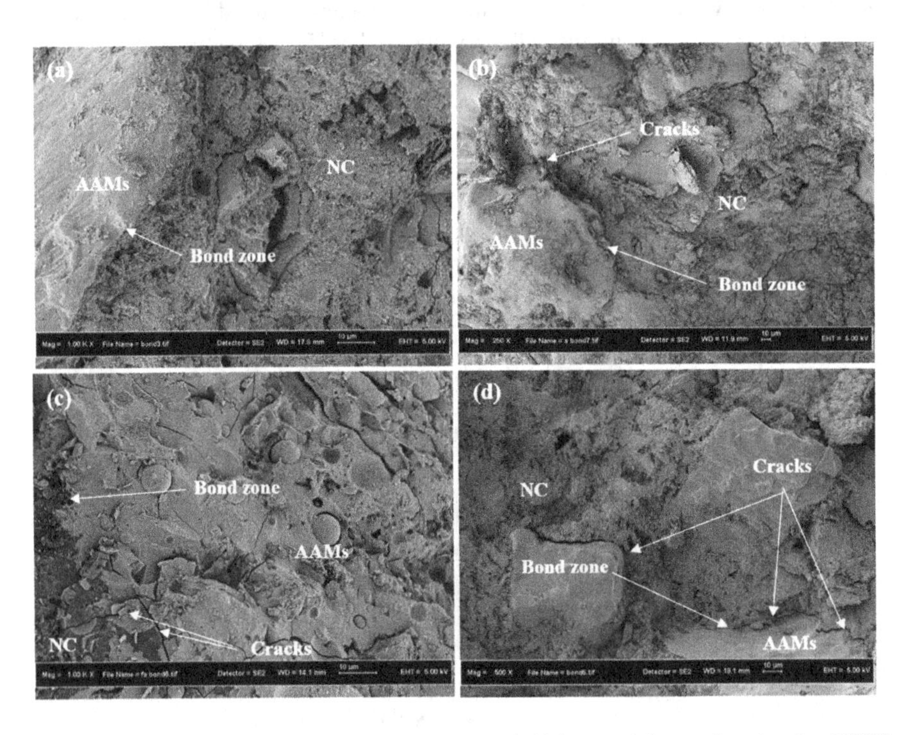

FIGURE 6.13 The SEM of the bond zone between AAMs containing various levels of WCP and normal concrete. (a) $AAMs_1$, (b) $AAMs_5$, (c) $AAMs_4$, and (d) $AAMs_8$.

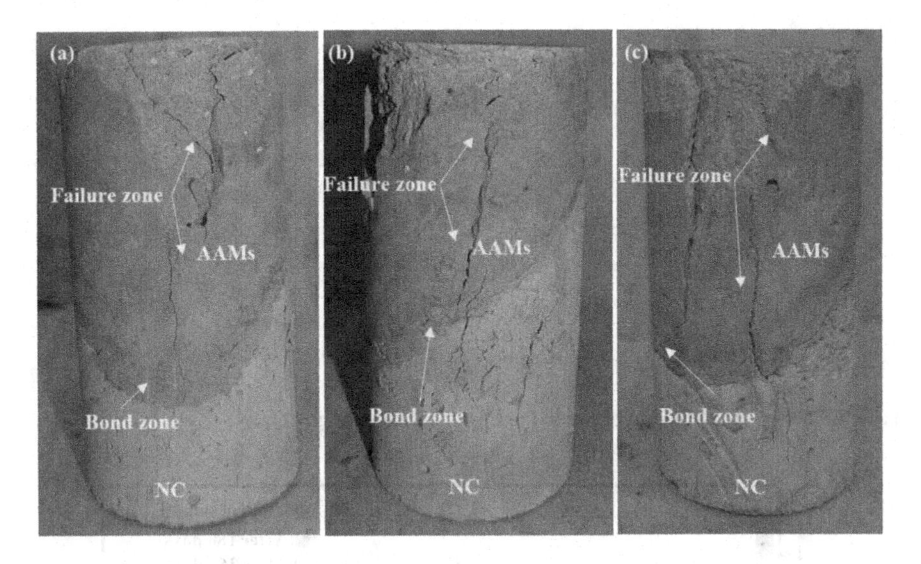

FIGURE 6.14 The failure zone under the slant shear bond test for the AAMs containing a high volume of WCP.

that the failure occurred in the alkali-activated and substrate concrete sides, which explained the excellent bond strength between the AAMs matrix and NC. However, the deterioration in the specimens increased with an increase in the WCP content in the matrix. The specimens prepared with 50% of WCP (Figure 6.14a) presented a higher performance compared to the specimens prepared with 60% of WCP (Figure 6.14b) and 70% of WCP (Figure 6.14c).

The results for the compressive strength and slant shear bond strength of the AAMs at 1, 3, 7, and 28 days are presented in Figure 6.15. The bond strength values were found to be correlated with their compressive strength. The achieved compressive strength values were used as a response factor with the bond strength values for the predictive parameters. The linear regression analyses were performed to relate the experimental data using Equation 6.1. For all the samples, the value of R^2 was found to be 0.76, signifying excellent correlation confidence. The linear regression relation can be written as

$$\text{Bond Strength} = 0.0411\text{CS} + 0.7091\left(R^2 = 0.7556\right) \tag{6.1}$$

6.5.4 Effect of Sulphuric Acid Attack

The results of the slant shear bond strength tests at 30° for the AAMs exposed to an acid environment for 180 and 365 days are illustrated in Figure 6.16. For all the samples, the bond strengths were decreased with the increased time duration of exposure in the acid solution. However, the loss in the bond strength of the AAMs were decreased with the rising content of the WCP replaced GBFS. The loss in the bond strength of the AAMs containing 50% of the WCP after 180 days of exposure

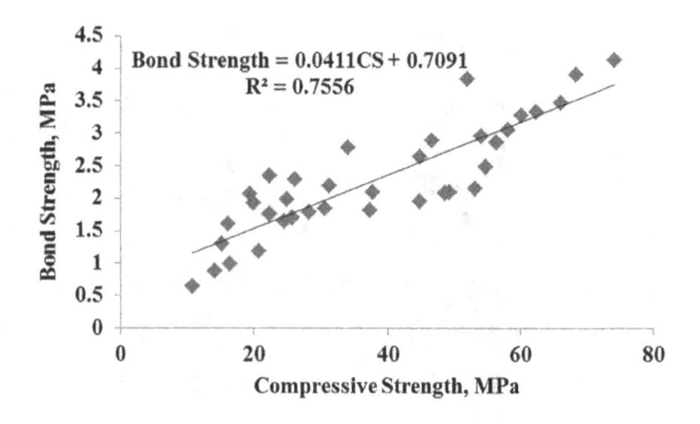

FIGURE 6.15 Relationship between the CS and bond strength.

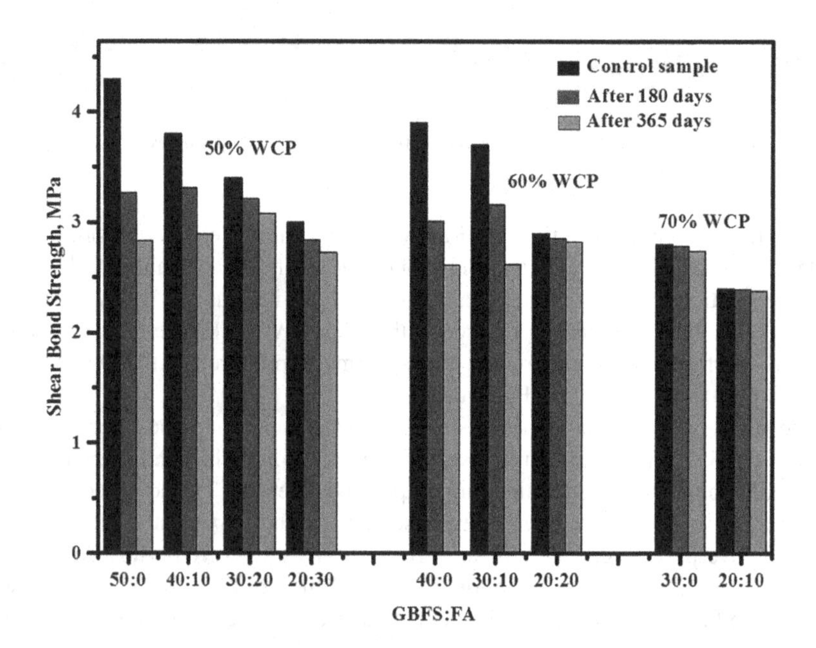

FIGURE 6.16 Effect of acid attack on the shear bond strength of the AAMs.

was 23.7%, which dropped from 19.8 to 1.3% with the respective increase of the WCP from 60 to 70%. Similarly, the loss in the bond strength of the AAMs after 365 days dropped from 34.1 to 3.1% with an increase of the WCP content from 50 to 70% replacement of the GBFS matrix, respectively. It is evident that the AAMs with high content of the WCP possessed better bonding character in the acid environment than the GBFS materials. The effect of FA substitution in the GBFS on the bond strength of the AAMs at various levels of the WCP was evaluated. It was demonstrated that a rise in the FA level to 30% for the mortar matrix prepared with 50% (high volume) of WCP can lower the loss in the bond strength around 18% compared to the specimen prepared without FA. However, the specimens prepared with the elevated quantity of WCP (70%) and 10% of FA presented an excellent performance and showed loss in the bond strength lower than 0.7% and 1.1% after 180 and 365 days, respectively. When the mortar specimens were exposed to the sulphuric acid, the $Ca(OH)_2$ reacted with the SO_4^{-2} and formed the gypsum ($CaSO_4.2H_2O$). This in turn resulted in the matrix expansion and extra cracks within the specimens [60] thereby leading to the bond strength weakening between the mortar and NC. The reduction in the GBFS content replaced by the WCP and FA in the mortar matrix led to restricting the amount of $Ca(OH)_2$ and reduced the formulation of the gypsum and cracks. This disclosure was attributed to the excellent performance of the AAMs prepared with high contents of WCP and FA.

Figure 6.17 displays the XRD patterns of the proposed AAMs under 10% of H_2SO_4 exposure (immersion) after 360 days. These AAMs showed the existence of the gypsum (calcium sulphate hydrate). The intensity of the gypsum peak at 2θ = 29.8° was reduced with the rise in the FA content. Also a gypsum peak at 11.8° and another at 20.9° near to the quartz peak (20.8°) appeared. Bellmann and Stark [61] also reported the appearance of the quartz and gypsum peaks at 20.8 and 20.9° 2θ, respectively. which were hard to differentiate. Nevertheless, double peaks that were evidenced upon the close scrutiny indicated the attendance of both quartz and gypsum. An increase in the FA level to 30% in place of GBFS in the mortar could restrict the gypsum formulation as indicated by the appeared peaks at 29.8°, thereby enhancing the resistance of the mortars against the H_2SO_4 attack. In addition, it suggested

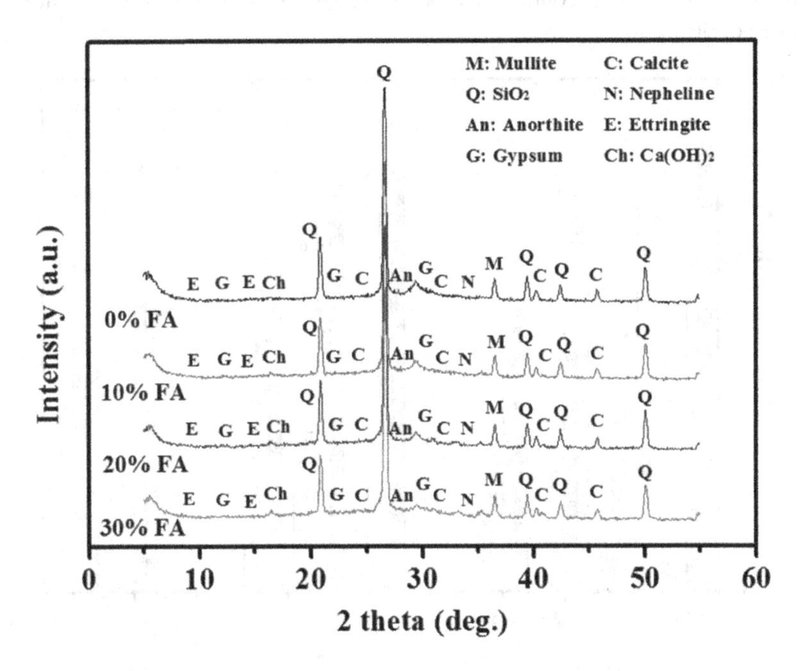

FIGURE 6.17 XRD patterns of the AAMs containing 50% of WCP and exposed to the H_2SO_4 environment for 12 months.

that the AAMs containing a high volume of GBFS can generate more C-A-S-H and C-N-A-S-H gels that are susceptible to H_2SO_4 attack.

6.5.5 Effect of Elevated Temperatures

Figure 6.18 shows the effect of high elevated temperatures on the slant shear bond strength of the AAM specimens to concrete substrate prepared with diagonal slanted at 30° angle. For all the AAMs, the loss in bond strength increased with the increase in temperature from 27 to 900°C. The results indicated an increase in the loss of the bond strength with the rise in temperatures from 27 to 900°C. The bonding strength at 27°C for all the mixtures was higher than after the exposure to various elevated temperatures. The loss in the bond strength values decreased with the addition of WCP and FA, where the loss percentage at 400°C dropped from 37 to 3.8% with the addition of WCP in place of GBFS from 50 to 70%, respectively. The effect of the FA substitution for GBFS in each level of high volume WCP on the slant shear bond strength was also assessed. The positive effect on the bond value was observed with the rise in the FA content. The percentage of loss on the bond strength dropped from 37 to 21% with the respective increase of FA substitution for GBFS from 0 to 30%. A similar trend was observed in the specimens exposed to 700 and 900°C. Those prepared with the high volume of WCP and FA showed lower loss in the bond strength compared to other samples. However, the highest loss on the bond strength (82%) was observed in the AAMs made from 50% of the WCP and GBFS compared to the mortar (41%) that contained 70% WCP, 20% GBFS and 10% FA under the exposure of 900°C.

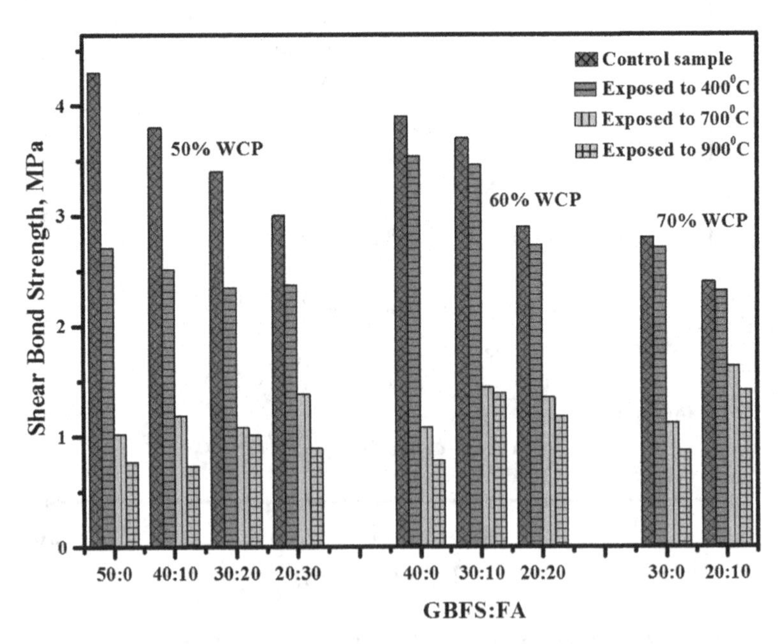

FIGURE 6.18 Effects of elevated temperatures on shear bond strength of AAMs.

The XRD patterns of synthesized AAMs under the exposure of the elevated temperatures up to 900°C are presented in Figure 6.19. The semi-crystalline alumina-silicates gel and quartz (Q) peak (Figure 6.19a) was observed after the exposure below 400°C. The appearance of the broad peak in all the AAMs in the range of $2\theta = 24.8 - 31°$ was ascribed to the formation of the crystalline zeolites as secondary reaction products after the completion of fire resistance test. The XRD patterns of the AAMs exposed to 700 and 900°C are depicted in Figure 6.19b and 6.19c, respectively. For WCP-based AAMs heated at 700°C, the emergence of the strong peaks can primarily be assigned to the presence of the mullite, quartz and nepheline phases. The mullite was the only stable crystalline structure of the Al_2O_3–SiO_2 coordination. Due to their high stability, low thermal expansion and excellent oxidation resistance against elevated temperature, mullite was retained in the mortars. After exposing to 700°C, the quartz peaks were steady but the peaks of mullite appeared increasingly intense. Furthermore, around 400°C a phase transformation from the goethite to hematite was observed due to the release of most of the constituent water molecules. The outward OH- flux and the concurrent diffusion flow in the grain structure might have caused a local accumulation of the internal stress, and thus produced hematite grains fractures. At this temperature, the shapes and sizes of the hematite grains more or less remained the same like the original goethite. However, the AAMs exposed to 900°C showed the disappearance of hematite peaks and appearance of the crystalline nepheline (sodium aluminium silicate, $AlNaSiO_4$) peaks even if the quartz and mullite phase were the dominant ones (Figure 6.19c). Compared to other mortars, the mortar prepared with 70% WCP and a high volume of FA showed some stable peaks at high temperatures.

Figure 6.20 illustrates the SEM micrographs of the AAMs prepared with high WCP content. The influence of the elevated temperature exposure (400, 700, and 900°C) on the microstructure of the AAMs prepared with 50 and 70% of the WCP were assessed. The structures of the AAMs were increasingly converted into the less dense networks with micro-cracks and bigger pores with the rise in temperatures. The SEM images of the AAMs prepared with 70% of WCP after the exposure to ambient temperatures (400, 700, and 900°C) were obtained from a crushed section. The Fe micro-cracks on the AAMs surface was observed when subjected to high temperatures where unreacted particles of the WCP, FA and some spherical holes were also evidenced. It is known that the WCP contains several hollow spheres and the partial dissolution of such spherical particles can generate the highly dispersed tiny pores in the matrix [1]. The voids of the hollow cavities left by the dissolved WCP and FA particles in the matrix were filled by such un-reacted tiny WCP spheres. In opposition, the mortar prepared with 70% of the WCP showed more stable surface at high temperatures than the one made from 50% of the WCP.

6.5.6 Effect of Freeze–Thaw Cycles

The influence of freeze–thaw cycles on the bond strength performance of the AAMs specimens are illustrated in Figure 6.21. The tested specimens showed an inverse relationship between the residual bond strength and freeze–thaw cycles as well as

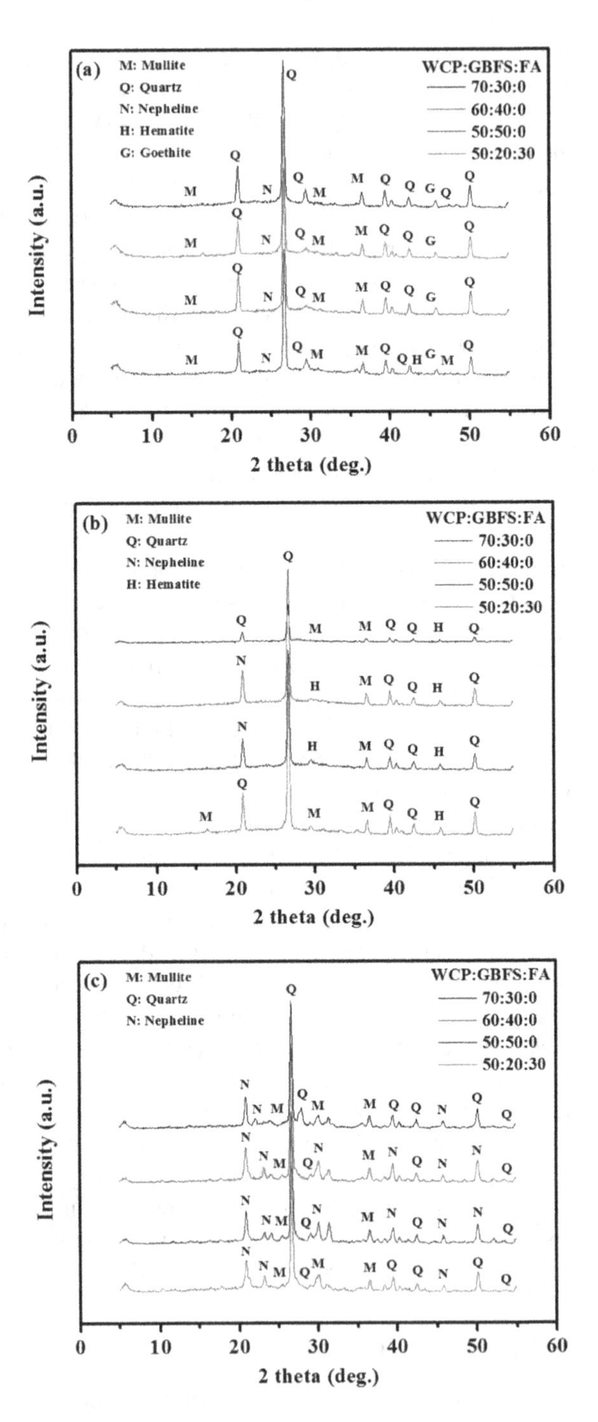

FIGURE 6.19 XRD analysis of the specimens after fire exposure to (a) 400°C, (b) 700°C, and (c) 900°C.

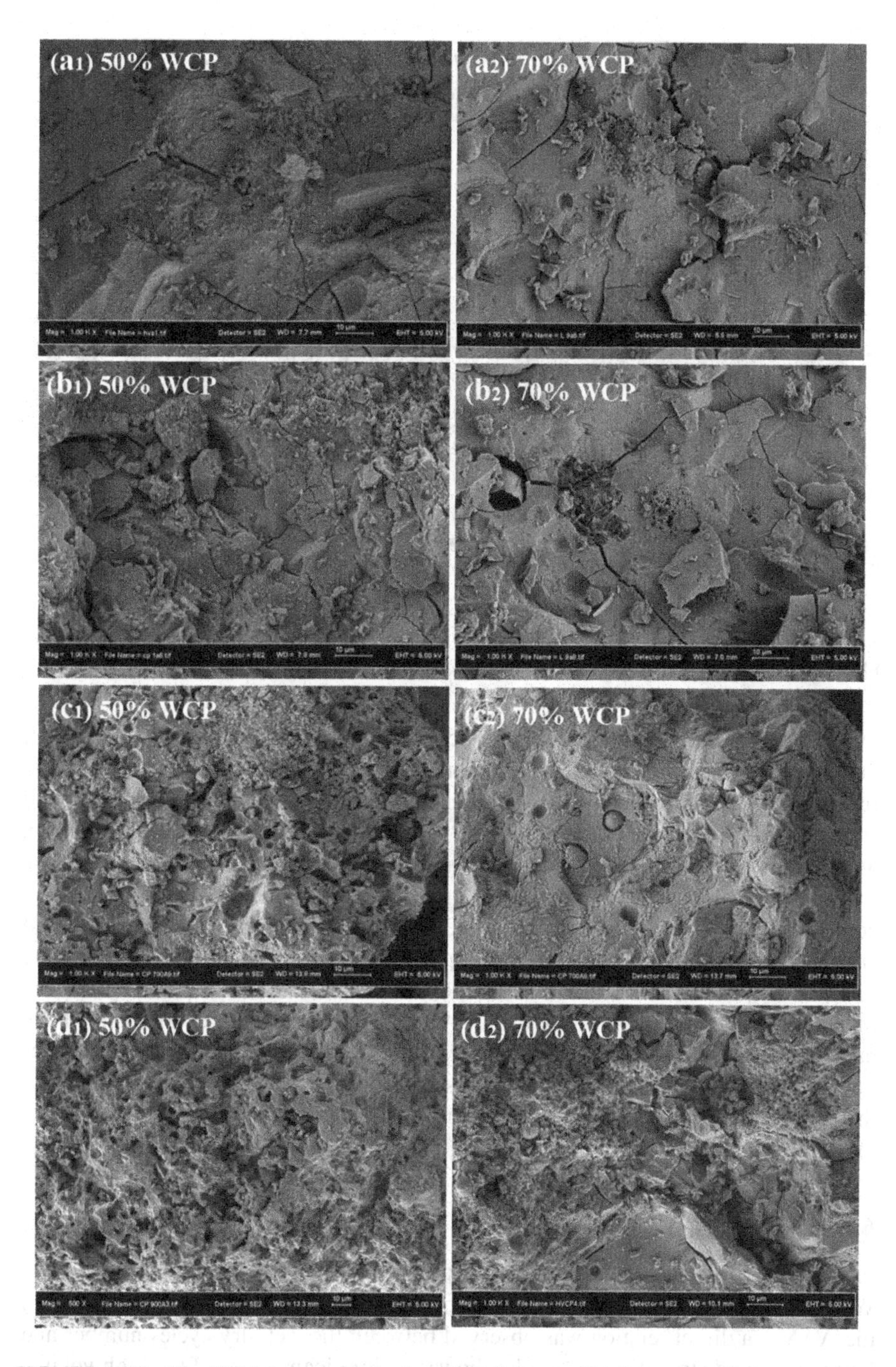

FIGURE 6.20 The SEM images displaying the influence of high temperatures on the morphologies of AAMs with different amounts of WCP. (a) 27°C, (b) 400°C, (c) 700°C, and (d) 900°C.

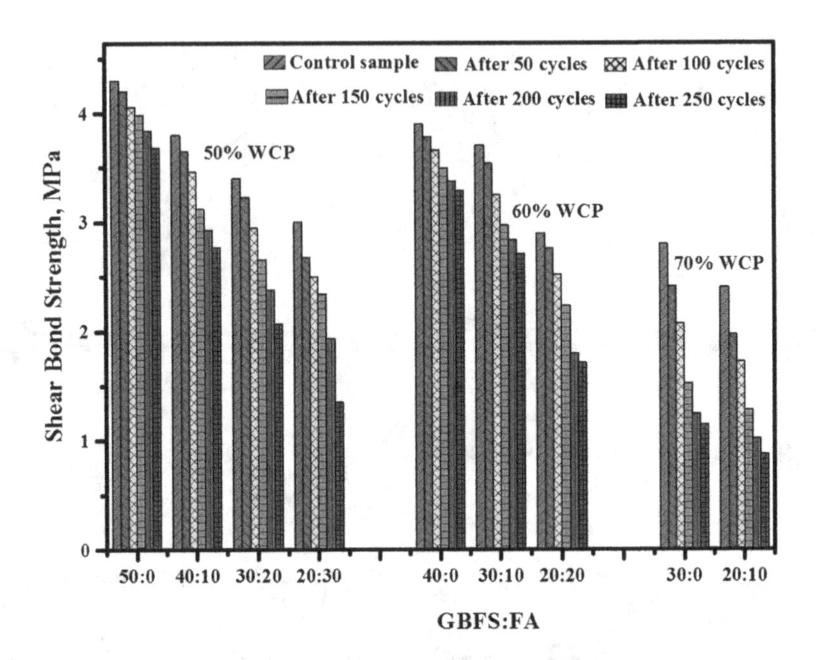

FIGURE 6.21 Effects of freeze–thaw cycles on shear bond strength of AAMs.

the WCP and FA content. The rise in the content of WCP from 50 to 70% caused an increase in the loss percentage on the bond strength from 13.8 to 58.4% after 250 freeze–thaw cycles, respectively. A similar trend was observed for the residual bond strength at a high amount of FA as GBFS replacement in each level of WCP matrix where the loss percentage on the bond strength was increased from 26.7 to 54.3% with the increase in the FA content from 10 to 30% compared to 13.8% with FA content of 0%. It was found that the increase in the content of WCP and FA could enhance the amount of non-reacted and partly reacted silicate and minimized the final C-S-H product. The amount of the porous structure in the prepared specimens increased with the increase of WCP and FA where a high quantity of non-reacted and partly reacted silicate was formed. The existence of the voids aided the intensification of the ice and destroyed the particles interlocking [62]. This was attributed to the reduction of the AAMs resistant to the freeze–thaw cycles, indicating a higher loss in the bond strength.

6.5.7 Effect of Wet–Dry Cycles

Figure 6.22 illustrates the residual bond strength of the AAMs with a high level of WCP plus GBFS and FA when exposed to wet–dry cycles at 28 days of age. For all the AAMs, a direct relation was observed between the wet–dry cycles number and loss in the bond strength. The results showed a significant effect of the high volume of WCP and FA on the strength loss of the AAM specimens. The percentage of the total loss in the bond strength increased from 6.9 to 21.7% with the increase in the WCP level from 50 to 70% replacement for the GBFS, respectively. Also,

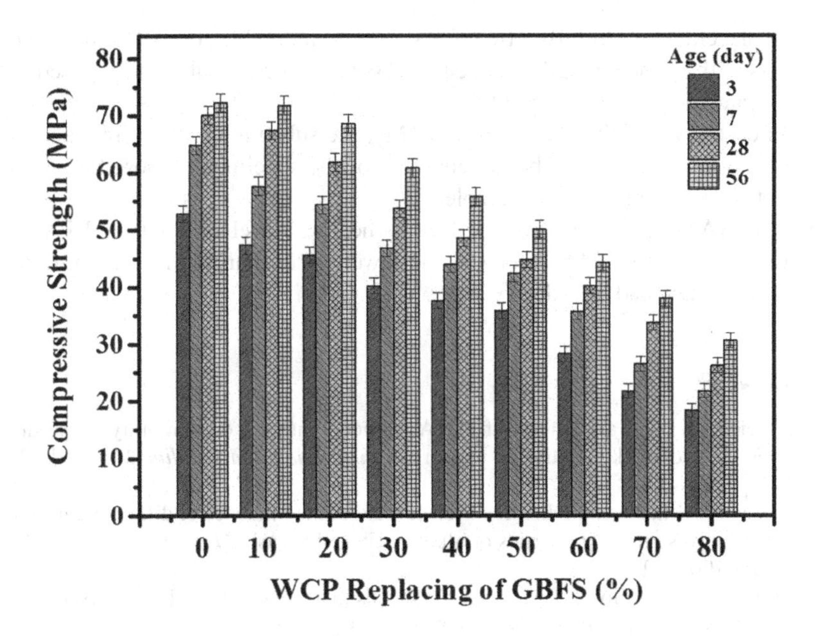

FIGURE 6.22 Effects of the wet–dry cycles on the shear bond strength of the AAMs.

the strength loss was influenced by the level of FA replaced GBFS in the high volume WCP mortar specimens. An increase in the FA level from 0 to 30% negatively affected the microstructures of the specimens and led to an increase in the strength loss from 6.9 to 25.3%, respectively. According to the previous findings [63], the main reason for the loss of the strength was due to the increase in overall porosity with the rise in WCP and FA contents, where no pores were available in the matrix. This greatly favoured the water entry into the matrix during the wet–dry cycles. The internal and external deterioration increased, leading to a loss in the bond strength and low durability over time.

6.6 CONCLUSIONS

This chapter reported the effects of aggressive environments (i.e. sulphuric acid attack, elevated temperatures, freeze–thaw cycles and wet–dry cycles) on the bond strength of AAMs containing various levels of WCP, GBFS, and FA. The experiments conducted to assess these parameters were bond tests performed in two cycles of 6 and 12 months for the acid attack, elevated temperatures from 400 to 900°C, 250 freeze–thaw cycles, and 150 wet–dry cycles tests. The XRD patterns and SEM images of the AAMs were also recorded to determine the changes in their structures under the aggressive conditions. The most important conclusions of the study are as follows:

i. Achievement of high durability performance of AAMs (containing a high amount of WCP and FA in place of GBFS) against the acid environment.

ii. Replacement of the GBFS by WCP showed remarkable ability to reduce the loss in the bond strength between AAMs and concrete substrate exposed to elevated temperatures up 900°C.
iii. In each level of the WCP-based AAMs, the substitution of FA in place of GBFS led to enhanced bond behaviour of the specimens exposed to up to 900°C compared to other samples.
iv. The AAM specimens prepared with the high level of WCP and FA as replacement to GBFS disclosed the lower performance when exposed to freeze–thaw and wet–dry cycles.

REFERENCES

1. Huseien, G. F., et al., Effects of POFA replaced with FA on durability properties of GBFS included alkali activated mortars. *Construction and Building Materials*, 2018. **175**: pp. 174–186.
2. Rashad, A. M., A comprehensive overview about the influence of different admixtures and additives on the properties of alkali-activated fly ash. *Materials & Design*, 2014. **53**: pp. 1005–1025.
3. Provis, J. L., A. Palomo, and C. Shi, Advances in understanding alkali-activated materials. *Cement and Concrete Research*, 2015. **78**: pp. 110–125.
4. Castel, A. and S. J. Foster, Bond strength between blended slag and Class F fly ash geopolymer concrete with steel reinforcement. *Cement and Concrete Research*, 2015. **72**: pp. 48–53.
5. Shekhawat, P., G. Sharma, and R. M. Singh, Strength behavior of alkaline activated eggshell powder and flyash geopolymer cured at ambient temperature. *Construction and Building Materials*, 2019. **223**: pp. 1112–1122.
6. Samadi, M., et al., Influence of glass silica waste nano powder on the mechanical and microstructure properties of alkali-activated mortars. *Nanomaterials*, 2020. **10**(2): p. 324.
7. Huseien, G. F., et al., Geopolymer mortars as sustainable repair material: A comprehensive review. *Renewable and Sustainable Energy Reviews*, 2017. **80**: p. 54–74.
8. Du, K., C. Xie, and X. Ouyang, A comparison of carbon dioxide (CO2) emission trends among provinces in China. *Renewable and Sustainable Energy Reviews*, 2017. **73**: pp. 19–25.
9. Huseien, G. F. and K. W. Shah, Durability and life cycle evaluation of self-compacting concrete containing fly ash as GBFS replacement with alkali activation. *Construction and Building Materials*, 2020. **235**: p. 117458.
10. Senthamarai, R. and P. D. Manoharan, Concrete with ceramic waste aggregate. *Cement and Concrete Composites*, 2005. **27**(9–10): pp. 910–913.
11. Huang, B., Q. Dong, and E. G. Burdette, Laboratory evaluation of incorporating waste ceramic materials into Portland cement and asphaltic concrete. *Construction and Building Materials*, 2009. **23**(12): pp. 3451–3456.
12. Senthamarai, R., P. D. Manoharan, and D. Gobinath, Concrete made from ceramic industry waste: Durability properties. *Construction and Building Materials*, 2011. **25**(5): pp. 2413–2419.
13. Hussein, A. A., et al., Performance of nanoceramic powder on the chemical and physical properties of bitumen. *Construction and Building Materials*, 2017. **156**: pp. 496–505.
14. Samadi, M., et al., Properties of mortar containing ceramic powder waste as cement replacement. *Jurnal Teknologi*, 2015. **77**(12): pp. 93–97.

15. Huseien, G. F., et al., Waste ceramic powder incorporated alkali activated mortars exposed to elevated temperatures: Performance evaluation. *Construction and Building Materials*, 2018. **187**: pp. 307–317.
16. Fernandes, M., A. Sousa, and A. Dias, Environmental impacts and emissions trading-ceramic industry: A case study. In *Coimbra: Technological Centre of Ceramics and Glass, Portuguese Association of Ceramic Industry* (in Portuguese), APICER, 2014: pp. 1–10.
17. Pacheco-Torgal, F. and S. Jalali, Reusing ceramic wastes in concrete. *Construction and Building Materials*, 2010. **24**(5): pp. 832–838.
18. Mohammadhosseini, H., et al., Enhanced performance of green mortar comprising high volume of ceramic waste in aggressive environments. *Construction and Building Materials*, 2019. **212**: pp. 607–617.
19. Limbachiya, M., M. S. Meddah, and Y. Ouchagour, Use of recycled concrete aggregate in fly-ash concrete. *Construction and Building Materials*, 2012. **27**(1): pp. 439–449.
20. Heidari, A. and D. Tavakoli, A study of the mechanical properties of ground ceramic powder concrete incorporating nano-SiO_2 particles. *Construction and Building Materials*, 2013. **38**: pp. 255–264.
21. Huseien, G. F., et al., The effect of sodium hydroxide molarity and other parameters on water absorption of geopolymer mortars. *Indian Journal of Science and Technology*, 2016. **9**(48): pp. 1–7.
22. Huseien, G. F., et al., Effect of metakaolin replaced granulated blast furnace slag on fresh and early strength properties of geopolymer mortar. *Ain Shams Engineering Journal*, 2016. **9**(4): pp. 1557–1566.
23. Zhou, W., et al., A comparative study of high-and low-Al 2 O 3 fly ash based-geopolymers: The role of mix proportion factors and curing temperature. *Materials & Design*, 2016. **95**: pp. 63–74.
24. Ranjbar, N., et al., Compressive strength and microstructural analysis of fly ash/palm oil fuel ash based geopolymer mortar. *Materials & Design*, 2014. **59**: pp. 532–539.
25. Rickard, W. D., et al., Assessing the suitability of three Australian fly ashes as an aluminosilicate source for geopolymers in high temperature applications. *Materials Science and Engineering: A*, 2011. **528**(9): pp. 3390–3397.
26. Chen, R., et al., Effect of particle size of fly ash on the properties of lightweight insulation materials. *Construction and Building Materials*, 2016. **123**: pp. 120–126.
27. Huseien, G. F., et al., Influence of different curing temperatures and alkali activators on properties of GBFS geopolymer mortars containing fly ash and palm-oil fuel ash. *Construction and Building Materials*, 2016. **125**: pp. 1229–1240.
28. Kumar, S., R. Kumar, and S. Mehrotra, Influence of granulated blast furnace slag on the reaction, structure and properties of fly ash based geopolymer. *Journal of Materials Science*, 2010. **45**(3): pp. 607–615.
29. Li, C., H. Sun, and L. Li, A review: The comparison between alkali-activated slag (Si+ Ca) and metakaolin (Si+ Al) cements. *Cement and Concrete Research*, 2010. **40**(9): pp. 1341–1349.
30. Deb, P. S., P. Nath, and P. K. Sarker, The effects of ground granulated blast-furnace slag blending with fly ash and activator content on the workability and strength properties of geopolymer concrete cured at ambient temperature. *Materials & Design (1980-2015)*, 2014. **62**: pp. 32–39.
31. Lee, N., E. Kim, and H. Lee, Mechanical properties and setting characteristics of geopolymer mortar using styrene-butadiene (SB) latex. *Construction and Building Materials*, 2016. **113**: pp. 264–272.
32. Yusuf, M. O., et al., Evolution of alkaline activated ground blast furnace slag–ultrafine palm oil fuel ash based concrete. *Materials & Design*, 2014. **55**: pp. 387–393.

33. Mirza, J., et al., Preferred test methods to select suitable surface repair materials in severe climates. *Construction and Building Materials*, 2014. **50**: pp. 692–698.
34. Alanazi, H., et al., Bond strength of PCC pavement repairs using metakaolin-based geopolymer mortar. *Cement and Concrete Composites*, 2016. **65**: pp. 75–82.
35. Mohammadhosseini, H., M. M. Tahir, and M. Sayyed, Strength and transport properties of concrete composites incorporating waste carpet fibres and palm oil fuel ash. *Journal of Building Engineering*, 2018. **20**: pp. 156–165.
36. Kubba, Z., et al., Impact of curing temperatures and alkaline activators on compressive strength and porosity of ternary blended geopolymer mortars. *Case Studies in Construction Materials*, 2018. **9**: p. e00205.
37. Mohammadhosseini, H., et al., Effects of waste ceramic as cement and fine aggregate on durability performance of sustainable mortar. *Arabian Journal for Science and Engineering*, 2019: pp. 1–12.
38. Lim, N. H. A. S., et al., The effects of high volume nano palm oil fuel ash on microstructure properties and hydration temperature of mortar. *Construction and Building Materials*, 2015. **93**: pp. 29–34.
39. Morgan, D., Compatibility of concrete repair materials and systems. *Construction and Building Materials*, 1996. **10**(1): pp. 57–67.
40. Geissert, D. G., et al., Splitting prism test method to evaluate concrete-to-concrete bond strength. *ACI Materials Journal*, 1999. **96**(3): pp. 359–366.
41. Momayez, A., et al., Comparison of methods for evaluating bond strength between concrete substrate and repair materials. *Cement and Concrete Research*, 2005. **35**(4): pp. 748–757.
42. Alyousef, R., H. Mohammadhosseini, F. Alrshoudi, H. Alabduljabbar, and A. M. Mohamed, Enhanced performance of concrete composites comprising waste metalised polypropylene fibres exposed to aggressive environments. *Crystals*, 2020. 10(8): pp. 696–716.
43. Fodil, D. and M. Mohamed, Compressive strength and corrosion evaluation of concretes containing pozzolana and perlite immersed in aggressive environments. *Construction and Building Materials*, 2018. **179**: pp. 25–34.
44. ASTM, Standard specification for coal fly ash and raw or calcined natural pozzolan for use in concrete. 2013: ASTM.
45. Temuujin, J., A. van Riessen, and K. MacKenzie, Preparation and characterisation of fly ash based geopolymer mortars. *Construction and Building Materials*, 2010. **24**(10): pp. 1906–1910.
46. Gunasekara, C., et al., Zeta potential, gel formation and compressive strength of low calcium fly ash geopolymers. *Construction and Building Materials*, 2015. **95**: pp. 592–599.
47. Noushini, A. and A. Castel, The effect of heat-curing on transport properties of low-calcium fly ash-based geopolymer concrete. *Construction and Building Materials*, 2016. **112**: pp. 464–477.
48. ASTM, Standard test method for bond strength of epoxy-resin systems used with concrete by slant shear. 2005: ASTM.
49. ASTM, Standard test methods for fire tests of building construction and materials. 2012: ASTM.
50. ISO, Fire-resistance Tests: Elements of Building Construction. Commentary on Test Method and Test Data Application. 1994: ISO.
51. Huseien, G. F., et al., Synergism between palm oil fuel ash and slag: Production of environmental-friendly alkali activated mortars with enhanced properties. *Construction and Building Materials*, 2018. **170**: pp. 235–244.

52. Huseien, G. F., J. Mirza, and M. Ismail, Effects of high volume ceramic binders on flexural strength of self-compacting geopolymer concrete. *Advanced Science Letters*, 2018. **24**(6): pp. 4097–4101.
53. Rashad, A. M., Properties of alkali-activated fly ash concrete blended with slag. *Iranian Journal of Materials Science and Engineering*, 2013. **10**(1): pp. 57–64.
54. Puertas, F., et al., Alkali-activated fly ash/slag cements: Strength behaviour and hydration products. *Cement and Concrete Research*, 2000. **30**(10): pp. 1625–1632.
55. Huseien, G. F., et al., Properties of ceramic tile waste based alkali-activated mortars incorporating GBFS and fly ash. *Construction and Building Materials*, 2019. **214**: pp. 355–368.
56. Huseien, G. F., et al., Effects of ceramic tile powder waste on properties of self-compacted alkali-activated concrete. *Construction and Building Materials*, 2020. **236**: p. 117574.
57. Dombrowski, K., A. Buchwald, and M. Weil, The influence of calcium content on the structure and thermal performance of fly ash based geopolymers. *Journal of Materials Science*, 2007. **42**(9): pp. 3033–3043.
58. Huseien, G. F. and K. W. Shah, Performance evaluation of alkali-activated mortars containing industrial wastes as surface repair materials. *Journal of Building Engineering*, 2020. **30**: p. 101234.
59. Pacheco-Torgal, F., J. Castro-Gomes, and S. Jalali, Adhesion characterization of tungsten mine waste geopolymeric binder. Influence of OPC concrete substrate surface treatment. *Construction and Building Materials*, 2008. **22**(3): pp. 154–161.
60. Chen, M.-C., K. Wang, and L. Xie, Deterioration mechanism of cementitious materials under acid rain attack. *Engineering Failure Analysis*, 2013. **27**: pp. 272–285.
61. Matschei, T., F. Bellmann, and J. Stark, Hydration behaviour of sulphate-activated slag cements. *Advances in Cement Research*, 2005. **17**(4): pp. 167–178.
62. Cai, L., H. Wang, and Y. Fu, Freeze–thaw resistance of alkali–slag concrete based on response surface methodology. *Construction and Building Materials*, 2013. **49**: pp. 70–76.
63. Chang, H., et al., Influence of pore structure and moisture distribution on chloride "maximum phenomenon" in surface layer of specimens exposed to cyclic drying-wetting condition. *Construction and Building Materials*, 2017. **131**: pp. 16–30.

7 Ceramic Waste-Based Self-Compacting Alkali-Activated Concrete

7.1 INTRODUCTION

Self-compacting concrete constitutes a major advancement to the construction industry. It is characterized by the flowability and consolidation under its own weight; it fully permeates the formwork by penetrating the spaces separating the reinforcement bars and keeps its stable composition at the same time [1–3]. Self-compacting concrete has many advantages over conventional concrete, including decreased construction time and labour cost, reduced noise pollution, improved filling capacity of highly congested structural members, an enhanced interfacial transitional zone between the cement paste and aggregate or reinforcement, decreased permeability, and improved durability of concrete [4–6]. In the construction industry, cement production alone is responsible for about 7% of the overall amount of CO_2 emissions at the global level [7–11]. One strategy to attenuate the negative environmental effects of cement production is to use industrial and agriculture wastes to partially substitute the cement [12–14]. Another strategy proposed recently is the use of geopolymers as new construction binder, because the CO_2 emissions they generate are far lower than those associated with standard Portland cement [15–17]. The production of aluminosilicate-based inorganic polymers, known as geopolymers, is achieved through synthesis of pozzolanic waste materials with highly alkaline hydroxide and/or alkaline silicate [18, 19]. Geopolymerization can be achieved by using industrial solid wastes and by-products enriched in silica and/or alumina contents (e.g. fly ash) as pozzolanic components [20–24].

Alkali-activated mortars and concrete have started to become distinguished from standard Portland cement as an effective construction binder without high CO_2 emissions [25–27]. An inorganic polymer material, alkali-activated concrete consists of calcium (Ca) and aluminosilicates (ASs). It is synthesized from alkaline-activator solution containing sodium silicate (NS) and sodium hydroxide (NH) and pozzolanic compounds [28, 29]. The production of alkali-activated concrete/mortars is based on the exploitation of a wide range of by-products and agricultural wastes comprising calcium, silica, and aluminium (Al) such as granulated blast furnace slag (GBFS), palm oil fuel ash (POFA), and fly ash (FA) with low and high calcium, etc. [30, 31].

Besides being non-recyclable, ceramic waste is also bulky and creates problems for landfill disposal. Hence, for the purposes of natural resource conservation and environmental protection, it is necessary to develop new products with ceramic waste to recycle and to use in construction projects. The pozzolanic properties of ceramic waste have been confirmed by earlier studies [32–34]. Because they benefit the mechanical and durability properties of concrete, natural pozzolanic wastes were employed as construction material in earlier times [35, 36]. However, industrial wastes are currently the preferred pozzolanic materials because of strict environmental policies [37–39]. The ceramics industry has a significant effect on the environment and relies on landfills to dispose of the massive volumes of its waste. In 2015, ceramic tile production at the global level was around 12.4 million square metres [40, 41], of which 10–30% is discarded as waste [32, 42]. Additionally, the greatest proportion of this waste is non-recyclable, which is highly problematic in terms of disposal. To address this issue of extensive amounts of ceramic waste with little usage, researchers started investigating its use in mortar or concrete instead of cement. This, in turn, is advantageous as it reduces cost and energy consumption, improves ecological equilibrium, and promotes more careful use of natural resources [43–46].

Due to the high concentration of calcium and silica, ground blast furnace slag (GBFS) carries both pozzolanic and cementing qualities, which makes it a popular material in the construction industry for making modified concrete more durable with enhanced mechanical properties [47]. It was reported that GBFS changed the microstructure and durability of alkali-activated concrete to which it was added [23, 48]. Meanwhile, Yusuf et al. [49, 50] observed that alkali-activated mortar was stronger when GBFS was added to it as binder. This was explained in terms of the fact that GBFS served to fill pores and led to the development of uniform microstructures with high arrangement and twin products comprising a highly polymerized alkaline unit and gel of calcium silicate hydrate (C-S-H). Moreover, evidence has also been put forward that GBFS made Ca progressively more soluble and improved products' inhomogeneity and amorphosity. Additionally, concrete compressive strength is improved by the development of (N(Ca)-A-S-H) and products of Na/Ca-aluminosilicate-hydrate [10, 51–55].

Self-compacting green concrete and high-performance concrete have led to the development of a new type of cement-free concrete, known as self-compacting high-performance alkali-activated concrete (SCAAC). This type of concrete displays not only self-compacting concrete properties, such as filling and penetration capability as well as the ability to withstand separation, but is also extremely strong and durable [3]. However, the mechanical properties of SCAAC on its own and in combination with ceramic waste have not been extensively studied. Therefore, to fill this gap in the literature, the present study used different concentrations of WCP to substitute GBFS as a cement-free binder, with minimal NH molarity and low viscosity of alkaline activator solution comprising sodium silicate (NS) and NH. The present study mainly aims to decrease the amount of solid waste in the environment, thus lowering CO_2 emissions. The other aims are to suggest feasible materials and physical requirements to produce alkali-activated GBFS–WCP self-compacting concrete, as well as to propose alternative materials for use in the construction industry.

7.2 RAW MATERIALS

Binders, aggregate and alkaline-activated solution were the three types of materials employed in the present study. WCP and GBFS were the waste materials used as binders in the production of SCAAC. Pure GBFS (Ipoh, Malaysia) was a resource material that was employed to produce cement-free binder, and its subsequent use did not require laboratory treatment. GBFS is characterized by its cementing and pozzolanic qualities. It is off-white in colour and results in a hydraulic reaction with water. With a composition of calcium silicate and alumina (around 90%), GBFS satisfies the pozzolanic material specification imposed by ASTM C618. The construction industry in Johor, Malaysia, supplied the ceramic waste employed in the current study. The first step upon acquisition of this material was to separate large particles by crushing followed by sieving through 600 μm mesh. After sieving, the material was ground for six hours, with hourly inspection to examine how fine the particles were. The subsequent step was to collect and to use the powder in the process of mixing with water. Figure 7.1 illustrates the manner in which the ceramic waste was treated.

To determine how GBFS and WCP were chemically constituted, X-ray fluorescence (XRF) spectroscopy was applied (Table 7.1). Results showed that GBFS and WCP primarily consisted of silica and aluminium in proportions of 41.7% and 84.8%, respectively. Furthermore, the concentration of calcium oxide was significantly higher in GBFS (51.8%) than in WCP. It is well known that the synthesis of alkali-activated mortars/concrete greatly depends on the concentration of silica, aluminium, and calcium oxide. Meanwhile, neither GBFS nor WCP had a high concentration of potassium oxide (K_2O). Additionally, both GBFS and WCP conformed to the ASTM C618 standard as they had extremely low loss on ignition contents.

FIGURE 7.1 Processing steps of WCP.

TABLE 7.1
Chemical Composition and Physical Characteristics of WCP and GBFS

Chemical Composition (% by Mass)									
Oxide	CaO	Al_2O_3	SiO_2	Fe_2O_3	MgO	K_2O	Na_2O	SO_3	LOI
WCP	0.02	12.6	72.6	0.56	0.99	0.03	13.5	0.01	0.13
GBFS	51.8	10.9	30.8	0.64	4.57	0.36	0.45	0.06	0.22

Physical Characteristics			
Item	Specific Gravity	Medium Particle Size (μm)	Specific Surface Area (m^2/g)
WCP	2.6	35	12.2
GBFS	35	12.8	13.6

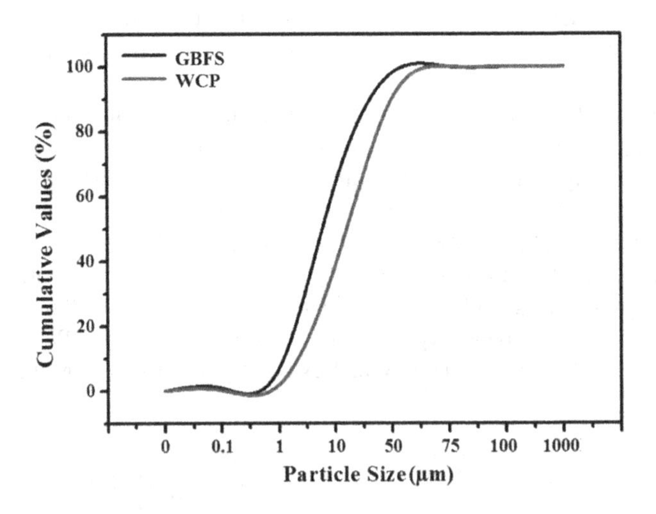

FIGURE 7.2 Particle size distribution of WCP and GBFS.

By using the particle size analyzer test, the particle size distributions of both GBFS and WCP were calculated and demonstrated in Figure 7.2. The median of particles for GBFS and WCP were estimated to be 12.8 and 35 μm, respectively. For physical properties of raw materials including GBFS and WCP, the surface areas were evaluated and are presented in Table 7.1.

The results of X-ray diffraction (XRD) of GBFS and WCP are shown in Figure 7.3. For WCP, the crystalline silica and alumina compounds were believed to be the reason for the marked diffraction peaks exhibited by the XRD pattern at about $2\theta =$ 16–30° [17, 56, 57], whereas the crystalline quartz and mullite phases were identified as the cause of other crystalline peaks. For GBFS, the extreme amorphous characteristic of this material was confirmed by the XRD pattern, which did not show any sharp peak. The reactive silica and calcium contents are the main contributors to the development of GBFS. In fact, the possibility of alkali-activated preparation is due to

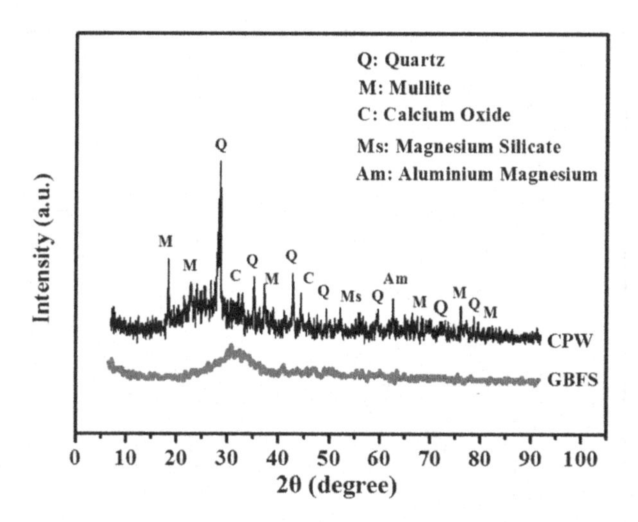

FIGURE 7.3 XRD patterns of GBFS and WCP.

FIGURE 7.4 Scanning electron microscopy (SEM) images of WCP and GBFS.

high levels of reactive amorphous silica and calcium in GBFS. Figure 7.4 illustrates the images of WCP and GBFS produced through scanning electron microscopy (SEM). It can be clearly seen that both materials contained irregular- and angular-shaped particles, corroborating earlier studies [20, 58].

Fourier transform infrared (FTIR) analysis provides a detailed signature of chemical bonding vibrations of the constituents and allows in making a comprehensive structural analysis. Transformation of GBFS and WCP from a crystalline to amorphous state with temperature was examined by recording their FTIR spectra in terms of wave number (cm^{-1}), which is equivalent to the frequency or energy of bonding vibrations (Figure 7.5). The stretching modes are sensitive to the Si–Al composition of the framework and may shift to a lower frequency with increasing number of tetrahedral aluminium atoms [59]. Sitarz and Handke [60] reported the occurrences of broad bands at 500–650 cm^{-1}, indicating the existence of silicate and aluminosilicate glass phases. It also possessed short-range structural order in the

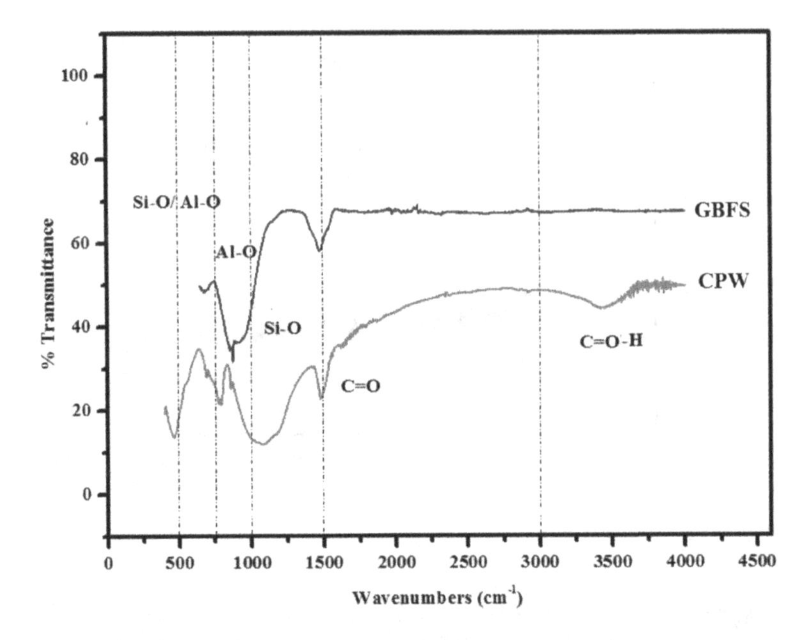

FIGURE 7.5 FTIR spectra of WCP and GBFS.

TABLE 7.2
Composition of Alkaline Activator Solutions

NaOH Solution (NH)			Na_2SiO_3 Solution (NS)					
Molarity (M)	Na_2 (Mass%)	H_2O (Mass%)	SiO_2 (Mass%)	Na_2O (Mass%)	H_2O (Mass%)	SiO_2:Na_2O (Mass%)	NS:NH (Mass%)	Final of Ms (SiO_2:Na_2O) (Mass%)
2	7.4	92.6	29.5	14.7	55.8	2.01	0.75	1.2

form of tetrahedral or octahedral rings. The spectral bands at 460 cm^{-1} were related to Al–O/Si–O in plane and bending modes, 730 cm^{-1} was assigned to octahedral site modes of Al, 820 cm^{-1} was allocated to tetrahedral Al–O stretching [61], and 1400 cm^{-1} was endorsed to asymmetric stretching vibrations of Al–O/Si–O bonds [62]. WCP presented a broad band with medium intensity (C=O) between 1460 and 1600 cm^{-1}. Furthermore, WCP displayed a higher intensity compared to GBFS. GBFS did not show any vibrational band in this region. In both waste materials, the bands appeared in the regions of 3500 cm^{-1} were assigned to bending vibrations of (C= – H) and stretching vibration of (–OH), respectively. Hydroxyl ions being characteristic of weakly bound ligaments of water were either adsorbed on the surface or trapped in large cavities.

Table 7.2 shows the composition of alkaline activator solution used in this study. To reduce the environmental effects of NaOH and Na_2SiO_3, a low molarity (2 M) of NaOH solution and Na_2SiO_3-to-NaOH ratio of 0.75 was adopted to prepare the alkaline activator solution. These were fixed for all solution dosages. Regarding the contents of Na_2O, SiO_2 and H_2O, the prepared alkaline solution is friendlier to the environment and is low in cost, energy consumption, and carbon dioxide emission. Sodium hydroxide solution with 2 M was prepared by using analytical grade NaOH (NH, 98% purity) in the form of pellets; then the NH pellets were disbanded in water. The prepared solution of sodium hydroxide was left for 1 day to cool and then added to a sodium silicate (NS) solution to achieve the final activator solution having the SiO_2:Na_2O ratio of 1.2. Analytical-grade NS solution comprising of Na_2O (14.70 wt%), SiO_2 (29.5 wt%), and H_2O (55.80 wt%) was utilized. The ratio of NS to NH was fixed to 0.75 for all alkaline activator solutions.

The physical properties of fine and coarse aggregates used to prepare SCAAC mixtures are presented in Table 7.3. River sand was obtained from a local supplier. The specific gravity of 2.62 was used as fine aggregate under saturated surface dry conditions for all mixes to guarantee that the alkaline solution binder (S:B) ratio remains unaffected. For coarse aggregate, the crushed granite, obtained from the quarry, was used for all the mixes with specific gravity 2.67 and water absorption 0.51%.

7.3 MIX DESIGN

WCP and GBFS were mixed in different concentrations to produce SCAACs. More specifically, the concentration of WCP ranged from 10 to 80% by weight, while 100% GBFS served as the control and its concentration ranged from 20 to 100% by weight. This binary blend in different concentrations was employed to conduct several tests to achieve the research aim of determining how SCAACs were affected by high WCP concentrations. All other compounds used in the mixtures did not vary in concentration as did WCP and GBFS. Table 7.4 provides an overview of SCAAC mixtures. NH and NS were rigorously blended for the preparation of a mixture measured by weight. This process released heat, so the mixture was left to reach room temperature prior to being used. For the preparation of SCAAC, a concrete mixer machine was employed to blend WCP and GBFS for a period of 3 minutes under dry conditions until a uniform mixture of fine and coarse aggregates was obtained. This was followed by the addition of the alkaline solution to the mixture, which was subjected to mixing for a further 6 minutes in the concrete mixer machine at moderate speed. After that, a number of techniques were applied to assess the

TABLE 7.3
Physical Characteristics of Fine and Coarse Aggregates

Material	Water Absorption (%)	Specific Gravity	Maximum Size (mm)
Fine aggregate	1.2	2.62	2.36
Coarse aggregate	0.51	2.67	10

TABLE 7.4
Self-Compacting Alkali-Activated Concrete Mix Design

	Binder (kg/m³)		Aggregate (kg/m³)		Alkaline Activator Solution (kg/m³)				
Mix Code	GBFS	WCP	Sand	Coarse	S:B	M	NS:NH	NS	NH
$SCAAC_1$	484	0							
$SCAAC_2$	338.8	145.2	844	756	0.50	2	0.75	104	138
$SCAAC_3$	290.4	193.6							
$SCAAC_4$	242	242							
$SCAAC_5$	193.6	290.4							
$SCAAC_6$	145.2	338.8							

S:B, ratio of alkaline activator solution to binder; M, molarity of sodium hydroxide; NS:NH, sodium silicate to sodium hydroxide; NS, sodium silicate; NH, sodium hydroxide.

prepared SCAAC in terms of main properties that SCAAC must fulfil to be classified as concrete mixtures, namely, filling and penetration capability as well as separation resistance.

7.4 TESTING OF FRESH AND HARDENED SCAAC

7.4.1 Workability Tests

Since a standard technique is yet to be devised to evaluate all three workability properties of SCAAC, multiple techniques were employed in this study to investigate those properties for every mix sample. Table 7.5 provides the techniques to assess the filling and penetration capabilities, while observation through visual stability is indicative of separation resistance. A range of test methods for the evaluation of SCAAC mixes have been put forth by the European guidelines of the European Federation of Specialist Construction Chemicals and Concrete Systems (EFNARC).

TABLE 7.5
Self-Compacting Concrete Acceptance Criteria as per the EFNARC

Test	Slump Flow (mm)	T_{50} Flow (sec)	V-Funnel (sec)	L-Box Ratio, H_2:H_1	J-Ring (mm)
		Acceptance Criteria as per EFNARC			
Min	650	2	6	0.8	0
Max	800	5	12	1.0	10

Source: EFNARC, *Specification and Guidelines for Self-Compacting Concrete*. Vol. 32, p. 34. 2002: Association House.

According to these guidelines, this study assessed the workability properties based on the methods of slump flow, V-funnel, L-box, T_{50}, and J-ring. In terms of setting time for the alkali-activated materials, a distinction was made between initial and final setting time. The mortar setting time was measured with the Vicat method, in keeping with the requirements of ASTM C191.

7.4.2 Hardened Concrete Testing

According to ASTM C109-109 M, the compressive strength test was conducted on samples after properly curing them in air for a period of 3, 7, 28, and 56 days. After every curing interval, testing was conducted on three specimens at each age, with every specimen being adequately prepared and placed between the superior and inferior metal bearing plates. The specimens were subjected to a fixed 2.5 kN/s loading rate until failure. Calculation of compressive strength was based on a standard deviation of 4.3 MPa and a coefficient of variation of 5.4. Assessment of tensile strength and flexural strength were carried out following ASTM C496/C496M-17 and ASTM C78, respectively. Both assessments were conducted after the specimens were cured for 28 days. The results were the average of three samples. Furthermore, comparison between the measured strengths and the control sample of 100% GBFS was also undertaken.

To assess how porous the generated SCAACs were, the vacuum saturation technique was applied as per ASTM C642-13. The saturated, suspended and oven-dry masses tests provided the required parameters. These tests were conducted on specimens cured for 28 days with saturated 100 ×100 mm surface. After breaking the alkali-activated specimens at 28 days, their middle portions were ground into a fine powder and subjected to microstructural assessment via XRD, SEM, and FTIR.

7.4.3 Sulphuric Acid Attack Test

The sulphuric acid solution (H_2SO_4) mainly attacks SCAAC by dissolving the binder paste matrix, leading to weakening of strength of affected concrete. The effect of sulphuric acid on concrete compositions under study is the 10% H_2SO_4 acid solution, which is prepared with deionized water. For each concrete mixture and after 28 days of curing, six specimens were selected and weighed before being immersed in H_2SO_4 solution for one year. Throughout the experiment and every 3 months, the H_2SO_4 solution was changed in order to maintain a continuity of the acid solution's pH level. After 12 months, the SCAAC specimens were checked while several different factors were considered during evaluating their performance, including remaining strength, loss in weight and ultrasonic pulse velocity according to ASTM C267 specifications [64].

7.5 RESULTS AND DISCUSSION

7.5.1 Filling Ability of Concrete

Three tests, namely slump flow, T_{50} and L-box tests, were adopted in this study to measure the impact of WCP content on the filling ability of concrete. The slump

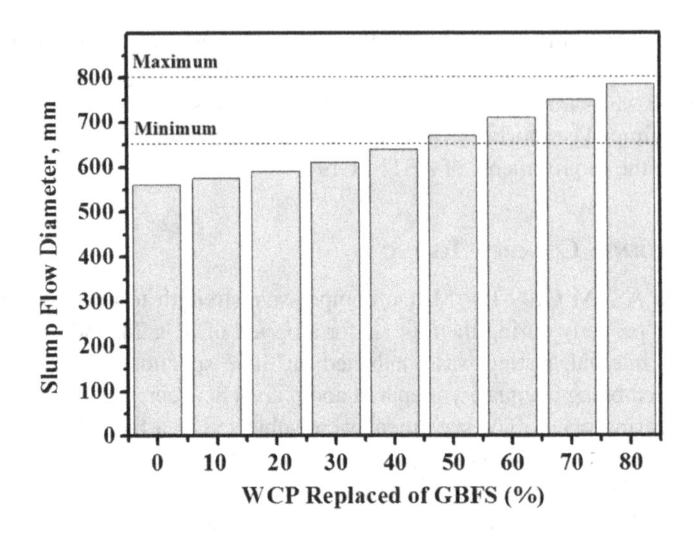

FIGURE 7.6 Slump flow of SCAACs containing different ratios of WCP replacing GBFS.

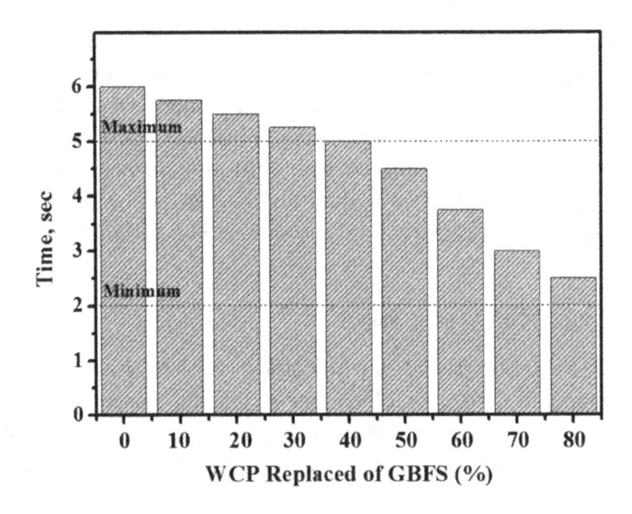

FIGURE 7.7 T_{50} flow of SCAACs containing different ratios of WCP replacing GBFS.

flow test results (Figure 7.6) showed a significant influence on the concrete workability with an increasing replacement level of GBFS by WCP in the alkali-activated matrix. With increasing WCP content replacing GBFS from 0 to 80%, the flowability enhanced and its flow diameter increased from 560 to 780 mm, respectively. Concrete mixes containing 50, 60, 70m and 80% of WCP as GBFS replacement, achieved high workability and are considered as self-compacting mixes according to EFNARC [63]. Figure 7.7 shows the effects of WCP levels on time flow (T_{50}) values of concrete mixes. The results showed a great enhancement on the workability of concrete with increased level of WCP replacing GBFS. The time dropped from 6 to 2.5 seconds with decreasing content of GBFS as the replacement level with WCP increased. Most mixes prepared with 40% WCP replacing GBFS or more achieved the EFNARC requirements for self-compacting concrete. The last test conducted to evaluate the filling ability of concrete was the L-box test. Its results revealed that all the samples passed the requirement for self-compacting concrete, except for samples containing 0 and 10% WCP which indicated the ratio of 0.79 and 0.77, respectively (Figure 7.8). The increments in flowability of concrete could majorly be influenced by the size distribution of large particles and the low specific surface area of WCP compared to GBFS (as presented in Table 7.1). Furthermore, the enhancement in concrete workability could be attributed to low water demand for WCP compared to GBFS [23, 36, 65]. As reported in previous studies [47, 66], the low water demand affects the chemical reaction rate and increases the plasticity of mixture which improves the concrete's workability.

7.5.2 Passing Ability of Concrete

Figure 7.9 shows the effect of WCP content on J-ring test results of concrete formulated from SCAACs at different WCP:GBFS ratios. It was observed that all mixtures containing 30% WCP and up achieved the criteria requirement of self-compacting concrete. As the WCP content increased from 0 to 10, 20, 30, 40, 50, 60, 70, and 80% to replace GBFS, the workability of concrete slightly tends to enhance and achieved the requirements. The highest value was obtained with the mixture containing 80% WCP, which presented 4.5 mm compared to 12 mm achieved with the control sample (100% GBFS). Several researchers [36, 46, 57] reported a reduction in CaO content, which enhanced the workability due to a slow pozzolanic chemical reaction.

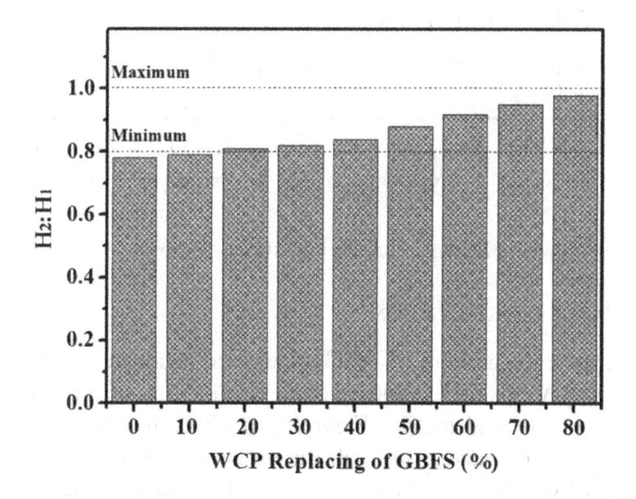

FIGURE 7.8 H_2:H_1 ratio of SCAACs containing different ratios of WCP replacing GBFS.

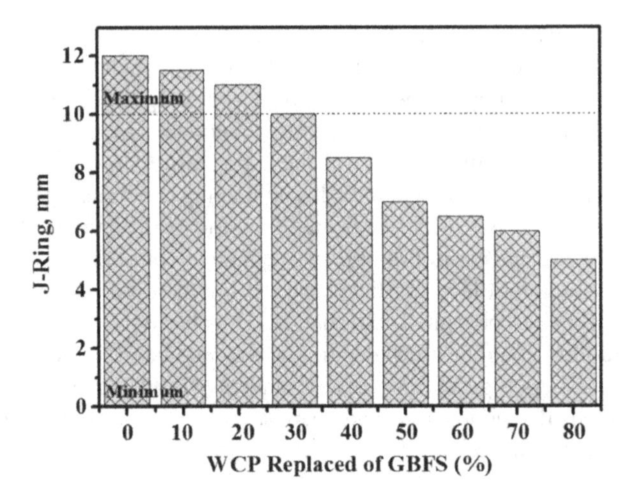

FIGURE 7.9 J-ring flow of SCAACs containing different ratios of WCP replacing GBFS.

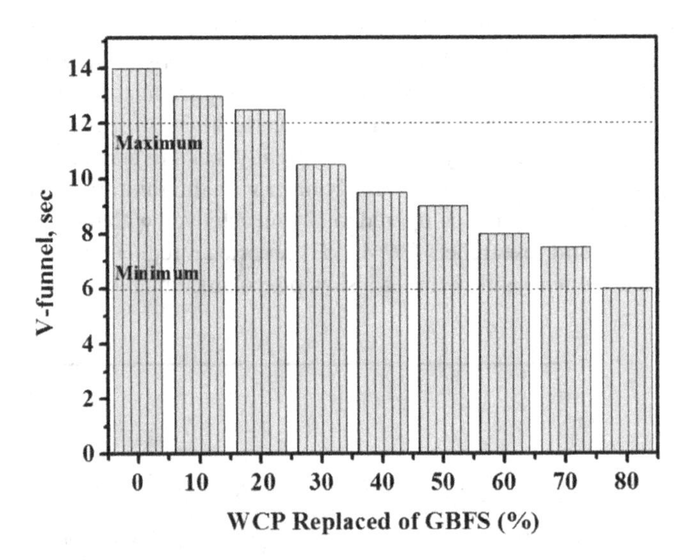

FIGURE 7.10 V-funnel of SCAACs containing different ratios of WCP replacing GBFS.

7.5.3 Resistance to Segregation

The resistance to segregation of SCAAC mixes were evaluated using the V-funnel test. The results showed a high influence on the workability of concrete by the WCP level in the alkali-activated matrix (Figure 7.10). Compared to the control sample and criteria for self-compacting mix, all the mixtures containing 30 to 70% WCP as replacement of GBFS achieved the requirements. The increment in workability of concrete including filling ability, passing and resistance to segregation could be attributed to the reduced CaO content while increasing the WCP level replacing

GBFS. The rapid rate of chemical reaction of high-volume GBFS content decreases the plasticity of mixture which is responsible for the reduction of the concrete's workability [57]. Also the low surface area of WCP compared to GBFS reduced the water demand from alkaline solution and enhanced the workability of alkali-activated matrixes [67].

7.5.4 Setting Time

Figure 7.11 illustrated the results of initial and final setting times of SCAAC prepared with varying levels of WCP replacing GBFS as binder. It shows that the initial and final setting times ranged from 6 to 64 minutes and 10 to 108 minutes, respectively. The control sample containing 100% GBFS recorded the lowest value for both the initial and final setting times. The initial and final setting times of alkali-activated pastes increased slightly with the increase in WCP content present in the alkali-activated matrix. The highest initial and final setting times recorded were 64 minutes and 108 minutes, respectively, in the SCAAC mix containing 80% WCP. This behaviour could be attributed to the lower alkali activation due to higher amounts of silica present in the higher WCP content in SCAAC which resulted in a lower setting time than with 100% GBFS. The difference in variation between the initial and final setting times is low. However, this difference in both setting times increased with the increasing WCP content in the matrix. This is due to slow a pozzolanic reaction between SiO_2 (from WCP) and $Ca(OH)_2$ evolved from GBFS hydration, which only occurs after the latter started to react with alkaline activator solution and water [36, 46, 47, 68].

In the mixture of $SCAAC_6$, the increased content of WCP to 80% and decreased content of GBFS to 20% led to a decrease in the CaO amount and increase in the SiO_2 content which resulted in a delayed chemical reaction rate and increased setting time. It also supported the fact that the higher the CaO content in the alkali-activated

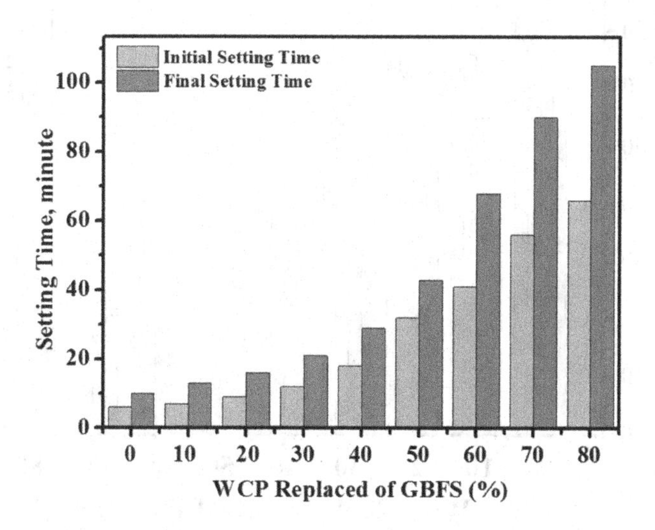

FIGURE 7.11 Setting times of SCAACs containing different ratios of WCP replacing GBFS.

matrix, the faster the setting rate [36, 69]. These results confirmed that WCP, as a part of the binary-blended alkali-activated binder, was effective in decelerating the setting time of concrete at ambient conditions.

7.5.5 Compressive Strength

Figure 7.12 summarizes the compressive strength test results of SCAACs. Each presented value consisted of the average of three measurements. The effects of WCP including the alkali-activated matrix as GBFS replacement were evaluated and are discussed in this section. Various levels of WCP replacing GBFS from 10 to 80% by mass were assessed and compared with control sample containing 100% GBFS. It is evident from Figure 7.12 that the use of WCP as a GBFS replacement decreased the early strength of the SCAAC mixes compared to the control at all the test ages. At the age of 3 days, the control sample (with 100% GBFS) provided the highest value of compression strength, i.e. 52.6 MPa. As soon as GBFS was replaced by 10% WCP in SCAAC, the compressive strength dropped to 47 MPa. This strength loss continued with the increased level of WCP from 10 to 20, 30, 40, 50, 60, 70, and 80%. The lowest compressive strength value of 18.6 MPa was recorded in a mixture containing 80% of WCP at 3 days of age. However, these results indicated that the development in compressive strength monotonically increased with the increase in age. After 28 days of age, the compressive strength gain rate continued to increase for all SCAAC mixtures and showed a percentage increase higher than 32%. After 56 days, the compressive strength increased between 3 and 16% with increasing hydration time as compared to specimens tested at 28 days. The reduction in compressive strength of concrete could be due to reduced CaO content, which led to increase

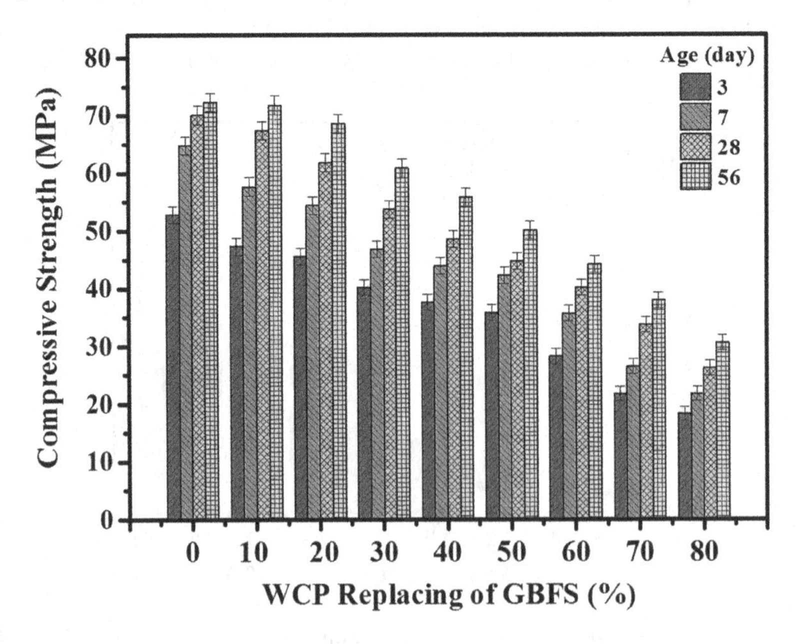

FIGURE 7.12 Effect of WCP replacing GBFS on compressive strength of SCAACs.

in the silicate-to-calcium ratio (SiO_2:CaO) and produced a negative effect on the compressive strength of SCAAC. Moreover, the production of C-(A)-S-H gels were significantly influenced by increased WCP content and reduced CaO content, which resulted in slower chemical reaction rates and lower strength. The occurrence of the low amount of C-(A)-S-H gels could be ascribed to low Ca content, which played a major role in loss of concrete strength. In a previous study Rashad [26] observed a decrease in compressive strength with increasing content of materials containing high amounts of ASs, such as POFA and FA, in the alkali-activated matrix. The lowering in compressive strength could be attributed to several factors. The first factor is the difference in chemical composition of raw materials used (WCP and GBFS), which significantly affected the alkali activation process of the binders. The second factor that affects compressive strength development is the lower reaction rates of WCP compared to GBFS which was partially dissolved [70]. The third factor could be associated with the lowering of compactness and density of concrete specimens with rising WCP amounts. The fourth factor is linked to the low NaOH concentration (2 M) where the compressive strength depended mainly on the level of CaO to replace the low amount of Na_2O. Thus, the formation of more C-S-H and C-A-S-H gels than N-A-S-H gel improved the prepared concretes' compressive strength development [17, 57].

7.5.6 X-Ray Diffraction Analysis

The X-ray diffraction (XRD) patterns of SCAAC in which GBFS was substituted with WCP to various degrees are illustrated in Figure 7.13. It is important to mention here that the glass content of fly ash consists of reactive and refractory glass, and

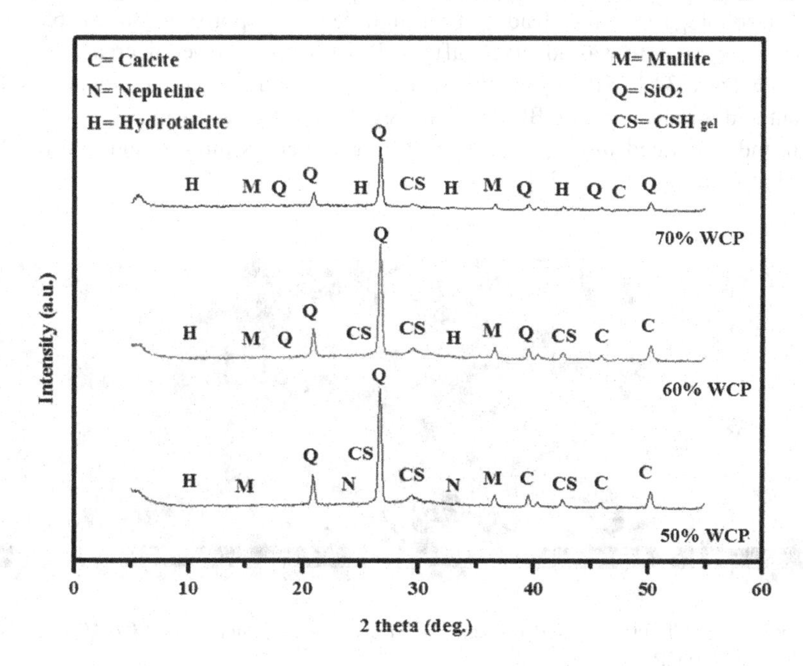

FIGURE 7.13 Effect of WCP replacing GBFS on XRD results of SCAACs.

only the former participates in geopolymerization. The broad and diffused background peak with maxima of around 10° emerged from the short-range order of the hydrotalcite ($Mg_6Al_2CO_3OH_{16}.4H_2O$) [69, 71]. Furthermore, the high content of MgO (4.56%) in GBFS (Table 7.1) formulated this peak during the geopolymerization process. An increase in WCP concentration showed a reduction in C-S-H peak intensity, particularly the 30° peak. At 70% level of WCP, a hydrotalcite peak substituted the C-S-H peak at 43°. Meanwhile, at a WCP level higher than 50%, a quartz peak substituted the calcite peak at 17°, 21°, 28°, 38.5°, and 51°. Moreover, the non-reacted silicate quantity increased and the C-S-H product decreased as the WCP level increased, which was due to the fact that the CS diminished from 52.8 to 34.1 MPa (at the age of 28 days). In previous studies [58, 72], it was reported that a reduction in CaO amount produced less C-S-H gel and restricted the calcite formulation. Furthermore, the N-A-S-H gel was found to co-exist with C-S-H gel and calcite [73].

7.5.7 Scanning Electron Microscopy (SEM)

SEM images of SCAAC incorporating WCP in various concentrations are shown in Figure 7.14. Examination at 28 days was focused on two specimens with WCP substituting GBFS in proportions of 50% and 70%, respectively. The specimen with 50% WCP exhibited a dense surface with the presence of a small quantity of non-reacted and partly reacted particles (Figure 7.14a). On the other hand, the other specimen with 70% WCP exhibited a larger amount of particles non-reacted and partly reacted (Figure 7.14b). The increase in WCP concentration from 50 to 70% was also accompanied by an increased quantity of non-reacted quartz (SiO_2) and poor morphology structure leading to a high degree of porosity [46, 57, 65]. The compressive strength was adversely affected by this and showed decreasing values of 44.8 MPa to 34.1 MPa, respectively. It is established that an increment in WCP content and a reduction in GBFS content negatively affects C-(A)-S-H gels formulation and generated more partially reacted gel such as mullite and non-reacted particles including quartz [27].

FIGURE 7.14 Effects of WCP replacing GBFS on microstructure of SCAACs. (a) 50% WCP, (b) 70% WCP.

7.5.8 Fourier-Transform Infrared Spectroscopy (FTIR)

The compressive strength outcomes indicated that specimens became weaker as the concentration of WCP substituting GBFS increased in the SCAAC matrix. As the concentration of GBFS declined, so did the levels of dissolved Ca and Al, which was the reason for the gradual reduction in strength. Furthermore, an earlier study observed that the extent of silicate polymerization was also diminished by the decrease in dissolved Al [23]. An overview of the FTIR spectral analysis is provided in Figure 7.15. This analysis focused on the bonding vibrations within SCAAC, which were the source of compressive strength. This observation was corroborated by the occurrence of an FTIR spectral band of Si-O-Al of greater width. Moreover, as the WCP concentration increased from 0 to 50 to 70% and the GBFS concentration declined, the band frequency increased from 956.1 to 965.4 to 982.6 cm^{-1}, respectively. The decrease in Al levels in the SCAAC matrix and, implicitly, the reduction in strength of the three-dimensional structure were the reasons for the rise in FTIR vibration frequency.

7.5.9 Tensile Strength

Splitting tensile strength (TS) values of the SCAAC specimens containing various levels of WCP as binder replacement for GBFS are enlisted in Figure 7.16. At 28 days of age, the average value of the three samples was considered to determine the TS results. The TS results revealed a lower strength value (2.9 MPa) at 28 days with an

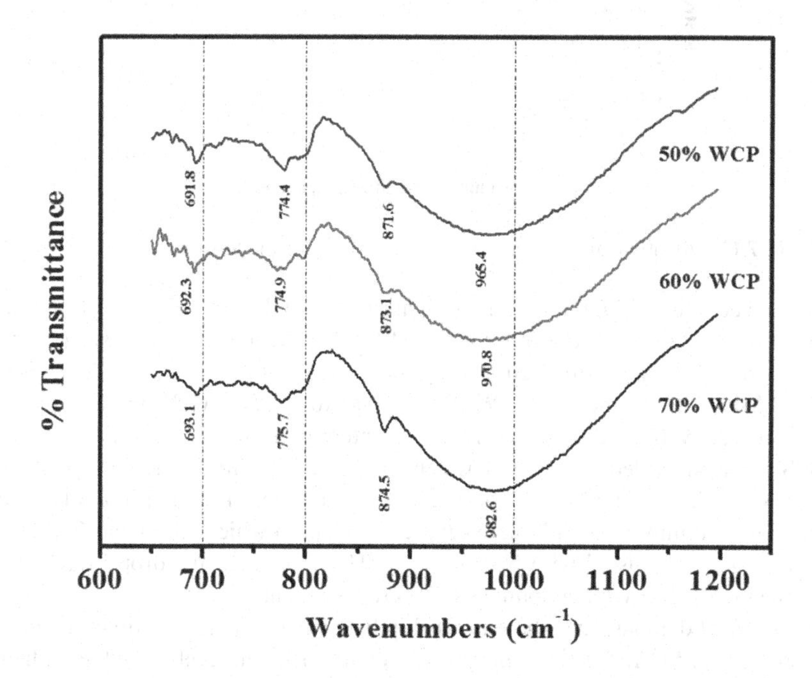

FIGURE 7.15 Effect of WCP replacing GBFS on FTIR results of SCAACs.

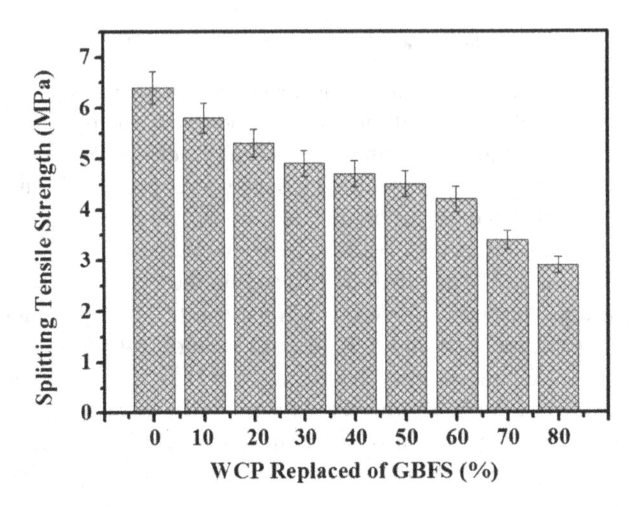

FIGURE 7.16 Effect of WCP replacing GBFS on tensile strength of SCAACs at 28 days of age.

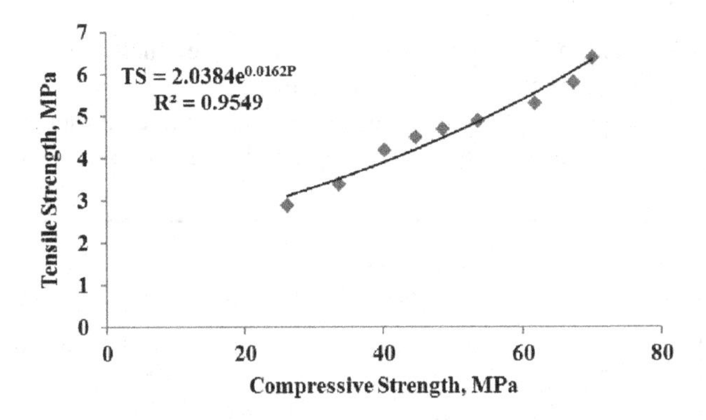

FIGURE 7.17 Relationship between compressive strength and tensile strength.

increased content of WCP (80%) as compared to 6.4 MPa achieved for 100% GBFS content (control sample). For other levels of WCP replacing GBFS, it was observed that the strength values dropped slightly, i.e. from 6.4 to 5.8, 5.3, 4.9, 4.7, 4.5, 4.2, and 3.4 MPa with the increase in WCP from 0 to 10, 20, 30, 40, 50, 60, and 70% (by mass), respectively. The reduction in CaO content with increasing WCP content as GBFS replacement led to a slow down in the rate of chemical reactions to produce C-(A)-S-H gel. Phoo-ngernkham et al. [53] reported that the strength of alkali-activated concrete improved with increasing Ca content in which additional C-S-H and C-A-S-H gels co-existed with N-A-S-H gel. This explained the drop in TS values with the increase in WCP amount as GBFS replacement.

Figure 7.17 demonstrates the relationship between compressive strength and tensile strength of SCAAC specimens prepared with different levels of WCP replacing

GBFS. An exponential relationship was found between compressive strength and tensile strength with coefficient of determination R^2 value up to 0.88 as given in Equation 6.1. This indicated a good confidence for the relationships.

$$TS = 2.0384e^{0.0162P}$$
$$R^2 = 0.9549 \tag{6.1}$$

where
TS = tensile strength 28 days of age, MPa
P = compressive strength at 28 days of age, MPa

7.5.10 Flexural Strength

Figure 7.18 shows the flexural strength development of SCAACs at 28 days of age containing different levels of WCP. The flexural strength was influenced by the WCP levels and dropped from 2.2 to 1.2 MPa as its contents increased from 0 to 80%, respectively. By adding WCP in SCAAC, the flexural strength was lower than the reference specimen (100% GBFS) for all replacement levels between 10 and 80% and showed similar trends to strength loss as in compressive and tensile strengths. These results were in agreement with the previous findings [21, 65], which reported that a decrease in GBFS content as FA replacement could negatively influence the formation of C-(A)-S-H product, thus reducing the strength. The aforementioned results were in agreement with the previous studies [23, 47, 74], which stated that the flexural strength value tend to drop as the GBFS content decreases in the alkali-activated matrix.

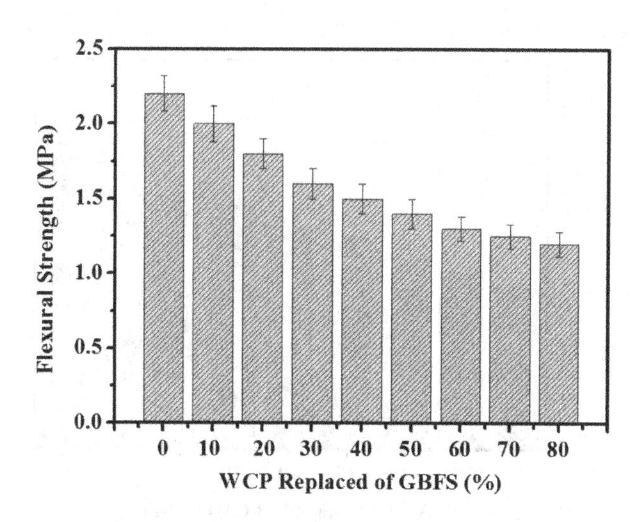

FIGURE 7.18 Effect of WCP replacing GBFS on FS of SCAACs at 28 days of age.

7.5.11 Water Absorption

Figure 7.19 displays the water absorption (WA) results for varying levels of WCP-containing SCAAC. The WA of SCAAC was significantly affected by the decrease in GBFS content replaced with WCP. The WA increased from 6.8 to 10.1 and 14.1% when the WCP levels increased from 0 to 50 and 80%, respectively. The reasons for such findings were attributed in Sections 7.5.5–7.5.7. As the WCP content increased, a less dense C-(A)-S-H gel was formed, which led to lower strength with high water absorption. It was reported [63] that a rise in WCP content could impact the partially and non-reacted particles, which produce a more porous structure and led to higher water absorption in SCAACs specimens [54].

7.5.12 Sulphuric Acid Attack

Figures 7.20, 7.21, and 7.22 illustrate the percentage loss in strength, weight, and ultrasonic pulse velocity (UPV), respectively, of SCAACs exposed to a 10% sulphuric acid solution attack. The resistance of specimens prepared with various WCP content as GBFS replacement (0, 10, 20, 30, 40, 50, 60, 70, and 80%) was examined. Figure 7.20 presents the strength loss percentages of SCAACs immersed in 10% H_2SO_4 solution for 12 months. The loss in strength decreased gradually with an increasing content WCP replacing GBFS in the alkali-activated matrix. As the WCP content increased from 0 to 80%, the loss in strength decreased from 74 to 13.3%, respectively. At each level of WCP replacing GBFS, the strength loss was assessed. It was observed that the strength dropped to 61.4, 54.8, 46.4, 24.9, and 15.2% as the WCP level increased from 10, 20, 50, 60, to 70%, respectively. Upon exposing the specimens in sulphuric acid, the presence of $Ca(OH)_2$ in concrete reacted with SO_4^{-2} ions and formed gypsum ($CaSO_4.2H_2O$). This product caused expansion in the alkali-activated matrix, which

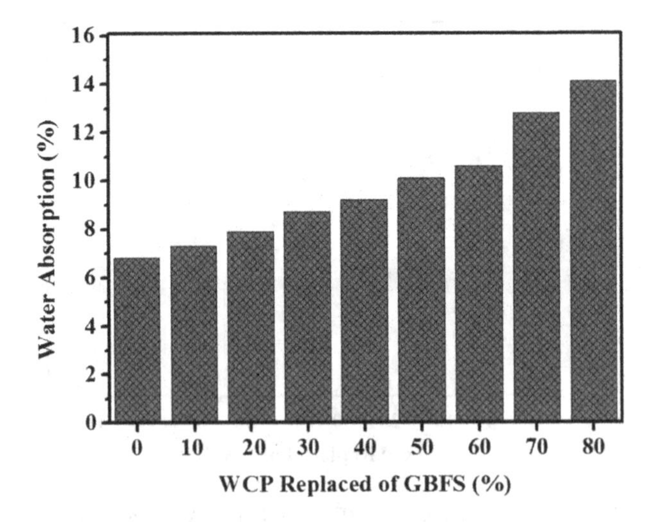

FIGURE 7.19 WA of SCAACs prepared with various ratios of WCP to GBFS.

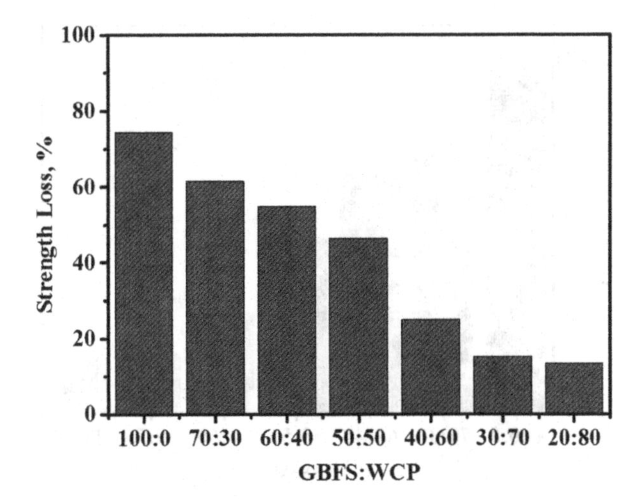

FIGURE 7.20 Strength loss percentage of SCAACs exposed to a 10% H_2SO_4 solution for 365 days.

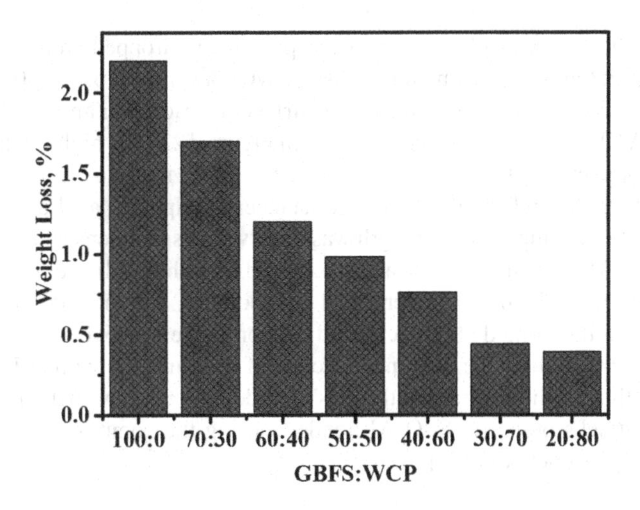

FIGURE 7.21 Weight loss percentage of SCAACs exposed to a 10% H_2SO_4 solution for 365 days.

created additional cracks in the interior of specimens and led to their further deterioration [17, 75]. The loss in weight percentage of specimens exposed to sulphuric acid was also evaluated and a similar trend was observed for all the specimens as in strength loss. The weight loss percentage decreased with increasing level of WCP replacing GBFS. The increase in WCP level from 0 to 30, 40, 50, 60, 70 and 80% led to drop in weight loss percentage from 2.2 to 1.7, 1.2, 0.98, 0.44 and 0.39, respectively. (Figure 7.21). The internal deterioration of specimens immersed for one year in acid solution was also evaluated by using the UPV test. Figure 7.22 shows the loss in UPV

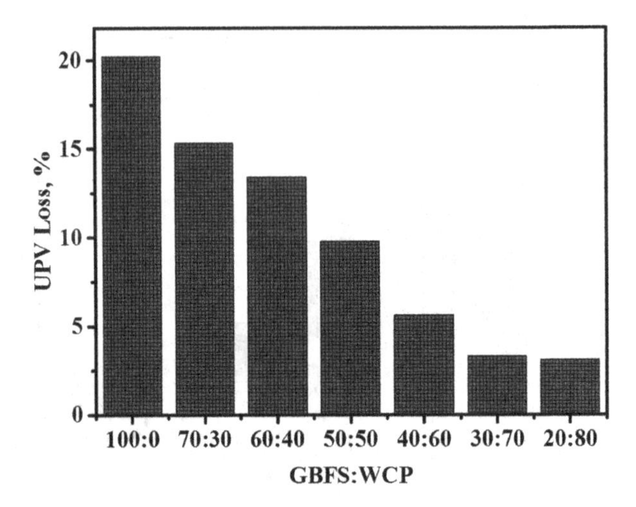

FIGURE 7.22 UPV loss percentage of SCAACs exposed to a 10% H_2SO_4 solution for 365 days.

percentage values. It was observed that the percentage dropped from 20.2 to 15.3, 13.4, 9.8, 5.6, 3.3 and 3.1% with an increase in WCP level from 0% to 30, 40, 50, 60, 70 and 80%, respectively. The decrease in surface deterioration and cracks with the increase in WCP contents can be clearly seen in Figure 7.23. The high calcium content in the control sample (100% GBFS) compared to other matrixes resulted in higher gypsum content. Therefore, the higher resistance to sulphuric acid attack and loss less than 13.3% of compressive strength was observed in specimens containing a high WCP content. Allahverdi and Skvara [76] reported on sodium silicate activated specimens containing 50% GBFS immersed in acid solution and indicated that the free Ca^{++} and Na^{+} in the formed alkali-activated network were negative to acid resistance. It was demonstrated that "the exchange reaction of Ca^{++} and Na^{+} with OH ions results in their loss, thus further degrading the gels". In addition, de-alumination could occur due to acid attacks prone on Si-O-Al bonds, causing the composition changes and structure of the AS network [77].

7.6 CONCLUSIONS

Based on the obtained results, the conclusions are as follows:

i. All SCAAC specimens prepared with 50% and higher WCP in this study achieved adequate flow, passing ability and resistance to segregation. These results indicated that the flow and passing ability of SCAAC increased with the increment of adding WCP in the matrix. Contrast to these results, the SCAAC's resistance to segregation decreased with increasing WCP content. However, incorporating GBFS and WCP led to achieve the aims of this study and produced high-performance SCAAC.

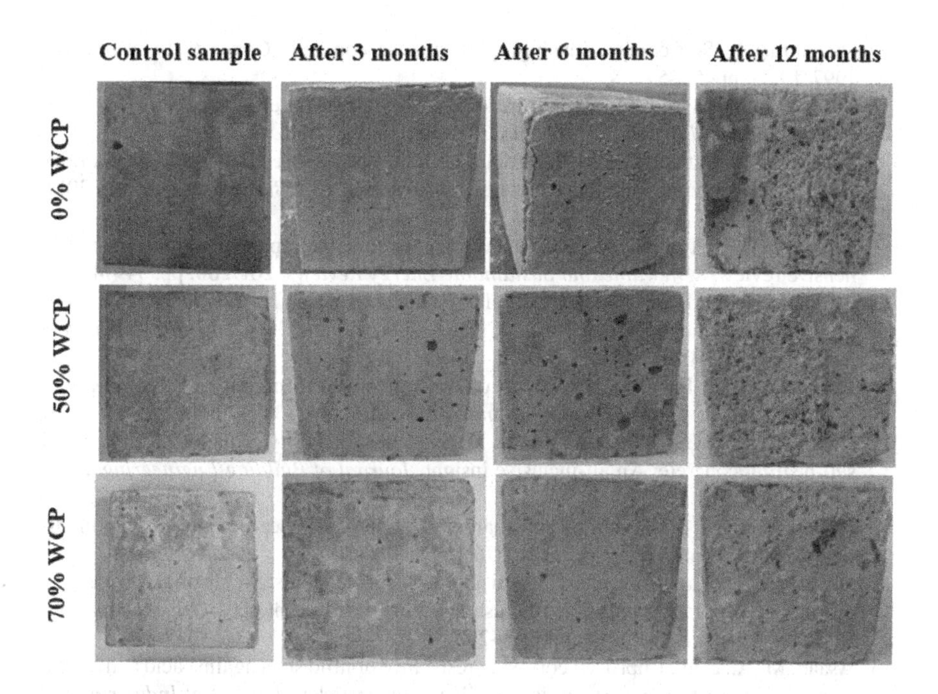

FIGURE 7.23 Surface texture of SCAACs exposed to 10% H_2SO_4.

ii. Inclusion of WCP enhanced both the initial and final setting times of SCAAC samples.
iii. Incorporating WCP as raw material in SCAAC showed excellent performance in the hardened state. All SCAAC specimens presented acceptable compressive strength between 30 and 50 MPa. However, the strength dropped as the WCP contents increased in the SCAAC matrix. Similar trends were also found with FS and TS.
iv. According to XRD, SEM, and FTIR microstructure results, the formation of C-S-H and C-A-S-H gels was influenced by high content of SiO_2 in WCP. Gel formation decreased as the WCP contents increased, which resulted in poor structure, low strength and high porosity.
v. Inclusion of WCP in the alkali-activated matrix as a GBFS replacement enhanced the durability of SCAACs exposed to sulphuric acid attack.

REFERENCES

1. Long, G., Y. Gao, and Y. Xie, Designing more sustainable and greener self-compacting concrete. *Construction and Building Materials*, 2015. **84**: pp. 301–306.
2. Domone, P., Self-compacting concrete: An analysis of 11 years of case studies. *Cement and Concrete Composites*, 2006. **28**(2): pp. 197–208.
3. Huseien, G. F., J. Mirza, and M. Ismail, Effects of high volume ceramic binders on flexural strength of self-compacting geopolymer concrete. *Advanced Science Letters*, 2018. **24**(6): pp. 4097–4101.

4. Okamura, H., Self-compacting high-performance concrete. *Concrete International*, 1997. **19**(7): pp. 50–54.
5. Shi, C., et al., A review on mixture design methods for self-compacting concrete. *Construction and Building Materials*, 2015. **84**: pp. 387–398.
6. Subaşı, S., H. Öztürk, and M. Emiroğlu, Utilizing of waste ceramic powders as filler material in self-consolidating concrete. *Construction and Building Materials*, 2017. **149**: pp. 567–574.
7. Huseien, G. F., et al., Geopolymer mortars as sustainable repair material: A comprehensive review. *Renewable and Sustainable Energy Reviews*, 2017. **80**: pp. 54–74.
8. García-Lodeiro, I., et al., Compatibility studies between NASH and CASH gels. Study in the ternary diagram Na2O–CaO–A12O3–SiO2–H2O. *Cement and Concrete Research*, 2011. **41**(9): pp. 923–931.
9. Abdel-Gawwad, H., et al., Recycling of concrete waste to produce ready-mix alkali activated cement. *Ceramics International*, 2018. **44**(6): pp. 7300–7304.
10. Huseien, G. F., K. W. Shah, and A. R. M. Sam, Sustainability of nanomaterials based self-healing concrete: An all-inclusive insight. *Journal of Building Engineering*, 2019. 23: pp. 155–171.
11. Huseien, G. F., et al., Synthesis and characterization of self-healing mortar with modified strength. *Jurnal Teknologi*, 2015. **76**(1): pp. 195–200.
12. Hassan, I. O., et al., Flow characteristics of ternary blended self-consolidating cement mortars incorporating palm oil fuel ash and pulverised burnt clay. *Construction and Building Materials*, 2014. **64**: pp. 253–260.
13. Asaad, M. A., et al., Improved corrosion resistance of mild steel against acid activation: Impact of novel Elaeis guineensis and silver nanoparticles. *Journal of Industrial and Engineering Chemistry*, 2018. **63**: pp. 139–148.
14. Huseien, G. F., et al., The effect of sodium hydroxide molarity and other parameters on water absorption of geopolymer mortars. *Indian Journal of Science and Technology*, 2016. **9**(48): pp. 1–7.
15. Ouellet-Plamondon, C. and G. Habert, Life cycle assessment (LCA) of alkali-activated cements and concretes. In *Handbook of Alkali-Activated Cements, Mortars and Concretes*. pp. 663–686. 2015: Elsevier.
16. McLellan, B. C., et al., Costs and carbon emissions for geopolymer pastes in comparison to ordinary Portland cement. *Journal of Cleaner Production*, 2011. **19**(9–10): pp. 1080–1090.
17. Huseien, G. F., et al., Evaluation of alkali-activated mortars containing high volume waste ceramic powder and fly ash replacing GBFS. *Construction and Building Materials*, 2019. **210**: pp. 78–92.
18. Hardjito, D., et al., Factors influencing the compressive strength of fly ash-based geopolymer concrete. *Civil Engineering Dimension*, 2004. **6**(2): pp. 88–93.
19. Soutsos, M., et al., Factors influencing the compressive strength of fly ash based geopolymers. *Construction and Building Materials*, 2016. **110**: pp. 355–368.
20. Huseien, G. F., M. Ismail, and J. Mirza, Influence of curing methods and sodium silicate content on compressive strength and microstructure of multi blend geopolymer mortars. *Advanced Science Letters*, 2018. **24**(6): pp. 4218–4222.
21. Huseiena, G. F., et al., Performance of sustainable alkali activated mortars containing solid waste ceramic powder. *Chemical Engineering*, 2018. **63**: pp. 673–678.
22. Duan, P., et al., Development of fly ash and iron ore tailing based porous geopolymer for removal of Cu (II) from wastewater. *Ceramics International*, 2016. **42**(12): pp. 13507–13518.
23. Huseien, G. F., et al., Synergism between palm oil fuel ash and slag: Production of environmental-friendly alkali activated mortars with enhanced properties. *Construction and Building Materials*, 2018. **170**: pp. 235–244.

24. Huseien, G. F., et al., Utilizing spend garnets as sand replacement in alkali-activated mortars containing fly ash and GBFS. *Construction and Building Materials*, 2019. **225**: pp. 132–145.
25. Yusuf, M. O., et al., Evolution of alkaline activated ground blast furnace slag–ultrafine palm oil fuel ash based concrete. *Materials & Design*, 2014. **55**: pp. 387–393.
26. Rashad, A. M., Properties of alkali-activated fly ash concrete blended with slag. *Iranian Journal of Materials Science and Engineering*, 2013. **10**(1): pp. 57–64.
27. Huseiena, G. F., et al., Effect of binder to fine aggregate content on performance of sustainable alkali activated mortars incorporating solid waste materials. *Chemical Engineering*, 2018. **63**: pp. 667–672.
28. Patankar, S. V., Y. M. Ghugal, and S. S. Jamkar, Effect of concentration of sodium hydroxide and degree of heat curing on fly ash-based geopolymer mortar. *Indian Journal of Materials Science*, 2014. **2014**: pp. 1–6.
29. Khale, D. and R. Chaudhary, Mechanism of geopolymerization and factors influencing its development: A review. *Journal of Materials Science*, 2007. **42**(3): pp. 729–746.
30. Ranjbar, N., et al., Compressive strength and microstructural analysis of fly ash/palm oil fuel ash based geopolymer mortar under elevated temperatures. *Construction and Building Materials*, 2014. **65**: pp. 114–121.
31. Salih, M. A., et al., Development of high strength alkali activated binder using palm oil fuel ash and GGBS at ambient temperature. *Construction and Building Materials*, 2015. **93**: pp. 289–300.
32. Pacheco-Torgal, F. and S. Jalali, Reusing ceramic wastes in concrete. *Construction and Building Materials*, 2010. **24**(5): pp. 832–838.
33. Huang, B., Q. Dong, and E. G. Burdette, Laboratory evaluation of incorporating waste ceramic materials into Portland cement and asphaltic concrete. *Construction and Building Materials*, 2009. **23**(12): pp. 3451–3456.
34. Pacheco-Torgal, F. and S. Jalali, Compressive strength and durability properties of ceramic wastes based concrete. *Materials and Structures*, 2011. **44**(1): pp. 155–167.
35. Abdollahnejad, Z., et al., Mix design, properties and cost analysis of fly ash-based geopolymer foam. *Construction and Building Materials*, 2015. **80**: pp. 18–30.
36. Nath, P. and P. K. Sarker, Effect of GGBFS on setting, workability and early strength properties of fly ash geopolymer concrete cured in ambient condition. *Construction and Building Materials*, 2014. **66**: pp. 163–171.
37. Huseien, G. F., et al., Effect of metakaolin replaced granulated blast furnace slag on fresh and early strength properties of geopolymer mortar. *Ain Shams Engineering Journal*, 2018. 9: pp. 1557–1566.
38. Roslan, N. H., et al., Performance of steel slag and steel sludge in concrete. *Construction and Building Materials*, 2016. **104**: pp. 16–24.
39. Çevik, A., et al., Effect of nano-silica on the chemical durability and mechanical performance of fly ash based geopolymer concrete. *Ceramics International*, 2018. **44**(11): pp. 12253–12264.
40. Baraldi, L. and MECS-Machinery Economics Studies by ACIMAC, World production and consumption of ceramic tiles. *AMERICA*, 2016. **1**(9.2): p. 7.7.
41. Hussein, A. A., et al., Performance of nanoceramic powder on the chemical and physical properties of bitumen. *Construction and Building Materials*, 2017. **156**: pp. 496–505.
42. Senthamarai, R., P. D. Manoharan, and D. Gobinath, Concrete made from ceramic industry waste: Durability properties. *Construction and Building Materials*, 2011. **25**(5): pp. 2413–2419.
43. Samadi, M., et al., Properties of mortar containing ceramic powder waste as cement replacement. *Jurnal Teknologi*, 2015. **77**(12): pp. 93–97.

44. Lim, N. H. A. S., et al., Microstructure and strength properties of mortar containing waste ceramic nanoparticles. *Arabian Journal for Science and Engineering*, 2018. **43**(10): pp. 5305–5313.
45. Mohit, M. and Y. Sharifi, Thermal and microstructure properties of cement mortar containing ceramic waste powder as alternative cementitious materials. *Construction and Building Materials*, 2019. **223**: pp. 643–656.
46. Mohammadhosseini, H., et al., Enhanced performance of green mortar comprising high volume of ceramic waste in aggressive environments. *Construction and Building Materials*, 2019. **212**: pp. 607–617.
47. Al-Majidi, M. H., et al., Development of geopolymer mortar under ambient temperature for in situ applications. *Construction and Building Materials*, 2016. **120**: pp. 198–211.
48. Lee, N., E. Kim, and H.-K. Lee, Mechanical properties and setting characteristics of geopolymer mortar using styrene-butadiene (SB) latex. *Construction and Building Materials*, 2016. **113**: pp. 264–272.
49. Yusuf, M. O., Performance of slag blended alkaline activated palm oil fuel ash mortar in sulfate environments. *Construction and Building Materials*, 2015. **98**: pp. 417–424.
50. Yusuf, M. O., et al., Evaluation of slag-blended alkaline-activated palm oil fuel ash mortar exposed to the sulfuric acid environment. *Journal of Materials in Civil Engineering*, 2015. **27**(12): p. 04015058.
51. Huseien, G. F., et al., Compressive strength and microstructure of assorted wastes incorporated geopolymer mortars: Effect of solution molarity. *Alexandria Engineering Journal*, 2018. **57**(4): pp. 3375–3386.
52. Karakoç, M. B., et al., Mechanical properties and setting time of ferrochrome slag based geopolymer paste and mortar. *Construction and Building Materials*, 2014. **72**: pp. 283–292.
53. Phoo-ngernkham, T., et al., High calcium fly ash geopolymer mortar containing Portland cement for use as repair material. *Construction and Building Materials*, 2015. **98**: pp. 482–488.
54. Kubba, Z., et al., Impact of curing temperatures and alkaline activators on compressive strength and porosity of ternary blended geopolymer mortars. *Case Studies in Construction Materials*, 2018. **9**: p. e00205.
55. Phoo-ngernkham, T., et al., Effects of sodium hydroxide and sodium silicate solutions on compressive and shear bond strengths of FA–GBFS geopolymer. *Construction and Building Materials*, 2015. **91**: pp. 1–8.
56. Rickard, W. D., et al., Assessing the suitability of three Australian fly ashes as an aluminosilicate source for geopolymers in high temperature applications. *Materials Science and Engineering: A*, 2011. **528**(9): pp. 3390–3397.
57. Huseien, G. F., et al., Properties of ceramic tile waste based alkali-activated mortars incorporating GBFS and fly ash. *Construction and Building Materials*, 2019. **214**: pp. 355–368.
58. Huseien, G. F., et al., Influence of different curing temperatures and alkali activators on properties of GBFS geopolymer mortars containing fly ash and palm-oil fuel ash. *Construction and Building Materials*, 2016. **125**: pp. 1229–1240.
59. Zhou, W., et al., A comparative study of high-and low-Al_2O_3 fly ash based-geopolymers: The role of mix proportion factors and curing temperature. *Materials & Design*, 2016. **95**: pp. 63–74.
60. Sitarz, M., W. Mozgawa, and M. Handke, Vibrational spectra of complex ring silicate anions—Method of recognition. *Journal of Molecular Structure*, 1997. **404**(1–2): pp. 193–197.

61. Phair, J. W., J. Van Deventer, and J. Smith, Mechanism of polysialation in the incorporation of zirconia into fly ash-based geopolymers. *Industrial & Engineering Chemistry Research*, 2000. **39**(8): pp. 2925–2934.
62. Guo, X., H. Shi, and W. A. Dick, Compressive strength and microstructural characteristics of class C fly ash geopolymer. *Cement and Concrete Composites*, 2010. **32**(2): pp. 142–147.
63. EFNARC, *Specification and Guidelines for Self-Compacting Concrete*. Vol. 32, p. 34. 2002: Association House.
64. ASTM, Standard test methods for chemical resistance of mortars, grouts, and monolithic surfacing and polymer concretes. 2012: ASTM.
65. Huseien, G. F., et al., Waste ceramic powder incorporated alkali activated mortars exposed to elevated Temperatures: Performance evaluation. *Construction and Building Materials*, 2018. **187**: pp. 307–317.
66. Sugama, T., L. Brothers, and T. Van de Putte, Acid-resistant cements for geothermal wells: sodium silicate activated slag/fly ash blends. *Advances in Cement Research*, 2005. **17**(2): pp. 65–75.
67. Alanazi, H., J. Hu, and Y.-R. Kim, Effect of slag, silica fume, and metakaolin on properties and performance of alkali-activated fly ash cured at ambient temperature. *Construction and Building Materials*, 2019. **197**: pp. 747–756.
68. Huseien, G. F., et al., Effects of POFA replaced with FA on durability properties of GBFS included alkali activated mortars. *Construction and Building Materials*, 2018. **175**: pp. 174–186.
69. Kumar, S., R. Kumar, and S. Mehrotra, Influence of granulated blast furnace slag on the reaction, structure and properties of fly ash based geopolymer. *Journal of Materials Science*, 2010. **45**(3): pp. 607–615.
70. Puertas, F., et al., Alkali-activated fly ash/slag cements: strength behaviour and hydration products. *Cement and Concrete Research*, 2000. **30**(10): pp. 1625–1632.
71. Mozgawa, W. and J. Deja, Spectroscopic studies of alkaline activated slag geopolymers. *Journal of Molecular Structure*, 2009. **924**: pp. 434–441.
72. Ravikumar, D., S. Peethamparan, and N. Neithalath, Structure and strength of NaOH activated concretes containing fly ash or GGBFS as the sole binder. *Cement and Concrete Composites*, 2010. **32**(6): pp. 399–410.
73. Ismail, I., et al., Modification of phase evolution in alkali-activated blast furnace slag by the incorporation of fly ash. *Cement and Concrete Composites*, 2014. **45**: pp. 125–135.
74. Song, S. and H. M. Jennings, Pore solution chemistry of alkali-activated ground granulated blast-furnace slag. *Cement and Concrete Research*, 1999. **29**(2): pp. 159–170.
75. Zhang, W., et al., The degradation mechanisms of alkali-activated fly ash/slag blend cements exposed to sulphuric acid. *Construction and Building Materials*, 2018. **186**: pp. 1177–1187.
76. Allahverdi, A. and F. Skvara, Sulfuric acid attack on hardened paste of geopolymer cements – Part 1. Mechanism of corrosion at relatively high concentrations. *Ceramics–Silikaty*, 2005. **49**(4): p. 225.
77. Gu, L., T. Bennett, and P. Visintin, Sulphuric acid exposure of conventional concrete and alkali-activated concrete: Assessment of test methodologies. *Construction and Building Materials*, 2019. **197**: pp. 681–692.

8 Durability Performance of Ceramic Waste–Based Alkali-Activated Mortars

8.1 INTRODUCTION

The production of cement and concrete is being explored by scientists in order to find more environmentally friendly methods of producing materials that will do the same job. This has occurred as people are becoming more aware of the harms to society that is caused by CO_2 emissions, a by-product of cement and concrete [1–5]. Around 10 billion metric tonnes of concrete are produced in the world each year, most of which contains ordinary Portland cement (OPC). Approximately one tonne of CO_2 is created with every tonne of OPC produced, which means that 7% of all the CO_2 emissions produced worldwide can be attributed to OPC factories [3, 6–8]. CO_2 is created due to the heat used during production and the energy expenditure of the raw materials [9–11]. With the production of OPC expected to quadruple over the next three decades, there is a real risk of serious environmental, economic and ecological damage [12, 13].

Recently, there has been a development in the area of alkali-activated concretes and mortars, which produce far less CO_2 and are therefore much better for the environment. Rather than the high levels of CO_2 created in the production of OPC, the alkaline method of making concrete using geopolymer binders [10, 14, 15] using wastes like palm oil fuel ash (POFA), metakaolin (MK), silica fume (SF), fly ash (FA), waste ceramic powder (WCP), and ground blast furnace slag (GBFS). This development in creating an alternative to OPC has caught the attention of professionals in commercial and academic sectors in recent years. The discussions have not only focused on the relatively similar functionality to OPC, but also on the natural features of the material, such as its good strength, resistance to fire and chemical attack, its ability to withstand heat, and the reduced level of CO_2 emissions it produces [16–18]. Some research has centred on several aluminosilicate materials and their response and microstructural classification when activated with various alkaline activators [19–22]. Research in the production of ecological concrete/mortar has stepped up in recent times [18, 23], but there have not been studies for the development of an outline for ecological concrete mix proportioning.

On the other hand, recent years have seen an increase in the constraints of landfill use due to the growing understanding of environmental issues in developing countries, such as Malaysia. As a result, the cost of managing waste will rise and manufacturers will need to seek alternative uses for it. Several recent studies have demonstrated how industrial waste can be used in the production of materials in the construction industry. For example, ceramic waste can be repurposed in the making of concrete, although most research has concentrated on the analysis of the durable and mechanical properties of ceramic as a binder or fine and coarse aggregates to partly take the place of the environmentally unfriendly OPC [24–28].

The process of using ceramic waste has been tested, but the levels of waste that are effectively used for this purpose are very small. The chemical qualities of ceramic with highly crystalline aluminosilicate make it possible for it to be used to manufacture good quality alkali-activated concrete and mortars.

In the past few years GBFS has shown great potential as a partial substitute for OPC or cement-free-based concrete. Unlike FA-based mortars which normally need a high level of alkaline solution (10–16 M of NaOH) and increased curing temperatures (between 40 and 100°C) [29, 30], GBFS-based mortars become strong and gain desirable mechanical properties at comparatively low temperatures and dosages of alkaline activator [31–33]. Several studies showed that more than 100 MPa strength values could be achieved when using GBFS-based, alkaline-activated concrete [32, 33]. However, there are limitations to the usage of GBFS-based alkali-activated mortars (AAMs), such as fast setting time, thus creating problems for its use in ready-mixed and pumped concretes [33–36]. The setting time can be controlled somewhat by altering the slag substance. This can be achieved by adding CaO in an alkali-activated SiO_2-Al_2O_3-CaO solid precursor system or by considering the design of the composition. Previous studies [18, 34, 37, 38] showed similar outcomes where concrete took a longer time to set when the FA levels were raised to 70% by weight of the complete binder (GBFS and FA), while there was a reduction in the compressive strength and the consistency of the material. A further obstacle in the use of GBFS-based AAMs, as shown in several studies, is the high (40–58% by mass) level of CaO, which reduces the mortar's ability to withstand an acidic attack. Despite its qualities, the high cost of GBFS, along with the energy consumption and CO_2 emissions created during production, make its use limited on a large scale and as a cement-free concrete.

Research into appropriate substances for geopolymers has identified low-calcium fly ash (class F) as a possibility, due to its widespread availability, low water demand, and relevant silica and alumina composition. Long- and short-term tests on heat-cured, low-calcium, fly ash–based geopolymer concrete have highlighted its exceptional qualities in mechanics and durability [39]. The need to cure fly ash–based AAMs at high temperatures, affecting the sustainability of FA, creates an obstacle for its widespread use. Attempts have been made to increase the reactiveness of FA in an alkaline atmosphere by adding elements containing calcium and by raising its fineness [40]. Adding calcium oxide (CaO) produces elements such as calcium silicate hydrate (C-S-H) as well as the alumina-silicate geopolymer network [31, 38]. Raised levels of CaO in the precursor materials were found to correlate to a lower setting time and greater strength in the hardened geopolymer [10, 34]. Further studies [41] showed that replacing FA with calcium hydroxide and CaO improved the mechanical properties for materials cured at ambient temperature, but had the opposite effect on materials cured at 70°C. The resilient qualities of fly ash–based geopolymer make it desirable as a cementing material [42] and some have identified comparable engineering properties of the material that make it also desirable for use in construction [43–45].

More research is being carried out into the use of alkali-activated materials as a sustainable substitute for OPC in order to reduce CO_2 emissions. However, issues surrounding cost and the ecological effects are standing in the way of mass production and introduction to the market. This research is focused on the analysis of the influence of ternary-blended alkali-activated materials with high levels of WCP, including GBFS and FA, on factors related to sustainability, including energy consumption, durability, CO_2 emissions, and cost. The resilience of AAMs under several ranges of GBFS replaced by FA and activated by low concentration alkaline solutions was extensively studied. Ambient heat used for curing in the production of AAMs can theoretically be used in more areas than just precast members, which in turn minimises energy consumption and costs that go hand in hand with heat curing. Consequently, this study sets out to create AAM compositions that will be effective in extreme environments, while also taking into account the need for low cost, energy consumption, and CO_2 emissions. Areas of the present research study included acid resistance, compressive strength, effects of elevated temperatures, freeze–thaw cycles, and sulphate attack resistance.

8.2 MATERIALS AND MIX DESIGN

8.2.1 MATERIALS

In this chapter, ceramic tile wastes were collected from White Horse ceramic manufacturer in Pasir Gudang Johor, Malaysia. Homogeneous ceramic tile wastes of the same thickness with no glassy coating were collected. They were crushed using a jaw crusher and then they were sieved with 600 μm mesh to remove big size particles. The ceramic waste particles that passed through 600 μm sieve were ground using a Los Angeles abrasion machine with 20 stainless steel balls of 40 mm in diameter for 6 hours; the final result is known as waste ceramic powder (WCP). The process for obtaining WCP is shown in Figure 8.1. The pure GBFS used in this study was obtained from Ipoh, Malaysia, by a supplier and used as received without any further treatment. From the Tanjung bin power station, Johor, Malaysia, low-calcium FA was collected and used as resource aluminosilicate material for making AAMs. The colours of WCP, GBFS, and FA were light grey, off-white and grey, respectively. Their physical properties are shown in Table 8.1. A lower specific gravity was observed with FA (2.2) compared to 2.6 and 2.9 for WCP and GBFS, respectively. The medium particle size of WCP, GBFS and FA were 35, 12.8, and 10 μm, respectively.

By using X-ray fluorescence spectroscopy (XRF), the chemical composition of WCP, GBFS and FA were determined (Table 8.1). It reveals that WCP, GBFS and FA have different characteristics based on chemical composition. It was found that the major compound in WCP and FA is SiO_2 (72.6 and 57.8%, respectively) and CaO (51.8%) in GBFS. Both WCP and FA presented very low CaO content (0.02 and 5.2%

FIGURE 8.1 WCP preparation process in the lab.

TABLE 8.1
Physical Properties and Chemical Composition of WCP, GBFS, and FA

Material	WCP	GBFS	FA
Physical Characteristics			
Specific gravity	2.6	2.9	2.2
Medium particle size (μm)	35	12.8	10
Chemical Composition (% by Mass)			
SiO_2	72.6	30.8	57.20
Al_2O_3	12.6	10.9	28.81
Fe_2O_3	0.56	0.64	3.67
CaO	0.02	51.8	5.16
MgO	0.99	4.57	1.48
K_2O	0.03	0.36	0.94
Na_2O	13.5	0.45	0.08
SO_3	0.01	0.06	0.10
Loss on ignition (LOI)	0.13	0.22	0.12

respectively) compared to a high content for GBFS. FA presented the highest content of Al_2O_3 (28.8%) compared to 12.6 and 10.9% for WCP and GBFS, respectively. As is well known, SiO_2, Al_2O_3 and CaO are very important oxides during the hydration process and produce calcium (aluminium) silicate hydrate (C-(A)-S-H) gels. The low content of Al_2O_3 and CaO in WCP require mixing with materials containing high amounts of Al_2O_3 (FA) and CaO (GBFS) to produce high-performance alkali-activated materials. Obviously, the presence of higher calcium in GBFS and aluminium in FA contents influence the pozzolanic reaction to produce extra C-(A)-S-H gels thus making the mortar more durable and denser. According to ASTM C618-15, WCP and FA are classified as class F pozzolans as the total percentage of (SiO_2+ $A12O_3$+ Fe_2O_3) is higher than 70%.

Mineralogical characterization was carried out by the X-ray diffraction (XRD) technique. The XRD patterns of WCP, GBFS, and FA are illustrated in Figure 8.2. The amorphous and crystalline quantity of the precursor WCP and FA has also been identified, which influence the reactivity and the compressive strength of the alkali-activated materials produced. Quartz (SiO_2) and mullite ($3Al_2O_3.2SiO_2$) of WCP and FA were the main crystalline components identified. The amorphous content (visible as the broad hump in the background of the XRD pattern centred at 20–30° 2θ) was quantified by adding an internal standard (corundum) and performing a Rietveld method-based analysis. Previous research studies have identified that the higher the amorphous content in the FA, the higher is its reactivity [46, 47]. GBFS is mainly composed of an amorphous phase as displayed by a halo around 29–40° with a small amount of magnetite. This observation is consistent with the findings of Ismail et al. [48].

Analytical-grade sodium hydroxide (NH) 98% purity pellets and sodium silicate (NS) solution made up of SiO_2 (29.5 wt%), Na_2O (14.70 wt%) and H_2O (55.80 wt%) were used in this study. The pellets were placed in water to dissolve in order to create the NS solution with a concentration of 4 M. This was cooled for a period of 24 hours before mixing it with NS solution to create the alkaline solution (S) with a SiO_2:Na_2O ratio of 1:02. All alkaline solutions maintained a low NS:NH ratio of 0.75 in order to minimize the adverse environmental effect of sodium silicate (Na_2SiO_3). Sieve analysis took place in line with ASTM C33-16 and used sand from local rivers. The bulk density used was 1614 kg/m^3 and 100% passed through a 2.36 mm sieve with specific gravity of 2.6. A fineness modulus of 2.8 was used (Figure 8.3).

8.2.2 Mix Design

The effects of FA replacing GBFS on durability performance of alkali-activated mortars containing a high amount of WCP were studied by preparing five mixtures (Table 8.2). For continuity, all mixtures had equal levels of NH molarity (4 M) with 50% WCP binder mass for each batch. Each mixture also had equal ratios of binder to fine aggregate (1.0); alkaline solution to binder (0.40); sodium silicate to sodium hydroxide (0.75) and modulus of solution (1.02). GBFS with a 50% content was chosen as the control batch and subsequent mixtures replaced 10, 20, 30, and 40%, by mass ratio of GBFS with FA. NH and NS were diligently mixed to create a weighed solution which was cooled to room temperature prior to using. In mixing the AAMs, fine aggregates were placed in the mortar mixer machine, before adding binders including FA, GBFS and WCP. They were mixed in dry conditions for 2 minutes to create a homogenous mixture before being activated with the alkaline solution. After a further 4 minutes of mixing, the

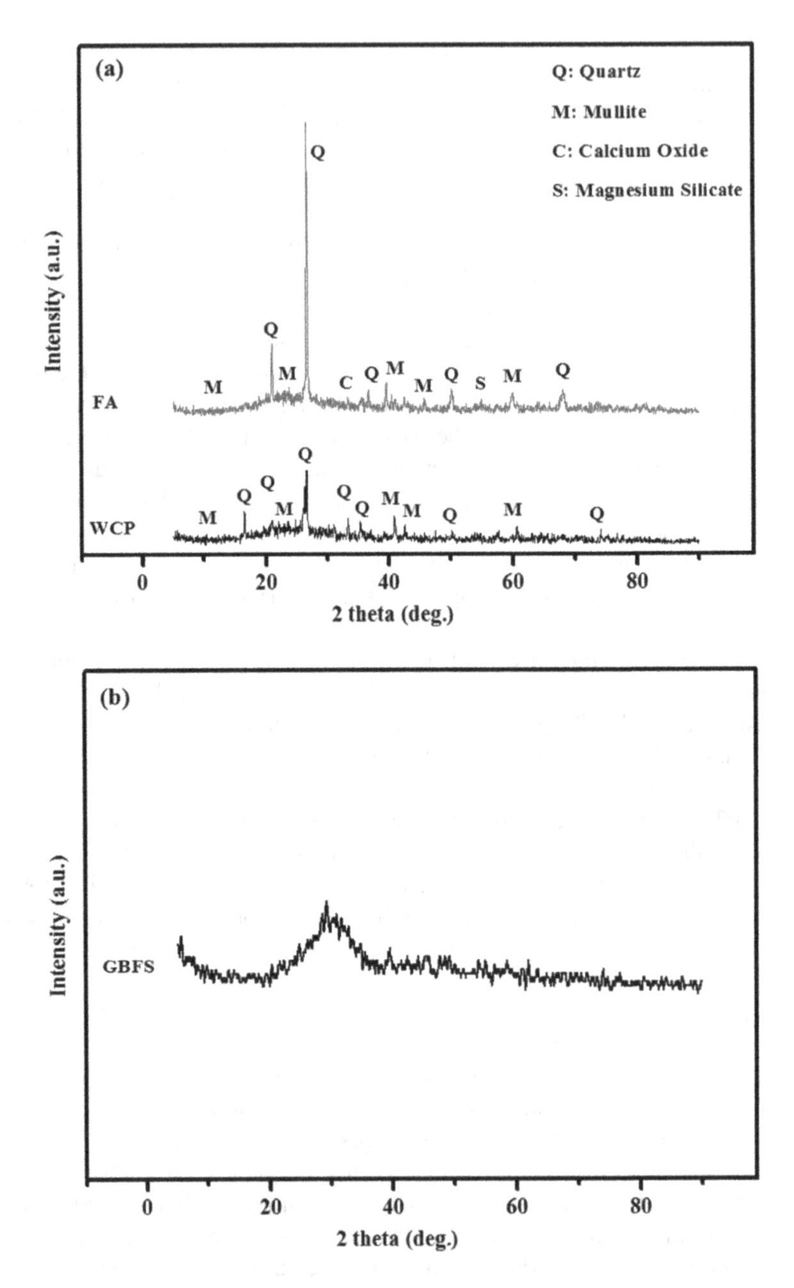

FIGURE 8.2 XRD patterns of (a) WCP and FA, and (b) GBFS.

final mixture was poured in the moulds as per ASTM C579-18 [49]. The mixture was poured in two layers into each mould, allowing 15 seconds on the vibration table after each layer to release any air bubbles. Keeping in mind the Malaysia weather, the AAM specimens were then left at a temperature of 27 ± 1.5°C and relative humidity of 75% for a period of 24 hours to allow them to cure.

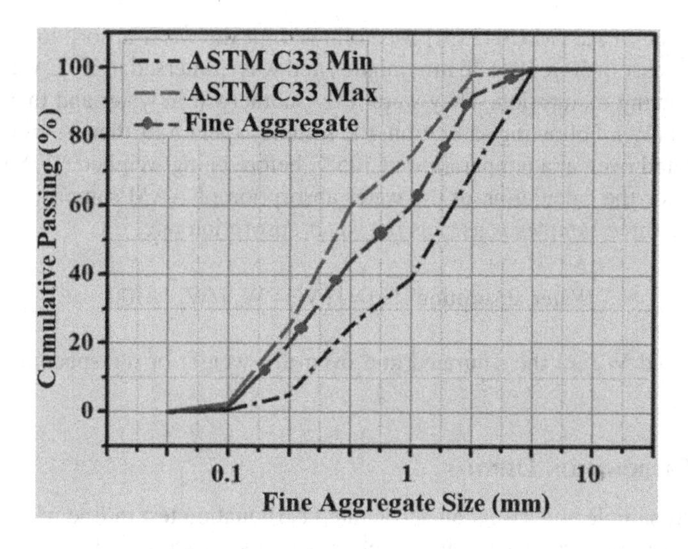

FIGURE 8.3 XRD patterns of GBFS and WCP.

TABLE 8.2
Mixture Design of Ternary-Blended (WCP, GBFS, and FA) Alkali-Activated Mortars

	Binder (kg/m³)				Alkaline Solution (kg/m³)	
Mix	WCP	GBFS	FA	Sand (kg/m³)	Na_2SiO_3	NaOH
$AAMs_1$	550	550	0	1100	188.6	251.4
$AAMs_2$	550	440	110	1100	188.6	251.4
$AAMs_3$	550	330	220	1100	188.6	251.4
$AAMs_4$	550	220	330	1100	188.6	251.4
$AAMs_5$	550	110	440	1100	188.6	251.4

Following that time, the specimens were opened and left in the same environment until the day of the test.

8.3 TESTING PROCEDURE

8.3.1 Strength and Water Absorption

The compression strength test was carried out after 1, 7, and 28 days, following ASTM C109-109M [50]. Three samples were tested for each age and, after preparation, each sample was placed precisely between the top and bottom metal-bearing plates, in line with the relevant standard specifications. A consistent loading rate of 2.5 kN/s was applied to the samples. Density and compressive strength figures, based on the weight and size of the samples, were automatically created due to the machine's construction.

The format of ASTM C642 [51] was followed for the water absorption test. Each sample was cast in 50 × 50 × 50 mm moulds and was immersed in 27°C water for 24 hours, following maturation. They were fully submerged in water and then hung to be weighed (W_s). Following saturation, the samples spent a further 24 hours drying in a ventilated oven at a temperature of 105°C before being weighed (W_d). Equation 8.1 allows for the calculation of the water absorption of AAM samples. The mean value of the three samples is used as the water absorption rate.

$$\text{Water absorption}\,(\%) = \left[\left(W_s - W_d\right)/W_d\right] \times 100 \tag{8.1}$$

where W_s and W_d are the saturated and oven-dry weight of the specimen (gram), respectively.

8.3.2 Carbonation Depth

Each AAM sample underwent an accelerated carbonation test in line with BS 1881-210:2013. An accelerated carbonation chamber was constructed by using pressure pipes to attach a CO_2 gas cylinder to a plastic tank, allowing the samples to be placed in a carbonated environment. Three AAM cylinders, each measuring 75 × 150 mm, were tested for each of the mixtures. After placing the samples in the chamber, a vacuum was created with a pressure of approximately 600 mm Hg at 60–65% relative humidity (RH), for 3 minutes. Twice each day for 90 days, CO_2 was passed through the tanks for 20 minutes, with a pressure of 1034 mm Hg. This was conducted at a temperature of 27 ± 3°C. After the 90 days, the samples were halved and the cross sections were sprayed with 1% phenolphthalein. This shows the un-carbonated areas which turn purple, while the carbonated parts stay the same. By measuring the distance between the purple boundary and the edge of the sample, the average carbonation depth could be calculated.

8.3.3 Sulphuric Acid and Sulphate Attack

The primary effect of sulphuric acid (H_2SO_4) on mortar is the weakening of the AAMs by disintegrating the binder paste solution. Deionized water was used to prepare the H_2SO_4 acid solution (10%) and its effects on the AAMs' composition was analyzed. Six 28-days-old specimens for each AAM mixture were weighed before being put into the solution for 365 days. In order to maintain a continuous pH level throughout the experiment, the solution was changed every 2 months. The AAMs' specimens were checked after 180 days and then after 365 days while several different factors were considered when evaluating their performance, including microstructure change, weight loss and remaining strength, based on ASTM C267 specifications [52]. The sulphate ions $(SO_4)^{2-}$ were the main cause of the sulphate attack on the AAMs. They were transferred to the mortar from a variety of concentrations in water along with magnesium, calcium or sodium cations. In a method that mirrored that of the sulphuric acid attack test, a magnesium sulphate solution was utilized to analyze the resistance of the AAM samples.

8.3.4 Freeze–Thaw Cycles

Method A in ASTM C666 [53] was carried out by subjecting each AAM sample to at least 300 cycles of freezing–thawing. To emulate the freeze–thaw cycle, in a 5-hour period of the cycle, the temperature of the samples can be reduced from 5 to –20°C for 75% of the time and increased back to 5°C for the remaining 25% of the time. This particular test was deemed the most significant for the AAM samples as there is no common method for alkali-activated materials and geopolymers. Therefore, this test was utilized in this study for the stated temperature range, temperature ramp and period of cycle. Two differing procedures, method A and method B, were recommended in ASTM C666. Method A stated that samples should be frozen and thawed in water, whereas method B said samples should be frozen in air and thawed in water. This study used the freezer and adopted method A due to its convenience. Two kinds of samples – six for each mixture – were made into cubes measuring 50 × 50 × 50 mm and prisms measuring 40 × 40 × 80 mm. Following a 28-day curing period the samples were thawed at room temperature to measure the mass. They were then placed in a vessel and immersed in water, with the water temperature automatically altered with a timer to create the freeze–thaw cycle. The specimens were subsequently put through 300 freeze–thaw cycles. Qualitative observations on weight loss, surface appearance and residual strength were used to analyze the performance of AAMs.

8.3.5 Wet–Dry Resistance

There is no common method for testing durability, especially in Malaysia where weather conditions are very changeable and can vary quickly; being hot and dry for only a few days then wet for a few days. This experiment was designed to simulate the Malaysian atmosphere and speed up the wet–dry cycle. Table 8.3 illustrates the test conditions of the wet–dry cycles. The experiment was run for 150 cycles. After every 50th cycle, the readings for the process were taken to highlight any alteration in the residual strength.

8.3.6 Elevated Temperatures

One of the aims of the study was to ascertain the capability of ternary-blended AAM samples to withstand high temperatures. Thus, this particular test was performed

TABLE 8.3
Test Conditions for Wet–Dry Cycles

Specimen size	**50 × 50 × 50 mm**
Wet condition	27°C
Dry condition	65°C
Cycle	3 days dry condition/1 day wet condition
Total cycles	150

using an automatic electric furnace on samples measuring 50 × 50 × 50 mm. Following a 28-day curing period, three samples of each mixture were put in temperatures at 400, 700 and 900°C for various lengths of time, then cooled in air. Weights of the samples were taken before and after the temperature changes in order to determine the extent of weight loss caused by the high temperatures. A compressive strength test was followed to determine the residual strength and percentage of strength. Finally, readings were taken to analyze the relative quality of the samples after heating. Furthermore, XRD measurements were used to analyze the microstructures of the samples.

8.4 MECHANICAL AND DURABLE PERFORMANCE

8.4.1 Compressive Strength

The developed compressive strength of various AAM specimens observed at 1, 7, and 28 days is illustrated in Figure 8.4. The results showed that all AAM specimens developed strength with the curing time. At 1 day of age, it was observed that the early strength dropped with an increase in replacement level of GBFS by FA. As the FA contents increased from 0 to 10, 20, 30, and 40%, the strength dropped from 30.6 to 28.2, 25.7, 22.5, and 17.8 MPa, respectively. Similar trends were also observed at 7 and 28 days of age for all AAM specimens with high content of FA compared to the control sample which was prepared with 0% FA. The results of the AAM specimens at 28 days showed a drop in strength to 66.2, 60.2, 56.5, and 45.9 MPa with 10, 20, 30, and 40% of FA content, respectively, compared to 72.1 MPa obtained with the control sample. The increased level of FA replacing GBFS from 0 to 40% affected total CaO content, which restricted the formation of C-(A)-S-H gels and led to lower strength. In previous studies, the authors reported a noticeable decrease in AAM

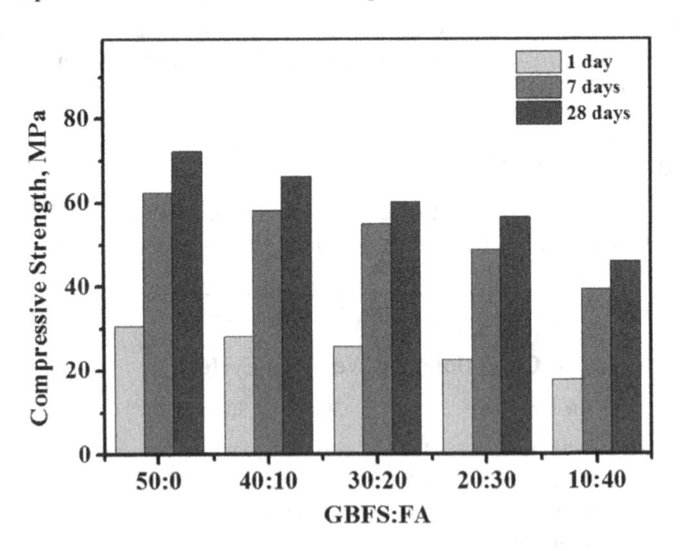

FIGURE 8.4 AAM compressive strength development at 1, 7, and 28 days of age.

strength that could be due to the decrease in the reaction products, but an improving strength of fly ash with increased calcium content where the additional C-S-H and C-A-S-H gels co-existed with N-A-S-H gel of AAMs [32, 34, 54]. However, the AAMs presented acceptable strength (45.9 MPa) even with a high amount of WCP (50%) and FA (40%) and could be used for several construction applications.

Figure 8.5 presents the XRD patterns displaying the effect of FA replacing GBFS on AAM structures containing 50% WCP. The intensity and the number of quartz peaks increased as the FA content increased. It was observed that the C-S-H peak intensity reduced as the FA content increased. This could be due to the presence of a low amount of calcium in the alkali-activated matrix. The calcite peak at 46° was replaced by mullite and then by quartz as the FA content increased from 20 to 30%, respectively. The formation of C-S-H gel was reduced with the increase in FA level where the C-S-H peak at 27° was replaced by hydrotalcite structure. An increase in the non-reacted or partially reacted silicate content was evidenced with the reduction in C-S-H product. This, in turn, led to a drop in AAM strength from 74.1 to 66.2 to 60.2 and to 56.5 MPa as the FA level increased from 0 to 10 to 20 to 30%, respectively.

8.4.2 Water Absorption

Figure 8.6 shows the effect of FA replacing GBFS on water absorption of high-volume WCP-based alkali-activated mortars at 28 days of age. The water absorption data revealed a direct proportionality with the FA content. The increase in FA

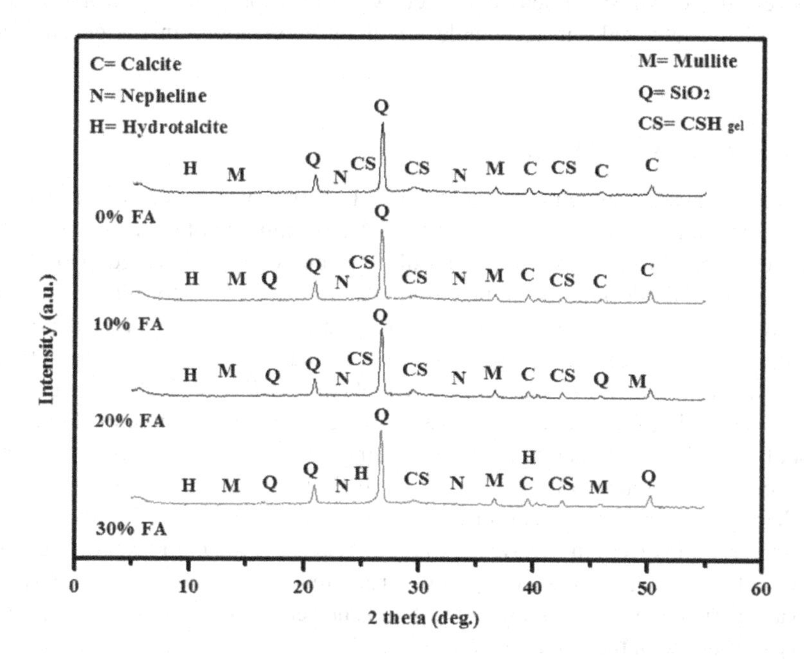

FIGURE 8.5 XRD patterns of FA content replacing GBFS in AAMs containing 50% WCP.

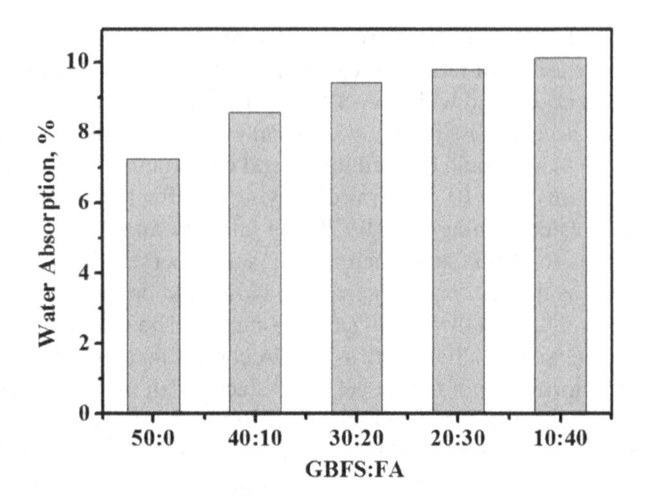

FIGURE 8.6 Water absorption of prepared AAMs.

level replacing GBFS increased the water absorption from 7.3 to 8.5, 9.4, 9.7, and 10.1% with the rising FA content from 0% to 10, 20, 30, and 40%, respectively. The water absorption of AAM specimens was influenced by the pore structure of mortar specimens prepared with high FA content. An increase in FA content could affect the non-reacted and partially reacted particles and lead to a more porous structure. However, all AAMs specimens displayed water absorption equal to or less than 10%, which could make it acceptable for many applications in the construction industry.

8.4.3 Carbonation Depth

Research work was devoted to the study of carbonation, which is considered as one of the major causes of concrete deterioration. A large amount of field data and accelerated carbonation in a laboratory control atmosphere were collected. This led to many formulas describing the carbonation rate. However, CO_2 concentration, environmental conditions, solution-to-binder ratios and time are the main parameters that affect carbonation. Also, carbonation is affected by binder contents and types. Hence, in this research, the effect of FA replacing GBFS on carbonation rate of AAMs containing high volume WCP is clarified through experiments. Figure 8.7 shows the AAMs' carbonation depth depending on FA to GBFS replacement ratio. The increase in the level of replacement of GBFS by FA from 0 to 10, 20, 30, and 40% led to an increase in the carbonation depth from 6.8 to 7.1, 7.3, 7.6, and 8.2 mm, respectively. This peculiar behaviour could be related to the high level of FA in the AAM matrix, which led to a more porous structure as the formulation of C-(A)-S-H gel was restricted by low calcium content and showed a higher porosity and permeability to water than the control sample [55–57].

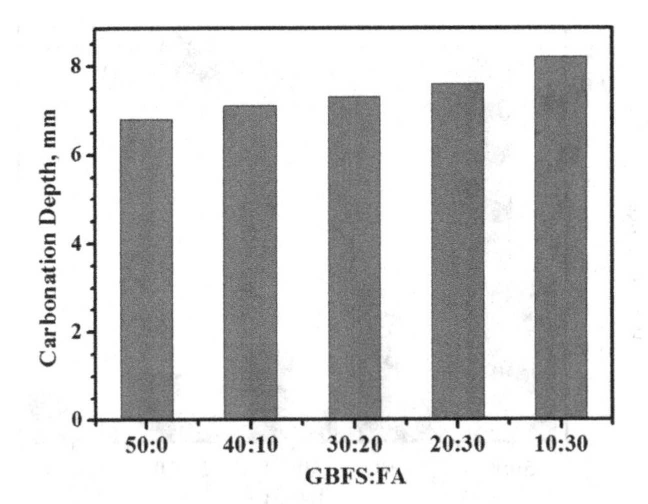

FIGURE 8.7 Porosity of prepared AAMs.

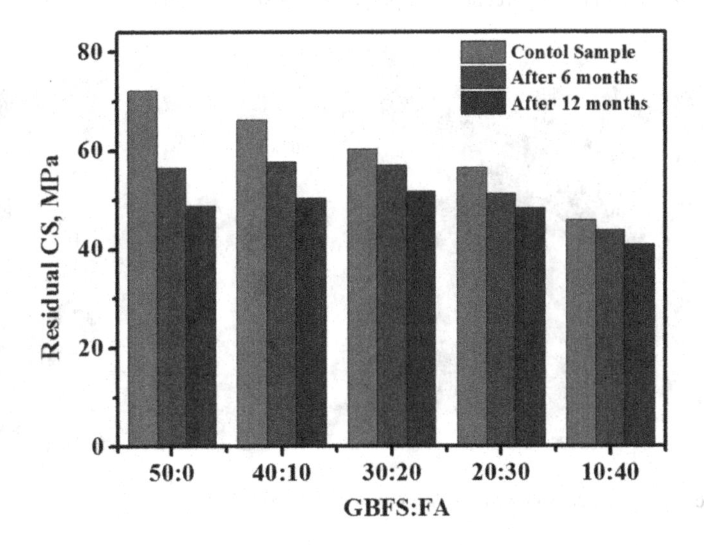

FIGURE 8.8 Residual strength of prepared AAMs exposed to 10% H_2SO_4.

8.4.4 Acid Attack Resistance

The effects of FA replacing GBFS were assessed on sulphuric acid attack resistance of AAMs in terms of residual compressive strength, external cracks and weight loss. Figure 8.8 shows the results of residual strength at 6 and 12 months of age after immersion in 10% H_2SO_4 solution. It was found that the residual strength is directly correlated with FA content (Figure 8.9). After immersion time, all the AAMs showed a reduction in residual strength compared to the control sample. An increase in FA

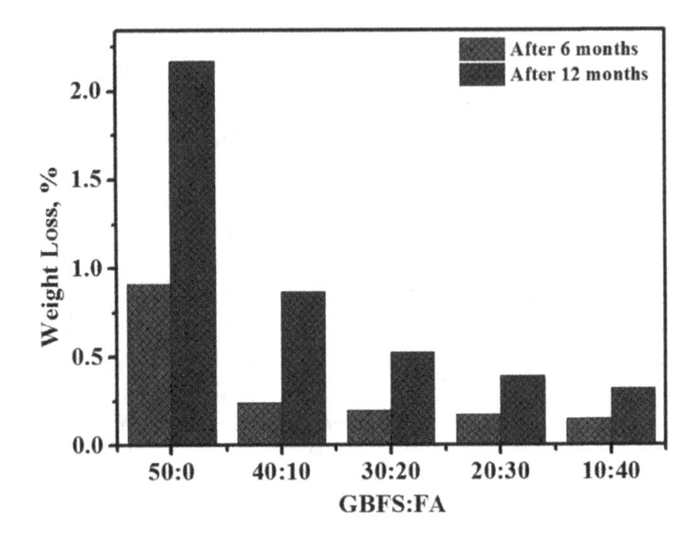

FIGURE 8.9 Weight loss percentage of prepared AAMs exposed to 10% H_2SO_4.

FIGURE 8.10 Surface texture of AAMs exposed to 10% H_2SO_4.

replacing GBFS from 0 to 10, 20, 30, and 40% led the increase in residual strength from 78.3 to 87.1, 93.8, 95.4, and 99%, respectively after 6 months of exposed duration. A similar trend in results was also found with FA replacing GBFS in a high level of WCP after 12 months. The residual strength increased from 67.8 to 76.1, 85.6, 85.7, and 89.2% with an increased FA content from 0 to 10, 20, 30, and 40% in AAM specimens. The increase in FA content from 0 to 10, 20, 30, and 40% also affected a reduction in weight loss percentage from 2.1, 0.86, 0.52, 0.39, and 0.32%, respectively. The surface deterioration and cracks decreased with the increase in FA content, which can be clearly seen in Figure 8.10. Upon exposing the AAMs to sulphuric acid, the $Ca(OH)_2$ compound in mortar reacted with SO_4^{-2} ion and formed gypsum ($CaSO_4.2H_2O$). This caused the expansion in the geopolymer matrix and

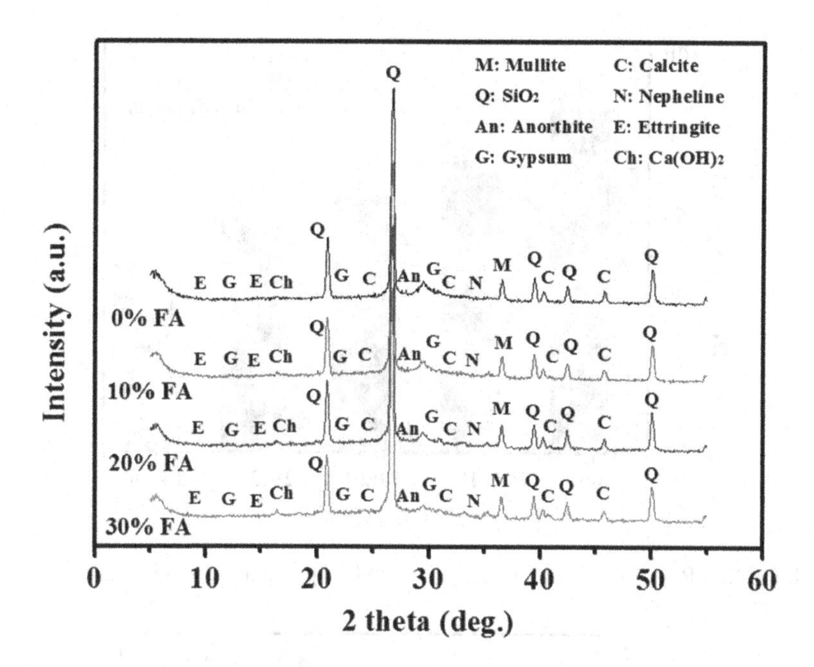

FIGURE 8.11 XRD of prepared AAMs exposed to 10% H_2SO_4.

additional cracking in the interior of specimens as reflected in the visual appearance results (Figure 8.10). Several researchers [58–61] reported that increased Si, Al, and Na contents reduced the gypsum formation, thus increasing the durability of AAMs.

Figure 8.11 shows the XRD pattern of prepared AAMs exposed to 10% H_2SO_4 after 360 days of immersion. The AAM samples demonstrated the presence of gypsum (calcium sulphate hydrate). A decreased intensity of the gypsum peaks at 29.8° 2θ was observed as the FA content increased. Also the gypsum peak grew at 11.8° 2θ and close to quartz (20.8° 2θ) at 20.9° 2θ. As reported by Bellmann and Stark [62], the quartz and gypsum peaks at 20.8 and 20.9° 2θ, respectively, can be difficult to distinguish. However, on close inspection, a double peak was observed indicating the presence of both quartz and gypsum. Nonetheless, the increased FA content to 40% as GBFS replacement restricted the gypsum formulation as can be clearly seen by the peak at 29.8° 2θ, which increased the AAMs' durability against acid attack. This suggested that the high content of GBFS in the AAM matrix could lead to formulate more C-A-S-H and C-N-A-S-H gels, which are vulnerable to sulphuric acid attack.

8.4.5 Sulphate Attack Resistance

Figure 8.12 shows the change in compressive strength of prepared ternary-blended AAMs due to 10% $MgSO_4$ solution attack. Upon sulphate solution exposure, all the mortars exhibited a continuous increase in compressive strength at the ages of 6 and 12 months. The observed increase in the compressive strength percentage was inversely proportional to the FA content. Moreover, the compressive strength of

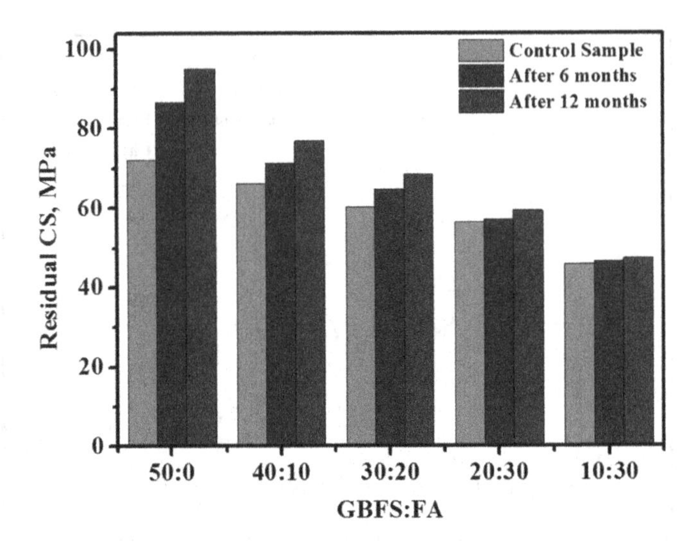

FIGURE 8.12 Residual strength of prepared AAMs exposed to 10% $MgSO_4$.

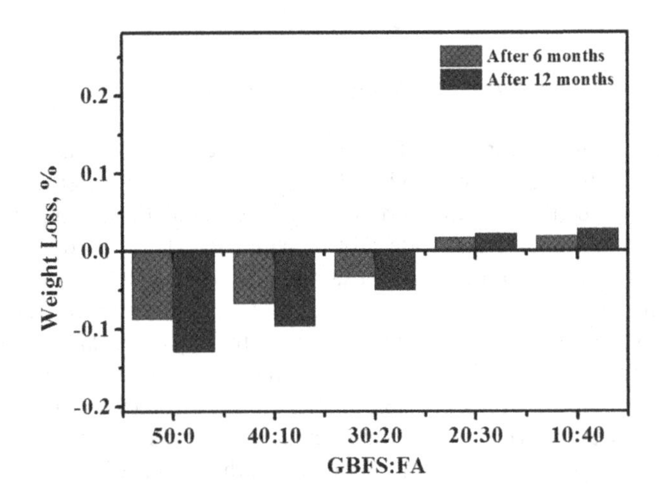

FIGURE 8.13 Weight loss percentage of prepared AAMs exposed to 10% $MgSO_4$.

AAMs increased by 24.1, 13.9, 12.2, 4.9, and 2.9% with the increase in FA content from 0 to 10, 20, 30, and 40%, respectively compared to control samples before immersion in sulphate solution. These results were somewhat unexpected because the strength normally reduces in the sulphate solution, but in these AAMs the strength revived excellently. The increase in the compressive strength could be due to high Na content in WCP composite (13.5%) as summarized in Table 8.1. This affects the formulation of $CaSO_4$ and stable behaviour against the sulphate environment.

The mass change of AAMs specimens containing FA and replacing GBFS is depicted in Figure 8.13. It was observed that all the specimens showed a very low

FIGURE 8.14 Surface texture of prepared AAMs exposed to 10% $MgSO_4$.

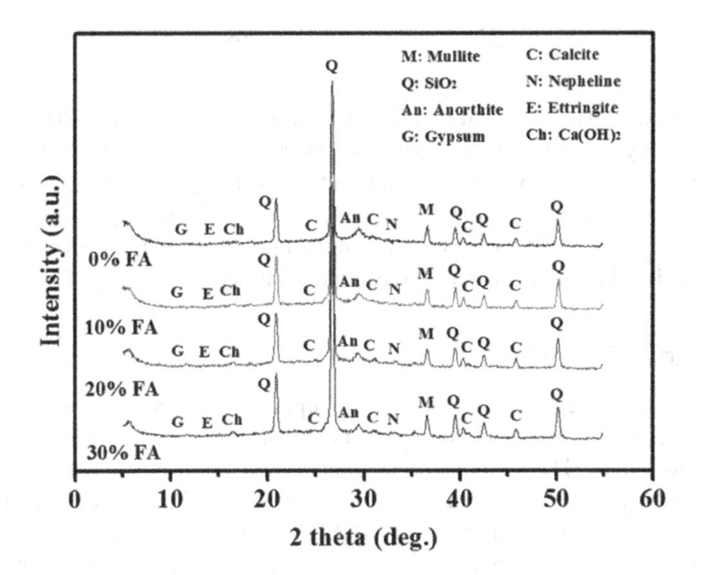

FIGURE 8.15 XRD patterns of prepared AAMs exposed to 10% $MgSO_4$.

change in mass influenced by sulphate solution. Prepared AAM specimens with 0, 10, and 20% of FA showed an increase in mass by 0.12, 0.09, and 0.04%, whereas the rising level of FA to 30 and 40% increased the mass loss from 0.02 to 0.03%, respectively. Figure 8.14 shows the surface texture of AAMs influenced by FA. All mortar specimens showed a very high resistance to sulphate attack. However, no deterioration was observed on all surfaces of AAMs specimens exposed to sulphate attack.

The XRD patterns of prepared AAMs exposed to 10% $MgSO_4$ for 360 days are presented in Figure 8.15. All AAM specimens showed a very high resistance to

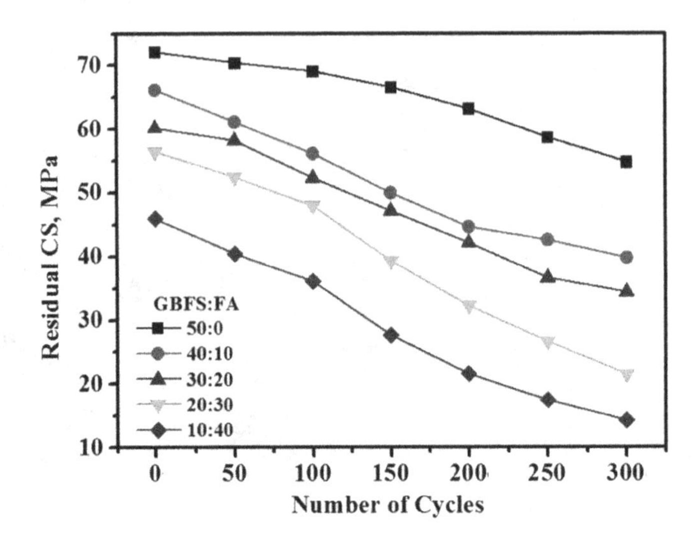

FIGURE 8.16 Residual strength of prepared AAMs exposed to 300 freeze–thaw cycles.

sulphate attack. It was observed that the gypsum peak grew in 11.8° 2θ only and its intensity is very small for all mixtures. However, the anorthite $CaAl_2Si_2O_8$ peak at 29.9° 2θ did not show any change in intensity or was replaced by a gypsum peak. The high stability of C, N-A-S-H gel could be explained by the high Na_2O content of WCP (more than 13%). It is well known that the Na is more reactive than Mg and Ca oxides, which led to high resistance to a sulphate attack.

8.4.6 Freeze–Thaw Cycles

Figure 8.16 illustrates the influence of FA replacing GBFS on the residual compressive strength and weight loss percentage of AAMs specimens exposed to freeze–thaw cycles. The results indicated an inverse relationship between residual strength and weight loss with increasing FA content. The increase in FA content from 0 to 10, 20, 30, and 40% dropped the strength after 300 freeze–thaw cycles for all mixtures. The strength values dropped from 72.1, 66.2, 60.2, 56.5, and 45.9 MPa to 56.8, 39.8, 34.4, 21.5, and 14.2 MPa of 0, 10, 20, 30, and 40% FA, respectively. A similar trend was also observed with residual weight which reduced from 94.5 to 92.7, 86.6, 80.4, and 79.6% with increasing FA content from 0 to 10, 20, 30 and 40%, respectively (Figure 8.17). The high amount of non-reacted silicate led to a more porous structure for 40% FA specimens as discussed in Section 8.4.1. The amount of voids may have allowed the growth of ice and destroyed the interlock between particles [63]. Figure 8.18 shows the deterioration of AAMs for 0 and 40% of FA replacing GBFS. It is clear that low durability and high surface scaling occurred with increasing content of FA to 40% as compared to the control sample.

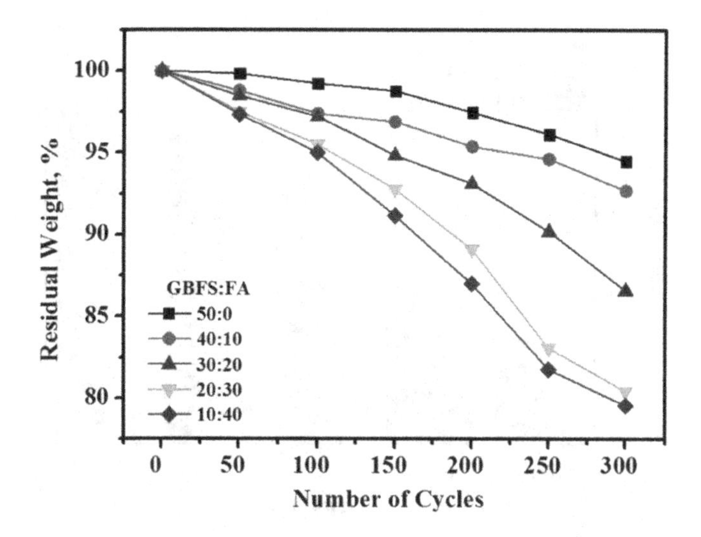

FIGURE 8.17 Residual weight of prepared AAMs exposed to 300 freeze–thaw cycles.

FIGURE 8.18 Surface texture of prepared AAMs exposed to 300 freeze–thaw cycles.

8.4.7 Wet–Dry Cycles

The influence of FA replacing GBFS on the residual compressive strength of AAMs was evaluated by exposing the specimens to wet–dry cycles. The results are displayed in Figure 8.19. The strength loss of AAMs was directly proportional to the number of wet–dry cycles and FA content. The strength loss increased from 6.9 to 11.3, 18.6, 20.3, and 22.1% with increasing FA as replacement of GBFS from 0 to 10, 20, 30, and 40%, respectively. The reason for this is because when total porosity increases, there will be more pores in the matrix, which provide more opportunities

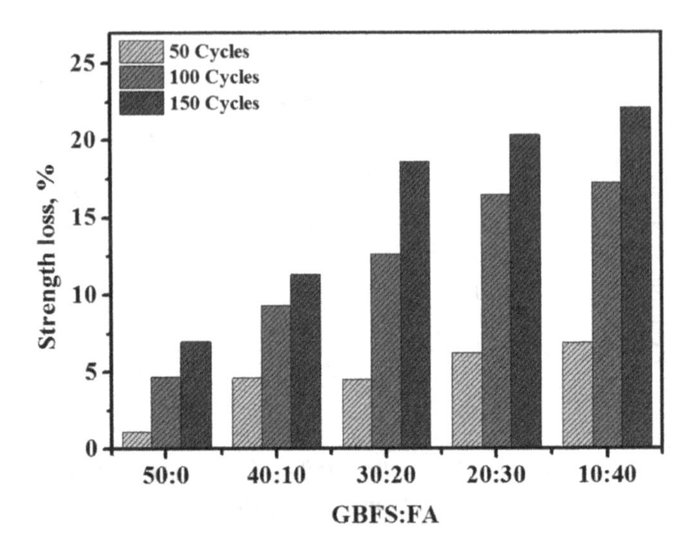

FIGURE 8.19 Residual strength of prepared AAMs exposed to 150 wet–dry cycles.

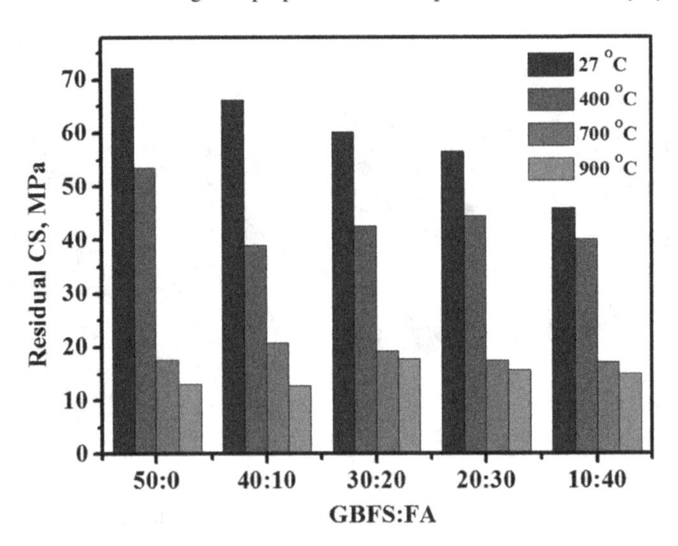

FIGURE 8.20 Residual strength of prepared AAMs exposed to elevated temperatures.

for water to enter the matrix during wetting and drying cycles. This, in turn, leads to an increase in the internal and external deterioration as well as loss of durability with time [64–66].

8.4.8 Elevated Temperatures

Figures 8.20 and 8.21 show the outcomes for the residual compressive strength and weight loss percentage of AAMs when GBFS was replaced by FA. All the AAM

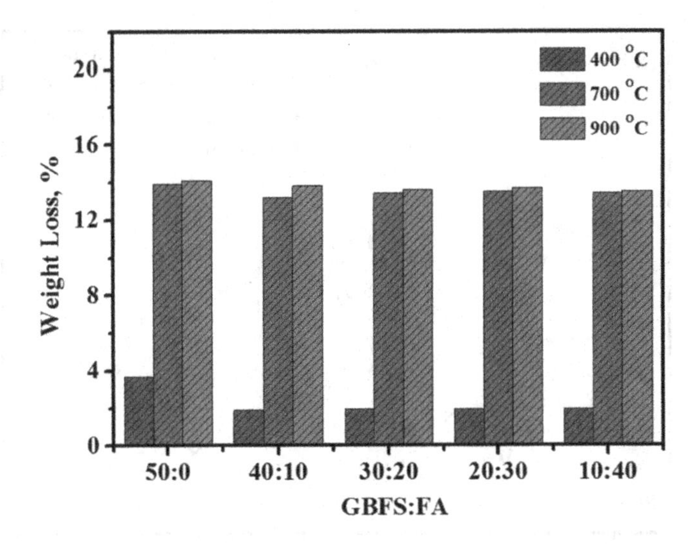

FIGURE 8.21 Weight loss percentage of prepared AAMs exposed to elevated temperatures.

samples showed a decrease in weight and strength when exposed to elevated temperatures, but the results differed somewhat. The residual compressive strength values were correlated to FA levels. With FA replacing GBFS at 0, 10, 20, 30, and 40%, the residual strength values increased in line with the rising percentages, i.e. at rates of 18.2, 19.2, 28.1, 29.6, and 32.6%, respectively. Conversely, the weight loss percentage dropped from 14.1 to 13.8, 13.7, 13.3, and 13.5% when the FA content increased from 0 to 10, 20, 30, and 40%.

The XRD patterns of AAMs mixed with different levels of FA replacing GBFS, before and after temperature increase from 27 to 900°C, are shown in Figure 8.22. The presence of semi-crystalline aluminosilicates gel and quartz (Q) were noticed in the original specimen both before and after exposure to high temperatures under 400°C. The broad limits of all alkali-activated compositions are observed between 25 and 30°C 2θ. Following the fire resistance test, the crystalline phase formed zeolites as a secondary reaction product. The firm peaks in the alkali-activated WCP base highlighted the existence of mullite, nepheline and quartz at temperatures of 700°C. Mullite was the only steady crystalline phase of the Al_2O_3–SiO_2 system. It maintained the strength it had at ambient temperatures even when the temperature was increased, and displayed low thermal expansion and oxidation resistance. Following exposure to 700°C heat, the quartz remained stable while the mullite grew in strength. At around 400°C, the majority of molecules in constitutional water were released and the phase changed from goethite to hematite. The release of the OH-groups at the same time when the grain structure changed could result in a build-up of internal pressure which may have caused the hematite grains to fracture. At this temperature, the grain shape and size of the freshly created hematite phase are much the same as the same elements of the original goethite. When the XRD pattern sample was exposed to a temperature of 900°C, the hematite phase started to disappear.

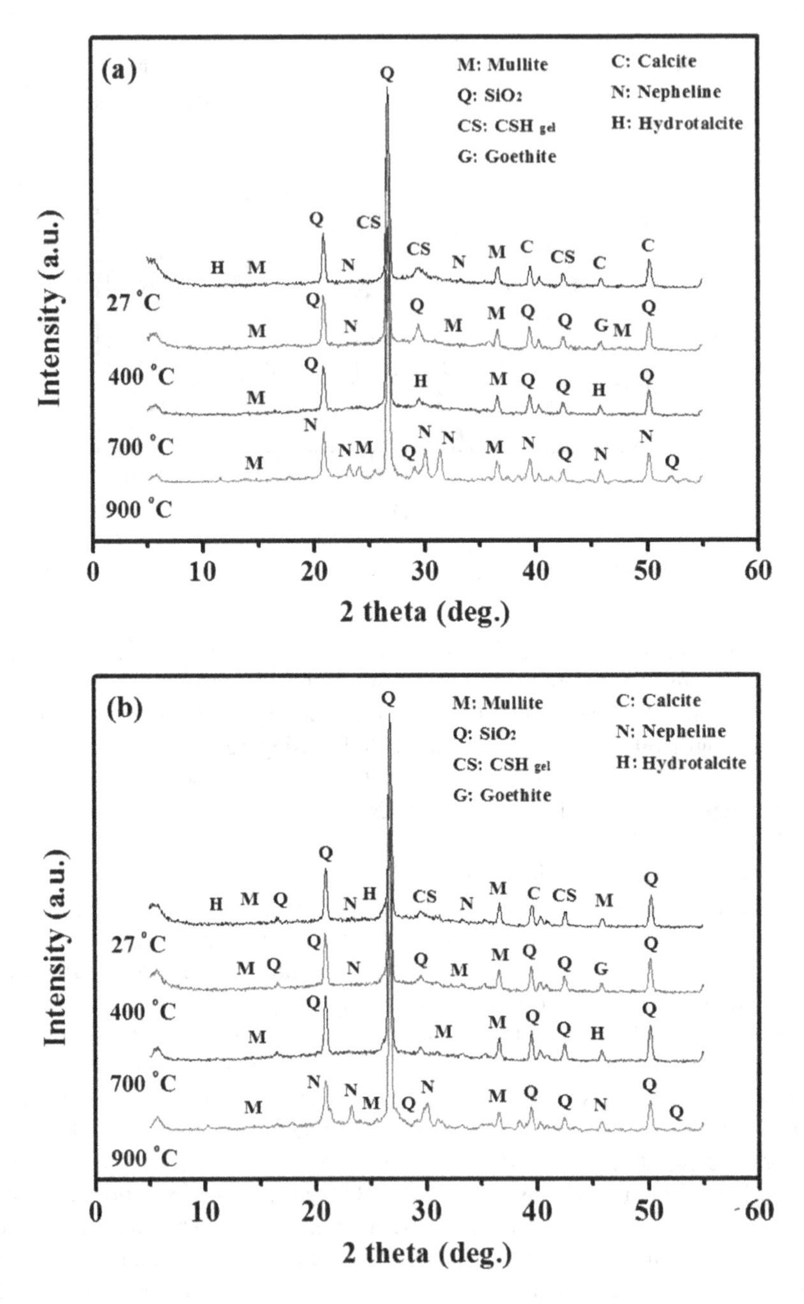

FIGURE 8.22 XRD of prepared AAMs exposed to elevated temperatures. (a) 0% FA, (b) 40% FA.

Quartz remained the primary phase, along with mullite, although the sample also showed the presence of crystalline nepheline $AlNaSiO_4$ (sodium aluminium silicate). When comparing samples prepared with no FA (Figure 8.22a) and those with 40% FA (Figure 8.22b), the peaks showed stability at high temperatures.

8.5 CONCLUSIONS

In response to the outlined objectives, the following conclusions have been made:

i. WCP-based AAMs incorporating FA influenced their compressive strength development. The strength gain dropped as the FA content increased from 0 to 40%. However, the AAMs prepared with 40% FA as replacement of GBFS achieved sufficiently high compressive strength (45.9 MPa), which could allow this product to be applied and used in several applications of construction industry.
ii. Replacing GBFS by FA in high-volume WCP-based AAMs produced high performance mortars in an aggressive environment. According to residual strength, mass loss and XRD patterns, the resistance of AAMs to acid and sulphate attack enhanced as the FA content in AAMs increased. AAMs prepared with 40% FA showed the highest resistance among all the mixtures. The low CaO content in FA mainly contributed to restrict the gypsum formation during the immersion duration.
iii. With an increasing replacement level of GBFS by FA, the drying shrinkage increased and showed low performance under freeze–thaw cycles. On the other hand, all AAMs presented very high performance under wet–dry cycle testing.
iv. All AAMs containing FA as a replacement of GBFS produced strength reduction and weight loss when subjected to high temperatures.
v. The results of this study suggested new compositions for AAMs that are cheaper and more environmental friendly than conventional OPC mortars. By replacing GBFS with FA in AAMs, the resulted product was more efficient, produced less CO_2, reduced costs and consumed lower fuel than OPC mortars.
vi. In order to decrease alkali-activated binder's cost by 42.9% (RM 157) compared to RM 275 with OPC binder, it is proposed to use 40% FA as a replacement of GBFS. This will make the new sustainable product a more suitable alternative to traditional OPC. Additionally, CO_2 emissions would decrease as the potentially new composition requires less than half the fuel for production than the OPC mortars.
vii. Aside from the positive impacts on the environment, the prepared AAMs also offered a superior product in terms of mechanical and durability properties; something that is likely of interest to many mortar and concrete manufacturers. This suggested substitute for OPC mortars and concrete has a widespread suitability and may also serve to fulfil sustainability aims for companies in the business of sustainable construction.

REFERENCES

1. Akeiber, H., et al., A review on phase change material (PCM) for sustainable passive cooling in building envelopes. *Renewable and Sustainable Energy Reviews*, 2016. **60**: pp. 1470–1497.
2. Nejat, P., et al., Iran's achievements in renewable energy during fourth development program in comparison with global trend. *Renewable and Sustainable Energy Reviews*, 2013. **22**: pp. 561–570.
3. Leung, D. Y., G. Caramanna, and M. M. Maroto-Valer, An overview of current status of carbon dioxide capture and storage technologies. *Renewable and Sustainable Energy Reviews*, 2014. **39**: pp. 426–443.
4. Mohammadhosseini, H., M. M. Tahir, and A. R. M. Sam, The feasibility of improving impact resistance and strength properties of sustainable concrete composites by adding waste metalized plastic fibres. *Construction and Building Materials*, 2018. **169**: pp. 223–236.
5. Mohammadhosseini, H. and M. M. Tahir, Durability performance of concrete incorporating waste metalized plastic fibres and palm oil fuel ash. *Construction and Building Materials*, 2018. **180**: pp. 92–102.
6. Scarlat, N., et al., Evaluation of energy potential of municipal solid waste from African urban areas. *Renewable and Sustainable Energy Reviews*, 2015. **50**: pp. 1269–1286.
7. Mohammadhosseini, H., et al., Enhanced performance for aggressive environments of green concrete composites reinforced with waste carpet fibers and palm oil fuel ash. *Journal of Cleaner Production*, 2018. **185**: pp. 252–265.
8. Huseien, G. F., K. W. Shah, and A. R. M. Sam, Sustainability of nanomaterials based self-healing concrete: An all-inclusive insight. *Journal of Building Engineering*, 2019. **23**: pp. 155–171.
9. Keyvanfar, A., et al., User satisfaction adaptive behaviors for assessing energy efficient building indoor cooling and lighting environment. *Renewable and Sustainable Energy Reviews*, 2014. **39**: pp. 277–295.
10. Huseien, G. F., et al., Geopolymer mortars as sustainable repair material: A comprehensive review. *Renewable and Sustainable Energy Reviews*, 2017. **80**: pp. 54–74.
11. Mohammadhosseini, H., et al., Effects of elevated temperatures on residual properties of concrete reinforced with waste polypropylene carpet fibres. *Arabian Journal for Science and Engineering*, 2018. **43**(4): pp. 1673–1686.
12. Asaad, M. A., et al., Enhanced corrosion resistance of reinforced concrete: Role of emerging eco-friendly *Elaeis guineensis*/silver nanoparticles inhibitor. *Construction and Building Materials*, 2018. **188**: pp. 555–568.
13. Asaad, M. A., et al., Improved corrosion resistance of mild steel against acid activation: Impact of novel *Elaeis guineensis* and silver nanoparticles. *Journal of Industrial and Engineering Chemistry*, 2018. **63**: pp. 139–148.
14. Li, N., et al., A mixture proportioning method for the development of performance-based alkali-activated slag-based concrete. *Cement and Concrete Composites*, 2018. **93**: pp. 163–174.
15. Kubba, Z., et al., Impact of curing temperatures and alkaline activators on compressive strength and porosity of ternary blended geopolymer mortars. *Case Studies in Construction Materials*, 2018. **9**: p. e00205.
16. Provis, J. L., A. Palomo, and C. Shi, Advances in understanding alkali-activated materials. *Cement and Concrete Research*, 2015. **78**: pp. 110–125.
17. Li, N., N. Farzadnia, and C. Shi, Microstructural changes in alkali-activated slag mortars induced by accelerated carbonation. *Cement and Concrete Research*, 2017. **100**: pp. 214–226.

18. Huseien, G. F., et al., Effects of POFA replaced with FA on durability properties of GBFS included alkali activated mortars. *Construction and Building Materials*, 2018. **175**: pp. 174–186.
19. Li, N., et al., Composition design and performance of alkali-activated cements. *Materials and Structures*, 2017. **50**(3): pp. 178.
20. Huseiena, G. F., et al., Potential use coconut milk as alternative to alkali solution for geopolymer production. *Jurnal Teknologi*, 2016. **78**(11): p. 133–139.
21. Huseien, G. F., et al., Effect of metakaolin replaced granulated blast furnace slag on fresh and early strength properties of geopolymer mortar. *Ain Shams Engineering Journal*, 2016. **9**(4): pp. 1557–1566.
22. Phoo-ngernkham, T., et al., Effects of sodium hydroxide and sodium silicate solutions on compressive and shear bond strengths of FA–GBFS geopolymer. *Construction and Building Materials*, 2015. **91**: pp. 1–8.
23. Huseien, G. F., et al., Waste ceramic powder incorporated alkali activated mortars exposed to elevated Temperatures: Performance evaluation. *Construction and Building Materials*, 2018. **187**: pp. 307–317.
24. Huang, B., Q. Dong, and E. G. Burdette, Laboratory evaluation of incorporating waste ceramic materials into Portland cement and asphaltic concrete. *Construction and Building Materials*, 2009. **23**(12): pp. 3451–3456.
25. Samadi, M., et al., Properties of mortar containing ceramic powder waste as cement replacement. *Jurnal Teknologi*, 2015. **77**(12): pp. 93–97.
26. Pacheco-Torgal, F. and S. Jalali, Reusing ceramic wastes in concrete. *Construction and Building Materials*, 2010. **24**(5): pp. 832–838.
27. Huseien, G. F., M. Y. Al-Fasih, and H. Hamzah, Performance of self-compacting concrete with different sizes of recycled ceramic aggregates. *International Journal of Innovative Research and Creative Technology*, 2015. **1**(3): pp. 264–269.
28. Hussein, A. A., et al., Performance of nanoceramic powder on the chemical and physical properties of bitumen. *Construction and Building Materials*, 2017. **156**: pp. 496–505.
29. Zhou, W., et al., A comparative study of high-and low-Al2O3 fly ash based-geopolymers: The role of mix proportion factors and curing temperature. *Materials & Design*, 2016. **95**: pp. 63–74.
30. Soutsos, M., et al., Factors influencing the compressive strength of fly ash based geopolymers. *Construction and Building Materials*, 2016. **110**: pp. 355–368.
31. Huseien, G. F., et al., Influence of different curing temperatures and alkali activators on properties of GBFS geopolymer mortars containing fly ash and palm-oil fuel ash. *Construction and Building Materials*, 2016. **125**: pp. 1229–1240.
32. Phoo-ngernkham, T., et al., High calcium fly ash geopolymer mortar containing Portland cement for use as repair material. *Construction and Building Materials*, 2015. **98**: pp. 482–488.
33. Huseien, G. F., et al., Synergism between palm oil fuel ash and slag: Production of environmental-friendly alkali activated mortars with enhanced properties. *Construction and Building Materials*, 2018. **170**: pp. 235–244.
34. Nath, P. and P. K. Sarker, Effect of GGBFS on setting, workability and early strength properties of fly ash geopolymer concrete cured in ambient condition. *Construction and Building Materials*, 2014. **66**: pp. 163–171.
35. Huseiena, G. F., et al., Performance of sustainable alkali activated mortars containing solid waste ceramic powder. *Chemical Engineering*, 2018. **63**, pp. 673–678.
36. Lee, N., E. Kim, and H.-K. Lee, Mechanical properties and setting characteristics of geopolymer mortar using styrene-butadiene (SB) latex. *Construction and Building Materials*, 2016. **113**: pp. 264–272.

37. Deb, P. S., P. Nath, and P. K. Sarker, The effects of ground granulated blast-furnace slag blending with fly ash and activator content on the workability and strength properties of geopolymer concrete cured at ambient temperature. *Materials & Design (1980-2015)*, 2014. **62**: pp. 32–39.
38. Kumar, S., R. Kumar, and S. Mehrotra, Influence of granulated blast furnace slag on the reaction, structure and properties of fly ash based geopolymer. *Journal of Materials Science*, 2010. **45**(3): pp. 607–615.
39. Fernández-Jiménez, A., I. García-Lodeiro, and A. Palomo, Durability of alkali-activated fly ash cementitious materials. *Journal of Materials Science*, 2007. **42**(9): pp. 3055–3065.
40. Somna, K., et al., NaOH-activated ground fly ash geopolymer cured at ambient temperature. *Fuel*, 2011. **90**(6): pp. 2118–2124.
41. Temuujin, J.V., A. Van Riessen, and R. Williams, Influence of calcium compounds on the mechanical properties of fly ash geopolymer pastes. *Journal of Hazardous Materials*, 2009. **167**(1–3): pp. 82–88.
42. Duan, P., et al., Fresh properties, mechanical strength and microstructure of fly ash geopolymer paste reinforced with sawdust. *Construction and Building Materials*, 2016. **111**: pp. 600–610.
43. Rashad, A. M., A comprehensive overview about the influence of different admixtures and additives on the properties of alkali-activated fly ash. *Materials & Design*, 2014. **53**: pp. 1005–1025.
44. Ramujee, K. and M. PothaRaju, Mechanical properties of geopolymer concrete composites. *Materials Today: Proceedings*, 2017. **4**(2): pp. 2937–2945.
45. Huseien, G. F., M. Ismail, and J. Mirza, Influence of curing methods and sodium silicate content on compressive strength and microstructure of multi blend geopolymer mortars. *Advanced Science Letters*, 2018. **24**(6): pp. 4218–4222.
46. Álvarez-Ayuso, E., et al., Environmental, physical and structural characterisation of geopolymer matrixes synthesised from coal (co-) combustion fly ashes. *Journal of Hazardous Materials*, 2008. **154**(1–3): pp. 175–183.
47. Gunasekara, C., et al., Zeta potential, gel formation and compressive strength of low calcium fly ash geopolymers. *Construction and Building Materials*, 2015. **95**: pp. 592–599.
48. Ismail, I., et al., Modification of phase evolution in alkali-activated blast furnace slag by the incorporation of fly ash. *Cement and Concrete Composites*, 2014. **45**: pp. 125–135.
49. ASTM, C579 Standard test methods for compressive strength of chemical-resistant mortars. In *Grouts, Monolithic Surfacings, and Polymer Concretes*. 2001: ASTM.
50. ASTM, Standard test method for compressive strength of hydraulic cement mortars (using 2-in. or [50-mm] cube specimens). 2013: ASTM.
51. ASTM, C642 Standard test method for density, absorption, and voids in hardened concrete. In *Annual Book of ASTM Standards*. Vol. 4, p. 2. 2006: ASTM.
52. ASTM, Standard test methods for chemical resistance of mortars, grouts, and monolithic surfacing and polymer concretes. 2012: ASTM.
53. ASTM, Standard test method for resistance of concrete to rapid freezing and thawing. 2008: ASTM.
54. Huseiena, G. F., et al., Effect of binder to fine aggregate content on performance of sustainable alkali activated mortars incorporating solid waste materials. *Chemical Engineering*, 2018. **63**: pp. 667–672.
55. Villain, G., M. Thiery, and G. Platret, Measurement methods of carbonation profiles in concrete: Thermogravimetry, chemical analysis and gammadensimetry. *Cement and Concrete Research*, 2007. **37**(8): pp. 1182–1192.

56. Morandeau, A., M. Thiery, and P. Dangla, Investigation of the carbonation mechanism of CH and CSH in terms of kinetics, microstructure changes and moisture properties. *Cement and Concrete Research*, 2014. **56**: pp. 153–170.
57. Huseien, G. F., et al., Compressive strength and microstructure of assorted wastes incorporated geopolymer mortars: Effect of solution molarity. *Alexandria Engineering Journal*, 2018. **57**(4): pp. 3375–3386.
58. Ahmed, M. A. B., et al., Performance of high strength POFA concrete in acidic environment. *Concrete Research Letters*, 2010. **1**(1): pp. 14–18.
59. Ariffin, M., et al., Sulfuric acid resistance of blended ash geopolymer concrete. *Construction and Building Materials*, 2013. **43**: pp. 80–86.
60. Bamaga, S., et al., Evaluation of sulfate resistance of mortar containing palm oil fuel ash from different sources. *Arabian Journal for Science and Engineering*, 2013. **38**(9): pp. 2293–2301.
61. Noruzman, A., et al. Strength and durability characteristics of polymer modified concrete incorporating vinyl acetate effluent. in *Advanced Materials Research*. 2013: Trans Tech Publ.
62. Matschei, T., F. Bellmann, and J. Stark, Hydration behaviour of sulphate-activated slag cements. *Advances in Cement Research*, 2005. **17**(4): pp. 167–178.
63. Cai, L., H. Wang, and Y. Fu, Freeze–thaw resistance of alkali–slag concrete based on response surface methodology. *Construction and Building Materials*, 2013. **49**: pp. 70–76.
64. Chang, H., et al., Influence of pore structure and moisture distribution on chloride “maximum phenomenon” in surface layer of specimens exposed to cyclic drying-wetting condition. *Construction and Building Materials*, 2017. **131**: pp. 16–30.
65. Huseien, G. F., et al., The effect of sodium hydroxide molarity and other parameters on water absorption of geopolymer mortars. *Indian Journal of Science and Technology*, 2016. **9**(48): pp. 1–7.
66. Huseien, G. F. and K. W. Shah, Durability and life cycle evaluation of self-compacting concrete containing fly ash as GBFS replacement with alkali activation. *Construction and Building Materials*, 2020. **235**: p. 117458.

9 Performance Evaluation Alkali-Activated Mortar Exposed to Elevated Temperatures

9.1 INTRODUCTION

For ages, ordinary Portland cement (OPC) has served as the primary structural material in the construction sector and widely used as concrete binder worldwide [1–3]. It is well known that large-scale manufacturing of OPC causes serious environmental pollution in terms of considerable amount of greenhouse gas emissions, but the right substitute to OPC has not emerged [4, 5]. OPC production alone is accountable for nearly 6–7% of total CO_2 emissions [6]. In recent times, alkali-activated mortars (AAMs)/concretes have been introduced as a new construction material to replace the traditional concrete in the construction industry [7–9]. Several notable merits of AAMs such as cheap production from abundant industrial wastes via recycling, reduction of the environmental pollution, enhanced durability, energy saving attributes, high early strength and great non-combustibility make them advantageous over other construction materials [10–12].

Concrete structures such as beam, column and slab used in the buildings must satisfy appropriate fire safety requirements specified to building codes [13, 14]. This is because fire represents one of the most severe environmental conditions to which structures may be subjected. Therefore, provision of appropriate fire safety measures for structural members is an important aspect of building design. Fire safety measures to structural members are measured in terms of fire resistance, which is the duration during which a structural member exhibits resistance with respect to structural integrity, stability and temperature transmission [15, 16]. Generally, AAMs provide the best fire resistance properties when applied as building material [17]. This excellent fire resistance attribute of AAMs is due to the chemical combination of the constituent materials that are essentially inert and have low thermal conductivity, high heat capacity, and slow strength degradation with temperature [18]. It is this slow rate of heat transfer and strength loss that enables concrete to act as an effective fire shield not only between adjacent spaces but also to protect itself from fire damage [19].

It is worth noting that millions of tons of natural, industrial, and agriculture wastes such as fly ash (FA), coal- and oil-burning by-products, bottom ash, palm oil fuel ash (POFA), bagasse ash (BA), used tires, cement dust, marble and crushed stone, and waste ceramic powders (WCP) are dumped every year in Malaysia [20–23]. These waste materials cause severe ecological setbacks such as air contamination and leaching out of hazardous substances. Several studies [24–28] revealed that many of these wastes may be potentially recycled in the form of innovative concrete materials as an alternative to OPC (often as much as 70%). Besides, these newly developed concretes, owing to their green chemical nature, are environmentally friendly, durable and inexpensive building materials.

Yet, the development of different AAMs as environmentally friendly construction materials by blending WCPs has rarely been explored. Ceramic powder is the principal waste of the ceramics industry and is generated as unwanted dust during the process of dressing and polishing. It has been estimated that 15–30% of the ceramic waste is produced from the total raw material used [22, 29, 30]. A portion of this waste is often utilized on-site for the excavation pit refilling. Ceramic wastes are not only non-biodegradable but also consume much space in a landfill. Thus, finding a new way to recycle this waste and subsequently using in the construction of infrastructures can be useful to preserve natural resources and the environment. Previous research [24] revealed that ceramic wastes have pozzolanic properties, which can be used to make concrete with improved strength and durability. According to Pacheco et al. [31], concrete combined with WCP has increased durability performance because of its pozzolanic properties. It was realized that by replacing the conventional sand with WCP, it is possible to produce mortar with superior strength and durability performance. This WCP-substituted conventional fine aggregate containing AAMs was effective with a bit lower water absorption performance. This water permeability of the AAMs implied that the substitution of conventional sand by WCP is an excellent option. The durability of WCP against sulphuric acid attack and sulphate attack were not evaluated.

This chapter was intended to determine the effects of WCPs on the thermal properties of the ternary AAMs containing FA and ground blast furnace slag (GBFS). Mortar properties such as residual compressive strength, weight loss and appearance colour are evaluated at various temperatures (400, 700 and 900°C). Microstructure tests including X-ray diffraction (XRD), scanning electron microscopy (SEM), energy dispersive X-ray spectroscopy (EDX), Fourier-transform infrared (FTIR), and thermogravimetric analysis (TGA) were used to explain the changes in physical and chemical properties of alkali-activated mortars.

9.2 MATERIALS AND MIX DESIGN

9.2.1 Materials

Three types of waste materials – ceramic, GBFS and fly ash – were obtained from a single source for each one and used to prepare the ternary-blended alkali-activated mortars. In the present study, the ceramic wastes were collected from the construction industry located in Johor, Malaysia. The materials were first crushed, grinded using a Los Angeles grinding machine (LAAM) and sieved through 600 μm mesh to isolate the large particles. The sieved ceramic was again ground for 6 hours and the finesse of particles was checked after every hour. Next, the powder was collected and used in the mixing process (Figure 9.1). The cement-free binder was made using pure GBFS (collected from Ipoh, Malaysia) as one of the resource materials, which was further utilized without any laboratory treatment. This slag possessed both cementitious and pozzolanic properties. GBFS shows off-white colour due to the development of hydraulic reaction during mixing with water. Low-calcium FA was acquired from Tanjung bin power station (Malaysia) as resource aluminosilicate materials for preparing AAMs. It satisfies the ASTM C618 [32] requirements and appears grey in colour. From the Brunauer–Emmett–Teller (BET) test, it was observed that the specific surface area of FA showed the highest value of 18.1 m^2/g compared to GBFS and WCP, which were 13.6 and 12.2 m^2/g, respectively. The fine size and light specific gravity of FA have influence on the surface area test. The lowest specific surface area was observed with WCP 12.2 m^2/g and its particle size was 35 μm.

X-ray fluorescence (XRF) spectroscopy was used to detect the chemical compositions of WCP, GBFS and FA as summarized in Table 9.1. The main oxide compositions were silica and aluminium in the presence of 84.8% GBFS, 41.7% GBFS, and 86% FA. GBFS presented very high content of calcium oxide (51.8%) compared to other materials. It is worth noting that the level of silicate, aluminium and calcium oxide played a significant role in the AAM synthesis by forming sodium aluminium silicate hydrate (N-A-S-H), calcium aluminium silicate hydrate (C-A-S-H), and calcium silicate hydrate (C-S-H) in the geopolymerization process. The content of potassium oxide (K_2O) was found to be very low in FA, GBFS and WCP. The content of sodium oxide (Na_2O) was observed at high proportions (13.5%) in the WCP chemical composite. It is known that both K_2O and Na_2O can strongly affect the activation of alkaline and the geopolymerization process.

The thermogravimetric analysis (TGA) and derivative thermogravimetric analysis (DTG) of WCP, GBFS and FA were carried out to examine the weight change

FIGURE 9.1 Processing steps of WCP.

TABLE 9.1
Chemical Composition of WCP, GBFS, and FA Obtained from XRF Analysis

Material	WCP	GBFS	FA
SiO_2	72.6	30.8	57.20
Al_2O_3	12.2	10.9	28.81
Fe_2O_3	0.56	0.64	3.67
CaO	0.02	51.8	5.16
MgO	0.99	4.57	1.48
K_2O	0.03	0.36	0.94
Na_2O	13.46	0.45	0.08
SO_3	0.01	0.06	0.10
$SiO_2:Al_2O_3$	5.95	2.82	1.98

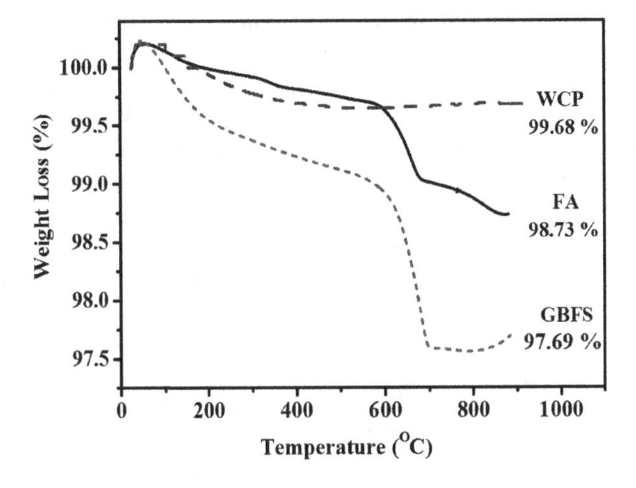

FIGURE 9.2 TGA curves of WCP, GBFS, and FA (weight loss up to 900°C).

as a function of temperature. Figures 9.2 and 9.3 display the TGA and DTG curves, respectively, of various minerals. The weight changes occurred at different temperatures. The results of the TGA test showed that the WCP was stable under high temperatures compared to other materials. Furthermore, WCP revealed a desorption/drying (rerun) stage only and lower weight loss (0.32%). Multi-stage decompositions (but no intermediates or fast heating rate) were observed for GBFS and FA with corresponding weight losses of 2.31 and 1.27%, respectively.

The DTG measurements were performed to assess mass loss during the pre-set heating of the samples to measure physical changes in materials. Figure 9.3 depicts the DTG curves for WCP, GBFS, and FA. Results revealed that WCP was more stable because it displayed a high bond strength against decomposition reaction. The

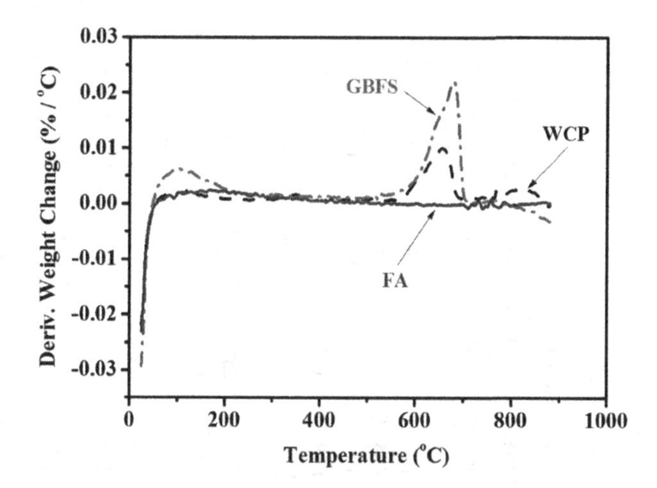

FIGURE 9.3 DTG curves for WCP, GBFS, and FA.

decomposition peak of FA started after 700°C, which displayed lower weight change (0.0078%/°C) when compared with GBFS. TGA curves also divulged that GBFS has the highest weight change of 0.02%/°C at temperature 718°C. Generally, WCP exhibited more stability compared to other materials under high temperature exposure.

For the proposed application, naturally occurring siliceous river sand was used as fine aggregate to make all mortar specimens. The sand was washed with water following the ASTM C117 standard [33] to reduce silts and impurity contents. Then, the sand was dried in the oven at 60°C for 24 hours to control the moisture content. The sand was graded to conform to ASTM C33-33M specifications [34]. The estimated fineness modulus of aggregate was 2.9 with specific gravity of 2.6. Analytical-grade sodium hydroxide (NaOH) (98% purity; from QREC, Malaysia) was used. The pellets were dissolved in water to prepare NaOH or NH solution of concentration 4. The solution was left for 24 hours to cool and then added to a sodium silicate solution (SiO_2 of 29.5 wt%), Na_2O of 14.70 wt% and H_2O of 55.80 wt%) to prepare the final alkaline solution with 1.02 ratio of SiO_2:Na_2O. The ratio of sodium silicate to sodium hydroxide (NS:NH) was fixed to 0.75 for all alkaline mixtures.

9.2.2 Mix Design and Sample Preparation

Nine mixtures containing a high amount of ceramic were prepared, as listed in Table 9.2. Three levels of ceramic-based AAMs of 50, 60 and 70% were adopted and replaced with GBFS and FA. The minimum content of GBFS was kept to 20% and the maximum as 50%. The effect of high volume of WCP and the level of GBFS and FA on the content of SiO_2, CaO and Al_2O_3 in AAMs are shown in Table 9.1 also. An increase in the content of WCP led to an enhancement of SiO_2 content. Meanwhile, the replacement of WCP by increasing the amount of GBFS led to an increase in the CaO content. The content of Al_2O_3 also increased with an increasing level of FA. The NH molarity, NS to NH, alkaline solution to binder and binder to fine aggregate

TABLE 9.2
Mix Proportions of Alkali-Activated Mortars

Materials		Mix Design Formulation of Alkali-Activated Mortars								
		AAM_1	AAM_2	AAM_3	AAM_4	AAM_5	AAM_6	AAM_7	AAM_8	AAM_9
Binder (B) kg/m^3	WCP	550	550	550	550	660	660	660	770	770
	GBFS	550	440	330	220	440	330	220	330	220
	FA	0	110	220	330	0	110	220	0	110
SiO_2:Al_2O_3 by mass%		4.48	4.08	3.77	3.53	4.79	4.35	4.01	5.09	4.62
CaO:SiO_2 by mass%		0.50	0.39	0.29	0.20	0.37	0.27	0.18	0.26	0.17
CaO:Al_2O_3 by mass%		2.24	1.59	1.09	0.70	1.77	1.19	0.74	1.31	0.79
Fine aggregate (A) kg/m^3		1100	1100	1100	1100	1100	1100	1100	1100	1100
Binder:aggregate (B:A)		1.0	1.0	1.0	1.0	1.0	1.0	1.0	1.0	1.0
Alkaline-activated	NH	251.4	251.4	251.4	251.4	251.4	251.4	251.4	251.4	251.4
solution (S) kg/m^3	NS	188.6	188.6	188.6	188.6	188.6	188.6	188.6	188.6	188.6
	NS:NH	0.75	0.75	0.75	0.75	0.75	0.75	0.75	0.75	0.75
Ms of solution by mass %		1.02	1.02	1.02	1.02	1.02	1.02	1.02	1.02	1.02
Solution:binder (S:B)		0.40	0.40	0.40	0.40	0.40	0.40	0.40	0.40	0.40

B:A, binder:aggregate by mass%; Ms, solution modulus SiO_2:Na_2O by mass%; S:B, solution:binder by mass%; NS:NH, Na_2SiO_3:NaOH by mass%.

were selected to be 4 M, 0.75, 0.40, and 1.0, respectively. These values were fixed for all mixes at this stage.

According to ASTM C579 [35], cubic moulds of 50 × 50 × 50 mm were prepared for the hardened tests (compressive strength). A mixture of NH and NS measured by weight was prepared by thoroughly mixing them together. The solution was allowed to cool to room temperature before use due to the production of generated heat in the process. Present AAMs were prepared by mixing WCP, GBFS, and FA for 2 minutes in dry conditions to achieve a homogeneous mixture of fine aggregates using a mortar mixer machine. Then, the acquired mixture was activated by adding the alkaline solution. The whole matrix was mixed in the machine operating at medium speed for another 4 minutes. Finally, the fresh mortar was casted in moulds in two layers, where each layer was consolidated using a vibration table for 15 seconds to allow the escape of air voids. After the casting, the specimens were left for 24 hours for curing at ambient temperature 27 ± 1.5°C under the relative humidity of 75%. Then, the specimens were opened and left in the same condition until testing.

9.3 TESTING PROCEDURES

Determination of the ability of ternary blended AAM specimens to resist exposure to elevated temperature was the main objective of the evaluation. The test was carried out using an automatic electric furnace. Specimens of 50 × 50 × 50 mm were prepared as described in Section 9.2.2. After 28 days of curing, three specimens for each mixture were subjected to temperatures of 400, 700, and 900°C with various time durations as shown in Figure 9.4. After heating the specimens, they were cooled using air cooling. The weight before and after the exposure to relevant temperatures were obtained to calculate the weight loss due to the exposure at elevated

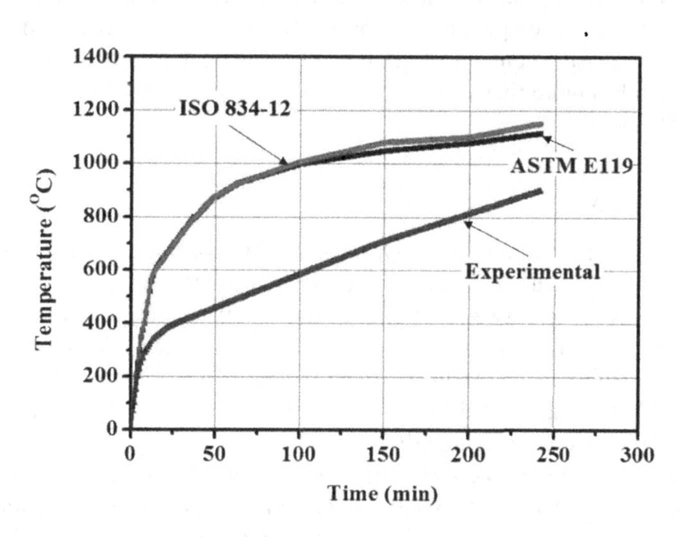

FIGURE 9.4 Experimental time-dependent temperature variation of ASTM E119 [37] and ISO 834-12 [38].

temperatures. Then, the specimens were subjected to compressive strength testing to obtain the residual strength and percentage of strength loss as well as ultrasonic pulse velocity (UPV) according to ASTM C597 [36]. Finally, the relative quality of specimens after heating was evaluated. Additionally, the microstructures of the specimens were analyzed using XRD, SEM, FTIR, DTG and TGA measurements.

9.4 STRENGTH AND MICROSTRUCTURE PROPERTIES

9.4.1 Residual Compressive Strength

Figure 9.5 shows the high volume WCP content dependent residual compressive strength percentages of AAMs. The residual compressive strength was directly proportional to the WCP content. With WCP increasingly replacing GBFS from 50 to 70% the residual strength values were enhanced from 17.7 to 40.1%, respectively. A similar trend was also found with FA replacing GBFS with 50, 60, and 70% of WCP containing alkali-activated specimens. The residual strength values increased from 17.7 to 29.6% with 0 to 30% of FA replacing GBFS in 50% WCP, respectively (Figure 9.5a). The values of residual strength obtained from 60 and 70% of WCP and a high level of FA replaced with GBFS indicated an increased resistance of AAMs to elevated temperatures with increasing level of WCP and FA in the alkali-activated matrix. It's known that high temperatures accelerate the process of bond breakage in the alkali-activated matrix [39]. The extent of damage and strength loss appeared higher with samples containing a high level of calcium – the product of calcium carbonate decomposition – resulting in the increase of volume, causing the development of cracks [40]. This phenomenon can be attributed to two contrasting processes occurring in the specimens as the temperature increases: the strength gain due to the continuation of geopolymerization and the strength loss because of thermal incompatibility to resist the temperature elevation. The enhancement in AAM resistance to heat with increasing contents of WCP and FA can be due to the increase of Al-Si content which has more thermal compatibility compared to the high level of GBFS (Figure 9.5b and c).

9.4.2 Mass Loss

The results for high-volume WCP (50, 60, and 70%) content-dependent weight loss percentages of AAMs are presented in Figures 9.6 and 9.7, respectively. An inverse relationship was found between weight loss and WCP content, with increased WCP content from 50 to 60 and 70%. The weight loss percentage dropped from 14.1 to 13.4 and 12% (Figure 9.6). A similar trend of results was found with FA replacing GBFS in 50% WCP alkali-activated matrixes. The weight loss values dropped from 14.1 to 13.6% with FA replacing GBFS from 0 and 30%, respectively (Figure 9.7a). This shows that the thermal compatibility was enhanced with the increase of WCP and FA content in 60 and 70% WCP groups and reduced the deterioration in samples. The weight loss percentage also dropped and showed values less than 12.8% (Figure 9.7b). The low deterioration was observed with the alkali-activated matrix

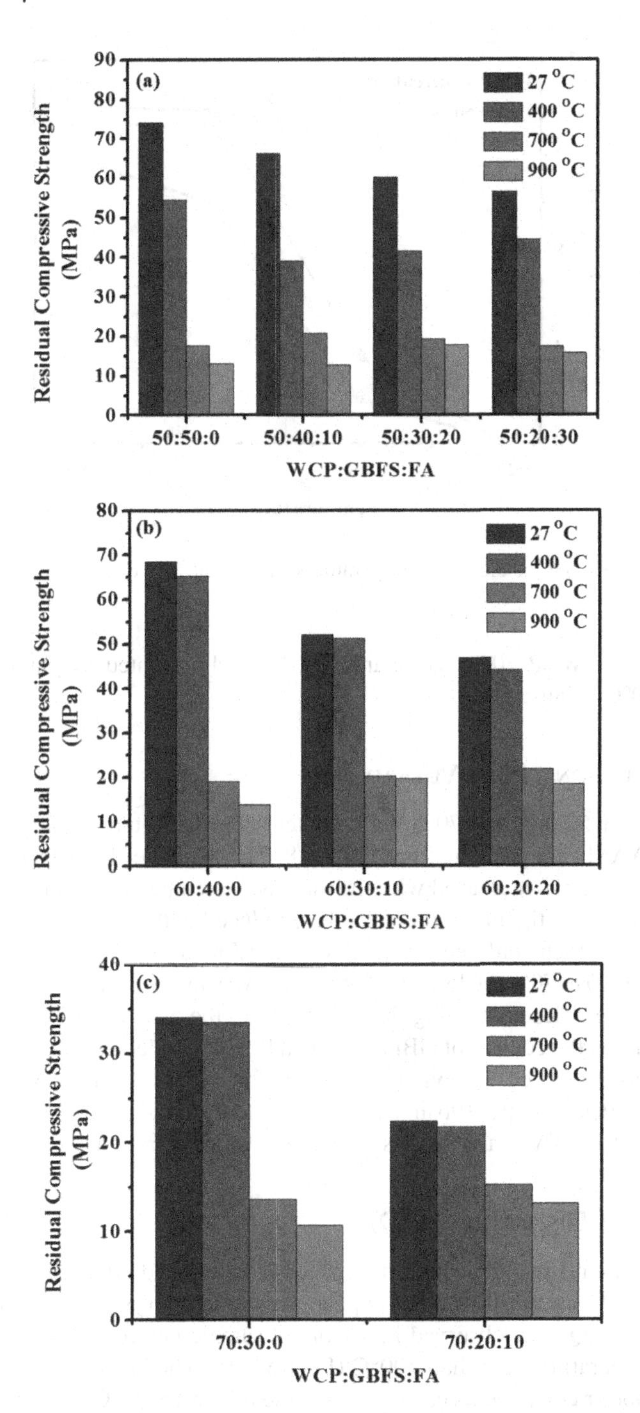

FIGURE 9.5 Effects of elevated temperatures on AAM residual strength with WCP content: (a) 50%, (b) 60%, and (c) 70%.

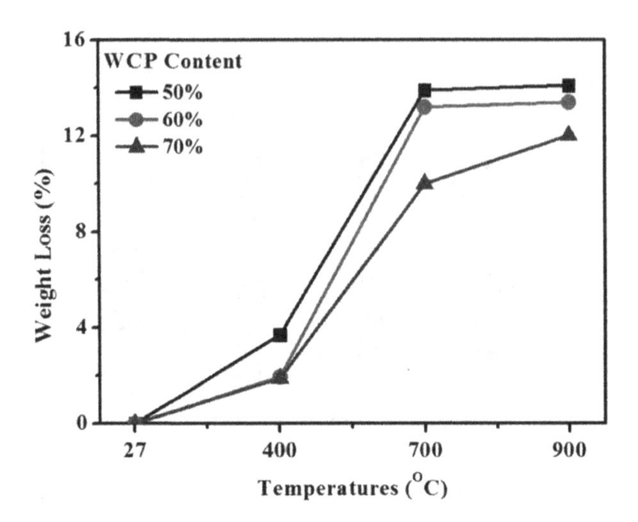

FIGURE 9.6 Effects of elevated temperatures on weight loss of 50, 60, and 70% WCP-content AAMs.

containing 70% WCP, 20% GBFS, and 10% FA and presented weight loss less than 8.5% at 900°C (Figure 9.7c).

9.4.3 Ultrasonic Pulse Velocity

The results for 50, 60, and 70% WCP content-dependent ultrasonic pulse velocity (UPV) of AAMs are presented in Figure 9.8. The results of UPV testing indicated that the deterioration increased with the increase of temperatures for all AAM mixtures. Furthermore, the internal cracks were reduced with the increasing content of WCP and FA, which could be due to slowed UPV reading. With increasing content of WCP from 50 to 70% the loss in UPV reading was lowered, indicting lesser cracks compared to samples containing 50% of WCP. In the 50% of WCP sample, it was observed that as the content of GBFS replaced by FA was decreased, the loss in UPV readings was decreased, implying occurrence few cracks due to elevated temperatures (Figure 9.9a). Figure 9.9b and c show the effects of increasing WCP content (60 and 70%) on the UPV of the AAMs revealing low deterioration.

9.4.4 X-Ray Diffraction (XRD)

Figure 9.9 presents the XRD patterns of AAMs before and after temperature elevation from 27°C up to 900°C. An appearance of semi-crystalline aluminosilicate gel and quartz (Q) was observed in the initial sample before and after exposing to elevated temperatures less than 400°C (Figure 9.9a). The broad peaks of all alkali-activated mortar components can be seen in the region 25–30°C 2θ. Zeolites formed as secondary reaction products, which were formed as the crystalline phase after the fire resistance test was completed. Figure 9.9b and c shows the XRD patterns of

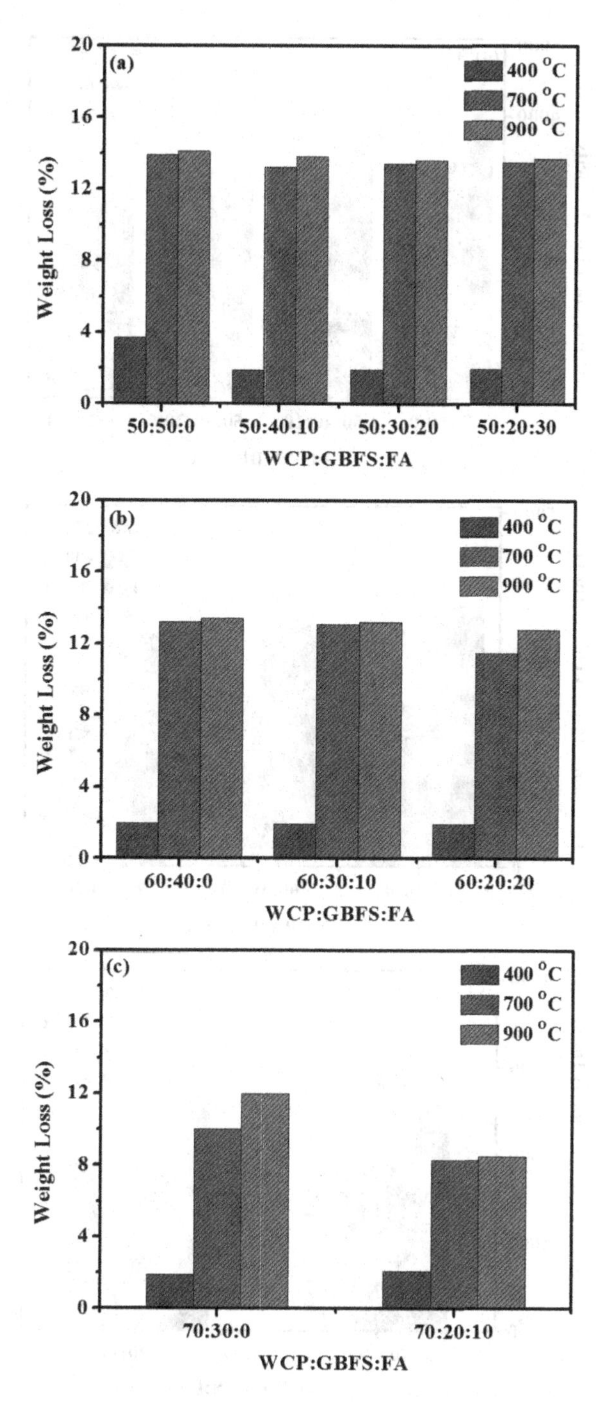

FIGURE 9.7 Effects of elevated temperatures on the weight loss of AAMs containing WCP of (a) 50%, (b) 60%, and (c) 70%.

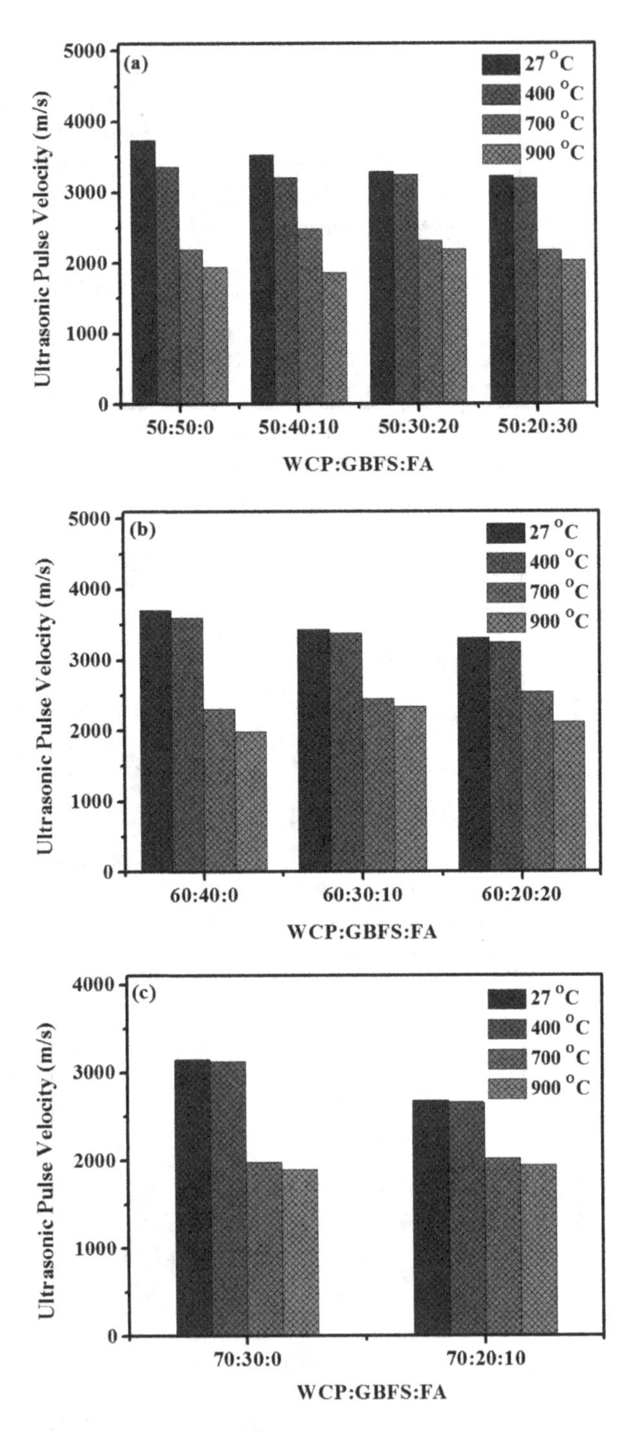

FIGURE 9.8 Effects of elevated temperatures on the UPV of AAMs containing WCP of (a) 50%, (b) 60%, and (c) 70%.

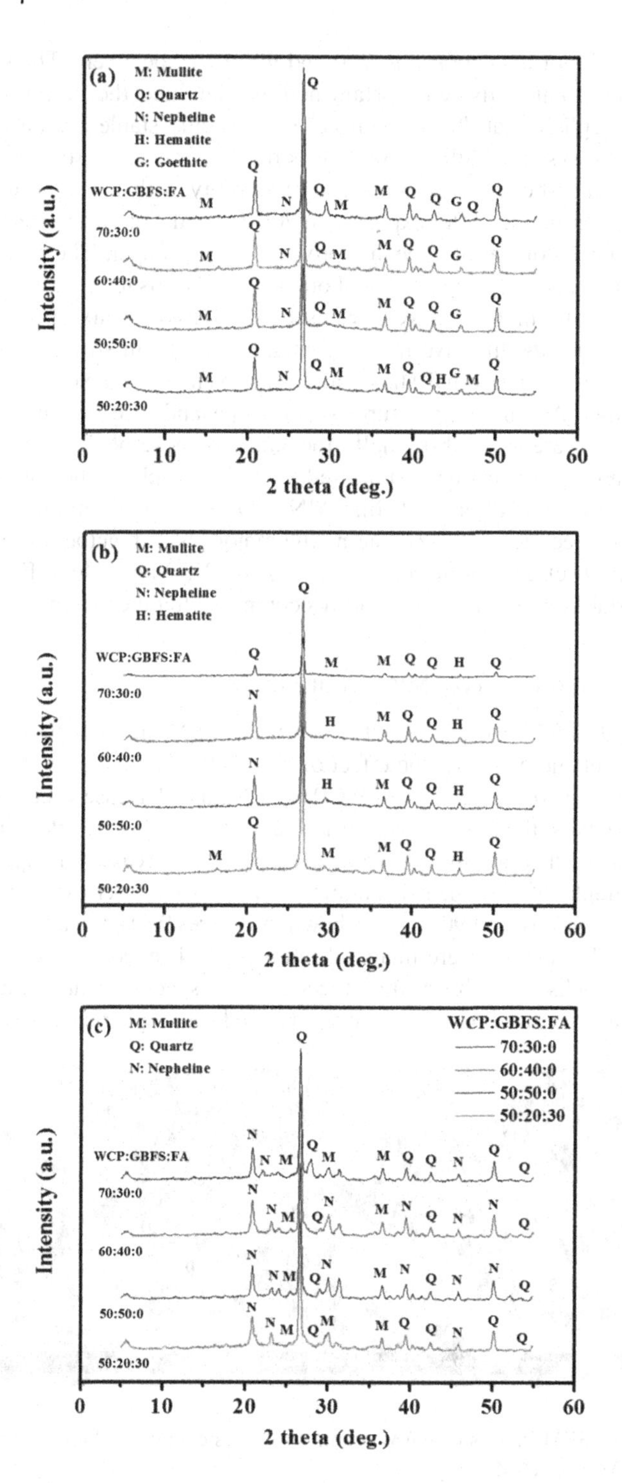

FIGURE 9.9 XRD of AAMs after fire exposure to (a) 400°C, (b) 700°C, and (c) 900°C.

alkali-activated mortars exposed to 700 and 900°C, respectively. The strong peaks in WCP-based alkali-activated mortars mainly identified the presence of quartz, mullite, and nepheline at 700°C. Mullite was the only stable crystalline phase of the Al_2O_3–SiO_2 system. Mullite retained its room temperature strength at elevated temperatures and showed high temperature stability with low thermal expansion and oxidation resistance. After exposure to 700°C, peaks of quartz were still stable, whereas the peaks of mullite became progressively stronger. The phase transition from goethite to hematite occurred at about 400°C. At this temperature, most of the constitutional water molecules were released. The outgoing flux of the OH- groups and the simultaneous diffusive rearrangement of the grain structure may cause a local accumulation of internal stress, and may even be capable of fracturing the hematite grains. At this temperature, grain shape and size of the newly formed hematite phase were still substantially the same as those of the original goethite. The XRD pattern of specimens subjected to 900°C disclosed hematite starting to disappear. While crystalline nepheline $AlNaSiO_4$ (sodium aluminium silicate) was present in the specimen, quartz remains the major phase together with the mullite phase (Figure 9.9c). In mortar containing 70% of WCP and a high FA amount the peaks were stable at elevated temperatures compared to other samples.

9.4.5 Scanning Electron Microscopy (SEM)

Figures 9.10, 9.11, 9.12 and 9.13 depict the results of SEM micrographs of AAMs containing high volume of WCP. The effect of 400, 700, and 900°C on the structure of AAMs containing 50, 60, and 70% of WCP were evaluated. Dense structures of AAMs were transformed to the less compact structure with a visible network of micro-cracks and large pores which are more pronounced with the increase in temperatures. The SEM micrographs of specimens containing 70% of WCP after exposure to ambient temperature, 400, 700, and 900°C are shown in Figures 9.10a, 9.11a, 9.12a, and 9.13a, respectively. The samples were imaged from a crashed section of the mortars. Only a few micro-cracks were observed on the surface of specimens at elevated temperature exposure. Unreacted particles of WCP, FA and some spherical holes were visible

FIGURE 9.10 SEM images of AAMs control samples containing high volume of WCP at 27°C. (a) AAM_9, (b) AAM_1.

FIGURE 9.11 SEM images showing the effect of elevated temperatures (400°C) on the morphology of AAMs containing a high volume of WCP. (a) AAM_9, (b) AAM_1.

FIGURE 9.12 SEM images showing the effect of elevated temperatures (700°C) on the morphology of AAMs containing a high volume of WCP. (a) AAM_9, (b) AAM_1.

(Figures 9.10–9.13). FA being known to contain a significant proportion of particles with hollow spheres, when these hollow spherical particles are partially dissolved they created highly dispersed small sized pores in the matrix [41]. These un-reacted particles were found in hollow cavities which might be due to the space left behind by dissolved FA particles. Conversely, specimens containing 70% of WCP exhibited a more stable surface at elevated temperatures compared to 50% of WCP specimens.

9.4.6 Fourier Transform Infrared Spectroscopy (FTIR)

The FTIR spectrum in Figure 9.14 exhibited major bands of 600 to 1200 cm^{-1} of alkali-activated samples exposed to 900°C. Table 9.3 presents the FTIR band positions and corresponding band assignments for AAMs containing high volume of WCP at 27°C. The changes in the band vibrations before and after exposure to elevated temperatures were clearly evidenced. The band frequency was found to increase with increasing temperatures and reduced content of WCP, FA and decreasing GBFS content. In specimens containing 50% of WCP, the frequency of bands

FIGURE 9.13 SEM images showing the effect of elevated temperatures (900°C) on the morphology of AAMs containing a high volume of WCP. (a) AAM_9, (b) AAM_1.

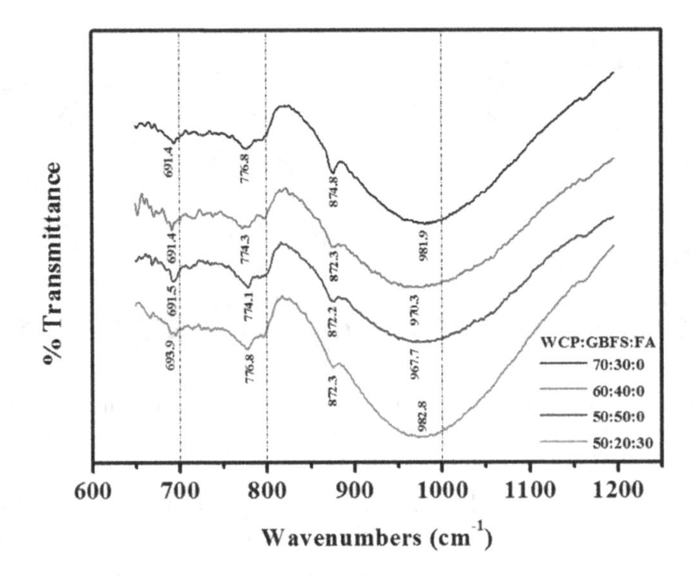

FIGURE 9.14 FTIR spectrums of AAMs after fire exposure to 900°C.

TABLE 9.3
FTIR Band Positions and Corresponding Band Assignments for AAMs Containing High Volume of WCP at 27°C

Mix				Band Positions (cm^{-1}) and Their Assignments				
WCP:GBFS:FA	Si/Al	Ca/Si	*fc* (MPa)	Al-O	Si-O	AlO_4	C-S-H	C(N)-A-S-H
70:30:0	5.09	0.26	34.02	669.8	695.9	778.1	875.7	994.2
60:40:0	4.79	0.37	68.44	666.6	693.9	774.6	874.5	966.3
50:50:0	4.48	0.50	74.12	659.1	691.1	773.5	873.9	943.1
50:20:30	3.53	0.20	56.47	660.2	692.7	775.2	875.1	971.5

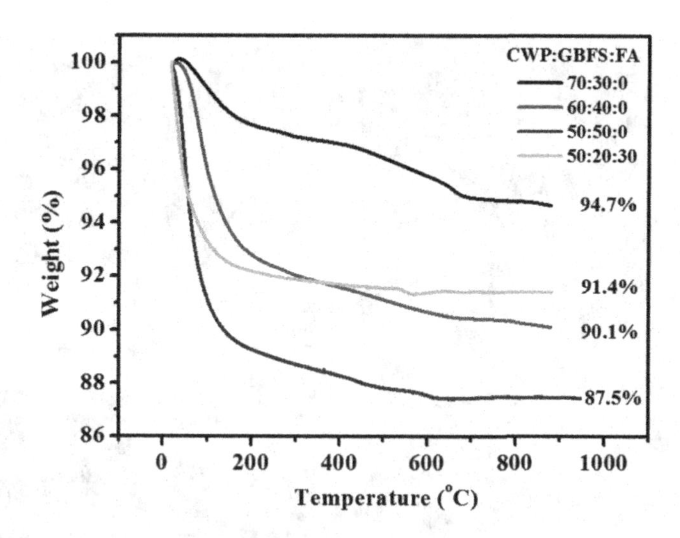

FIGURE 9.15 Effects of elevated temperatures on the weight loss of AAMs containing a high volume of WCP.

was increased from 943.1 to 967.7 cm^{-1} with increasing temperature from 27 to 900°C, respectively. This observed increase in the frequency of IR vibration was attributed to the reduction of C(N)-A-S-H gel in the AAMs network and thereby weakening of the 3D structure. A similar trend was also observed for C-S-H gel and the lost was increased with the increase in heat temperatures. Increased contents of WCP and FA in the AAM matrix have reduced the lost in C(N)-A-S-H and C-S-H gels and displayed high performance under elevated temperatures.

9.4.7 Thermogravimetric and Differential Thermal Analysis

Figure 9.15 illustrates the results of TGA curves of AAMs containing a high volume of WCP. The effect of 400, 700, and 900°C on structures of 50, 60, and 70% of WCP specimens was evaluated. Dense structures of AAMs were transformed to the less compact structure with a visible network of micro-cracks and large pores which are more pronounced with increasing temperatures. TGA results show weight loss in alkali-activated specimens, where it decreased with increasing WCP replaced GBFS contents of AAMs. The values of weight loss were decreased from 12.5 to 5.3% with increasing WCP replacing GBFS from 50 to 70%, respectively. A similar trend was observed in TGA, as the FA replacing GBFS content increased the weight loss decreased. The increased FA content as a replacement of GBFS in 50% of WCP specimens from 0 to 30% led to a reduction in weight loss from 12.5 to 8.6%.

9.5 Visual Appearance

9.5.1 Impact of Temperature Rise on Cracks

Figure 9.16 displays the effect of heating (900°C) on the surface of AAMs. Mortar prepared with 50% of WCP showed more cracks (Figure 9.16a) than those with

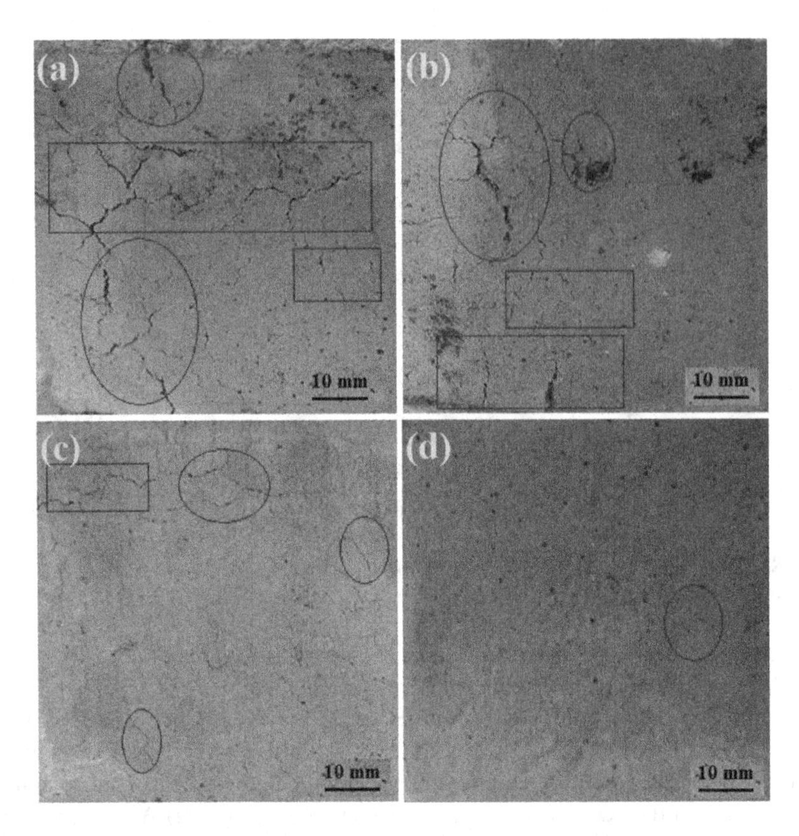

FIGURE 9.16 Effects of elevated temperatures at 900°C on crack development of (a) AAM_1, (b) AAM_5, (c) AAM_8, and (d) AAM_9.

60 and 70% of WCP. The deterioration was slightly reduced at high contents of WCP (Figure 9.16b and c). The addition of FA to AAMs enhanced the resistance of specimens to elevated temperatures and manifested a low number of cracks (Figure 9.16d). The heating process was normally accompanied by several transformations, moisture evaporation, internal vapour pressure, fine aggregate expansion, alkali-activated paste contraction, chemical decompositions, etc. During the early stage of heating these transformations may not have sufficient ability to cause any cracks. Except in situations where the rate of heating is quite high and the concrete is either dense or contains enough moisture content in which case spalling may occur within the first 30 minutes of exposure. The shapes in Figure 9.16 highlight the surface cracks of the AAMs with various WCP contents after exposure to 900°C.

9.5.2 Impact of Temperature Rise on Discolouration

The temperature-dependent colour changes of the proposed AAMs were evaluated. Alkali-activated samples showed normal grey colour at room temperature (27°C). Furthermore, this colour became slightly darker with the increased level of replacement of GBFS by FA, which was maintained till the end of first level of heating (400°C). With the increase of heat level up to 400°C, the colour was slightly changed to vanilla and at 700°C it appeared light grey. At 900°C, the AAMs looked whitish in colour for all levels of WCP. The burning effect was more severe at external planes, which was further diminished towards the interiors and caused most of these changes. In fact, close observations on crushed specimens revealed that the surface colour at a particular temperature penetrates into the mortar and this depended upon the intensity of the inward heat. While clear surface discolouration was seen to last up to about 10 mm deep, particularly on specimens exposed to 700°C or less, this penetration differs in the case of specimens subjected to 900°C. These observations were common to all alkali-activated specimens.

9.6 CONCLUSIONS

The influence of a high volume of WCP contents on the properties of ternary AAMs were evaluated in terms of resistance to elevated temperatures. Based on the experimental results the following conclusions were drawn:

i. An increase in content of WCP from 50 to 70% has enhanced the resistance of AAMs to elevated temperatures up to 900°C.
ii. Incorporating WCP with FA to replace GBFS caused a reduction in strength and weight loss of specimens when exposed to elevated temperatures.
iii. Specimens prepared with 70% of WCP, 20% of GBFS, and 10% of FA disclosed the optimum resistance at elevated temperatures.
iv. Results of XRD, SEM, FTIR, and TGA tests explained the thermal stability of AAMs containing a high level of WCP (70%) when exposed to heat.
v. A practical advantage of studying surface discolouration of fired AAMs in the preliminary assessment of damage caused by fire hazards is that the intensity of fire can be comprehended.
vi. It was established that by recycling the industrial WCP good-quality AAMs could be produced. This could be beneficial in terms of economy, enhanced durability performance and environmental friendliness.

REFERENCES

1. Mohammadhosseini, H., et al., Durability performance of green concrete composites containing waste carpet fibers and palm oil fuel ash. *Journal of Cleaner Production*, 2017. **144**: pp. 448–458.
2. Huseien, G. F., et al., Synthesis and characterization of self-healing mortar with modified strength. *Jurnal Teknologi*, 2015. **76**(1): pp. 195–200.
3. Asaad, M. A., et al., Improved corrosion resistance of mild steel against acid activation: Impact of novel *Elaeis guineensis* and silver nanoparticles. *Journal of Industrial and Engineering Chemistry*, 2018. **63**: pp. 139–148.
4. Khankhaje, E., et al., On blended cement and geopolymer concretes containing palm oil fuel ash. *Materials & Design*, 2016. **89**: pp. 385–398.
5. Huseiena, G. F., et al., Performance of sustainable alkali activated mortars containing solid waste ceramic powder. *Chemical Engineering*, 2018. **63**: pp. 673–678.

6. Yang, L., et al., Effects of nano-TiO 2 on strength, shrinkage and microstructure of alkali activated slag pastes. *Cement and Concrete Composites*, 2015. **57**: pp. 1–7.
7. Yusuf, M. O., et al., Evolution of alkaline activated ground blast furnace slag–ultrafine palm oil fuel ash based concrete. *Materials & Design*, 2014. **55**: pp. 387–393.
8. Huseien, G. F., et al., Effect of metakaolin replaced granulated blast furnace slag on fresh and early strength properties of geopolymer mortar. *Ain Shams Engineering Journal*, 2016. 93: pp. 1557–1566.
9. Huseiena, G. F., et al., Effect of binder to fine aggregate content on performance of sustainable alkali activated mortars incorporating solid waste materials. *Chemical Engineering*, 2018. **63**: pp. 667–672.
10. Al-Majidi, M. H., et al., Development of geopolymer mortar under ambient temperature for in situ applications. *Construction and Building Materials*, 2016. **120**: pp. 198–211.
11. Huseien, G. F., et al., Geopolymer mortars as sustainable repair material: A comprehensive review. *Renewable and Sustainable Energy Reviews*, 2017. **80**: pp. 54–74.
12. Karakoç, M. B., et al., Mechanical properties and setting time of ferrochrome slag based geopolymer paste and mortar. *Construction and Building Materials*, 2014. **72**: pp. 283–292.
13. Kodur, V., Properties of concrete at elevated temperatures. *ISRN Civil Engineering*, 2014. **2014**: pp. 1–14.
14. Novak, J. and A. Kohoutkova, Mechanical properties of concrete composites subject to elevated temperature. *Fire Safety Journal*, 2018. **95**: pp. 66–76.
15. Buchanan, A. H. and A. K. Abu, *Structural Design for Fire Safety*. 2017: Wiley.
16. Awal, A. A. and I. Shehu, Performance evaluation of concrete containing high volume palm oil fuel ash exposed to elevated temperature. *Construction and Building Materials*, 2015. **76**: pp. 214–220.
17. Saavedra, W. G. V. and R. M. de Gutiérrez, Performance of geopolymer concrete composed of fly ash after exposure to elevated temperatures. *Construction and Building Materials*, 2017. **154**: pp. 229–235.
18. Omer, S. A., R. Demirboga, and W. H. Khushefati, Relationship between compressive strength and UPV of GGBFS based geopolymer mortars exposed to elevated temperatures. *Construction and Building Materials*, 2015. **94**: pp. 189–195.
19. Ranjbar, N., et al., Compressive strength and microstructural analysis of fly ash/palm oil fuel ash based geopolymer mortar under elevated temperatures. *Construction and Building Materials*, 2014. **65**: pp. 114–121.
20. Huseien, G. F., et al., Influence of different curing temperatures and alkali activators on properties of GBFS geopolymer mortars containing fly ash and palm-oil fuel ash. *Construction and Building Materials*, 2016. **125**: pp. 1229–1240.
21. Hossain, M., et al., Durability of mortar and concrete containing alkali-activated binder with pozzolans: A review. *Construction and Building Materials*, 2015. **93**: pp. 95–109.
22. Hussein, A. A., et al., Performance of nanoceramic powder on the chemical and physical properties of bitumen. *Construction and Building Materials*, 2017. **156**: pp. 496–505.
23. Huseiena, G. F., et al., Potential use coconut milk as alternative to alkali solution for geopolymer production. *Jurnal Teknologi*, 2016. **78**(11): pp. 133–139.
24. Samadi, M., et al., Properties of mortar containing ceramic powder waste as cement replacement. *Jurnal Teknologi*, 2015. **77**(12): pp. 93–97.
25. Zhou, W., et al., A comparative study of high-and low-Al 2 O 3 fly ash based-geopolymers: The role of mix proportion factors and curing temperature. *Materials & Design*, 2016. **95**: pp. 63–74.
26. Yusuf, M. O., Performance of slag blended alkaline activated palm oil fuel ash mortar in sulfate environments. *Construction and Building Materials*, 2015. **98**: pp. 417–424.

27. Huseien, G. F., et al., The effect of sodium hydroxide molarity and other parameters on water absorption of geopolymer mortars. *Indian Journal of Science and Technology*, 2016. **9**(48): pp. 1–7.
28. Huseien, G. F., et al., Synergism between palm oil fuel ash and slag: Production of environmental-friendly alkali activated mortars with enhanced properties. *Construction and Building Materials*, 2018. **170**: pp. 235–244.
29. Huseien, G. F., J. Mirza, and M. Ismail, Effects of high volume ceramic binders on flexural strength of self-compacting geopolymer concrete. *Advanced Science Letters*, 2018. **24**(6): pp. 4097–4101.
30. Huseien, G. F., M. Ismail, and J. Mirza, Influence of curing methods and sodium silicate content on compressive strength and microstructure of multi blend geopolymer mortars. *Advanced Science Letters*, 2018. **24**(6): pp. 4218–4222.
31. Pacheco-Torgal, F. and S. Jalali, Reusing ceramic wastes in concrete. *Construction and Building Materials*, 2010. **24**(5): pp. 832–838.
32. ASTM, Standard specification for coal fly ash and raw or calcined natural pozzolan for use in concrete, 2013: ASTM.
33. ASTM, C117-13: Standard test method for materials finer than 75-µm (No. 200) sieve in mineral aggregates by washing, 2013: ASTM.
34. ASTM, C33: Standard specification for concrete aggregates. In *Annual Book of ASTM Standards*. 2004: ASTM.
35. ASTM, C579: Standard test methods for compressive strength of chemical-resistant mortars. In *Grouts, Monolithic Surfacings, and Polymer Concretes*. 2001: ASTM.
36. ASTM, C597: Standard test method for pulse velocity through concrete, 2009: ASTM.
37. ASTM, Standard test methods for fire tests of building construction and materials, 2012: ASTM.
38. ISO, *Fire-Resistance Tests: Elements of Building Construction. Commentary on Test Method and Test Data Application*. 1994: ISO.
39. Roviello, G., et al., Fire resistant melamine based organic-geopolymer hybrid composites. *Cement and Concrete Composites*, 2015. **59**: pp. 89–99.
40. Ismail, M., M. E. Ismail, and B. Muhammad, Influence of elevated temperatures on physical and compressive strength properties of concrete containing palm oil fuel ash. *Construction and Building Materials*, 2011. **25**(5): pp. 2358–2364.
41. Huseien, G. F., et al., Effects of POFA replaced with FA on durability properties of GBFS included alkali activated mortars. *Construction and Building Materials*, 2018. **175**: pp. 174–186.

10 Sustainability of Ceramic Waste in the Concrete Industry

10.1 INTRODUCTION

The primary goal of sustainability is to maintain life on earth for the foreseeable future without disturbing the ecological balance [1]. Sustainability is based upon three components: environment, economy and society. Sustainable development must protect these three components to preserve biodiversity and balance in the ecosystem. In the era of the industrial revolution, engineers, or architects ingeniously used the sustainability principle to minimize the net negative impact on the environment. Thus, in the context of construction materials the word *sustainability* appears to be synonymous with *environmentally sound* or *affable* and *green* [2–4].

Millions of tons of industrial wastes are generated every year worldwide, compounding the issue of sustainability. Those wastes become an environmental liability due to lack of storage and safety disposal. Subsequently, such wastes (fly ash, slag, ceramic waste powder, palm oil fuel ash, etc.) pollute the land and water in the vicinity of the factories [5]. In the recent past, several studies [5, 6] revealed that the cementless (geopolymer) production is the best way to recycle those wastes. Moreover, it is feasible to recycle the construction wastes including concrete and ceramic as aggregate in the geopolymer industries [7, 8]. It has become evident that large-scale production of geopolymers will not only save energy (fuel and electricity) consumed during the processing (heating and grinding) of cement but also remarkably reduce the pollution, especially carbon dioxide (CO_2) emissions. Geopolymers thus provide a viable solution to landfill problems and simultaneously minimize the overexploitation of natural resources [5].

Low carbon emissions and energy-saving building material incorporated with smart material (in self-healing technology) is a well-known candidate of energy technology for enhancing energy efficiency and sustainability of building. In the present industrial uprising era, engineers, scientists, policymakers, and architects are attempting to resourcefully use the sustainable model to reduce the negative impact on our ecosystem. Therefore, in the perspective of building materials the phrase *sustainability* is used synonymously with *robust* or *friendly* and *green environment* [2, 3]. In this viewpoint, self-healing materials have attracted increasing interest due to their potential to lessen the degradation, prolong the functional lifespan and suppress the maintenance costs of materials [9, 10]. However, the self-healing technology contributes directly to enhance the environment and reduce the pollution from increasing concrete lifespan and reducing the demand and consumption of ordinary Portland cement (OPC) as well as saving energy and increasing the sustainability of concrete.

During the past decade, the self-healing technique had been extensively studied and introduced as an innovative technique for concrete crack self-repair. However, several innovative strategies to self-healing for cementitious materials have been proposed and developed. Nowadays, it is known as the life cycle assessment (LCA) methodology, which has been standardized in ISO 14040-14044. Thus far, it has been used for assessing the environmental impact of all kinds of products and services preferably from cradle to grave. The goal of this LCA was a quantification of the environmental impact reduction that could be achieved by using the proposed self-healing concrete instead of a more traditional concrete. Some direct benefits of self-healing concrete include the reduction of the rate of deterioration, extension of service life, and reduction of repair frequency and cost over the life cycle of a concrete infrastructure. These direct benefits may be expected to lead to enhanced environmental sustainability since fewer repairs implies lower rate of material resource usage and reduction in energy consumption and pollutant emission in material production and transport, as well as that associated with traffic alterations in transportation infrastructure during repairs/reconstruction [11]. Van Belleghem et al. [12] reported that the self-healing of cracks with encapsulated polyurethane precursor formed a partial barrier against immediate ingress of chlorides through the cracks. Application of self-healing concrete was able to reduce the chloride concentration in a cracked zone by 75% or more. Service life of self-healing concrete in marine environments could amount to 60–94 years as opposed to only 7 years for ordinary (cracked) concrete. However, life cycle assessment calculations indicated important environmental benefits (56–75%) and achievable service life extension.

A review of literature revealed that modified cement, geopolymers and alkali-activated binders possess excellent mechanical strength and durability properties. These notable attributes include high early strength, low porosity, elevated temperature resistance, and high performance in acidic and sulphate environments. All these properties make these binders potential candidates for diverse industrial applications including the field of civil engineering, automotive and aerospace, non-ferrous foundries and metallurgy, plastics, waste management, art and decoration, and retrofitting of buildings.

10.2 SUSTAINABILITY OF MODIFIED CEMENT-BASED CONCRETE

10.2.1 Life Cycle Calculation

By utilizing and recycling solid waste for the manufacturing of construction materials, it is possible to develop more sustainable and greener products. However, these materials must have a competitive cost with the production of existing materials on the market or should have ecological benefits to the construction project. In order to compare the sustainability aspect of recycled ceramic mortar (RCM) with conventional mortar, parameters including greenhouse emissions, the cost of production, and the energy consumption associated with the manufacturing of the mortar were chosen. These parameters are considered the major reasons for utilizing RCM despite other key indicators that play important roles. These include the technical action, leaching, use of water, harmful material content, emission of other non-environmentally friendly gases during production, and the quantity of the waste mass that can be circumvented by using various industrial wastes in concrete.

In the environmental benefits analysis process, it is essential to examine the impact produced by the required feedstock and binder manufacturing, as well as the transport. The mixing, laying, curing and emission characteristics of different mortars over their working lifetime are not considered due to their similar nature. Thus, this technique may provide analogous life cycle impacts instead of an absolute one. This is a helpful strategy for comparable products because it reduces the time needed for the evaluation. The energy consumption, production cost and emission of greenhouse gas were calculated for the preparation of each material. The preparation stages chosen for ground ceramic (GC), ceramic fine aggregate (CFA), and RS are shown in Table 10.1.

In the preparation of the materials, the transportation fees were embedded in the final cost and for all materials a transportation fee of 1 t/km was used (Table 10.2). The price of ceramic waste in this study was assumed to be zero Ringgit Malaysia (RM), as this material was waste and obtained for free from the ceramics industry. Depending on the life cycle of each material, the total cost of the electrical consumption of the equipment was calculated according to the capacity of the engine and time of operation (Table 10.2). Table 10.3 depicts the electricity price rates in Malaysia for March 2020, upon which the electricity cost calculation for the different types of machinery was based on. The CO_2 emission equivalent for the production of 1 t of cement was considered as the total greenhouse gas emission. The total amount of greenhouse gas emissions was calculated following the McLellan method using the following relation:

$$\text{Total greenhouse gases released: } \sum_{1=1}^{n} mi\left(\left(di * ei\right) + pi\right) \tag{10.1}$$

where *di* is the transportation distance, which is dependent on the direction of the transport; *ei* is the emission factor for the different methods of transportation; and *pi* is the emission per unit mass of each material.

TABLE 10.1
Chosen Preparation Stages for the WCP, OPC, CFA, and RS

Material	Type	Collection	Transportation	Crush	Sieve	Dry	Grind
WCP	Waste	√	√	√	√	—	√
OPC							
CFA	Waste	√	√	√	√	—	—
RS	Natural	√	√	—	√	—	—

TABLE 10.2
Details of the Machinery and Materials in Malaysia

Items	Amount
Truck volume, m^3	12
Truck speed, km/hr	80
Diesel consumption, L/km	0.09
Diesel price, RM/L	2.18
Transport charge of 1 m^3, RM/km	0.75
GC density, kg/m^3	1470
CFA density, kg/m^3	1440
Transport distance of ceramic waste, km	35
RS density, kg/m^3	1750
Transport distance of river sand, km	62
Crushing machine power, watt	435
Crushing machine capacity, m^3	0.08
Sieving machine power, watt	250
Sieving machine capacity, m^3	0.05
Oven power, watt	1200
Oven capacity, m^3	0.18
Grinding machine power, watt	750
Grinding machine capacity, m^3	0.45
CO_2 release for 1 kwh electricity, ton	0.00013
CO_2 release for 1 L diesel, ton	0.0027
Energy consumption for 1 L diesel, GJ	0.0384
Energy consumption for 1 kWh electricity, GJ	0.0036
Portland cement CO_2 release, tonne/tonne	0.904
Energy consumption, GJ/tonne	5.13

TABLE 10.3
Electricity Cost in Accordance with the Consumption Rate (March 2020)

Tariff (Watt)	The unit (RM/kWh)
0 to 199	0.218
200 to 299	0.334
300 to 599	0.516
600 to 899	0.546
900+	0.571

The total energy consumption and cost of production for each batch was calculated according to the related data (Table 10.1–10.3). The cost, energy consumption, cost of production and greenhouse gas emissions of sample 40GC100CFA were compared with the control sample 0GC0CFA to achieve a CS of 30 MPa with a requirement of Portland cement of 460 kg/m^3.

$$\text{Total } CO_2\text{emissions } = \sum_{i=1}^{n} mi\left[\left(di \times Di \times k1i\right) + \left(Ei \times k2i\right)\right] \tag{10.2}$$

where mi is the mass of component i (t/m^3), di is the transport distance (km), Di is the diesel consumption (L/km), $k1i$ is the CO_2 emission for 1 L of diesel (t), Ei is the total electricity consumption (kwh), and $k2i$ is the CO_2 emission for 1 kwh electricity (t).

$$\text{Total energy consumption} = \sum_{i=1}^{n} mi\left[\left(di \times Di \times k3i\right) + \left(Ei \times k4i\right)\right] \tag{10.3}$$

where $k3i$ is the energy consumption for 1 L of diesel (GJ), Ei is the total electricity consumption (kwh), and $k4i$ is the energy consumption for 1 kwh of electricity (GJ).

$$\text{Total cost} = \sum_{i=1}^{n} mi\left[\left(di \times Di \times DPi\right) + Ti + \left(Ei \times EPi\right)\right] \tag{10.4}$$

where DPi is the diesel cost (RM/L), Ti is the transport charge for 1 m^3 (RM/km), and EPi is the electricity cost (RM/kwh).

$$\text{Electricity consumption of component } I\ (Ei) = \sum_{i=1}^{n} \left(MEi \times MPi\right) \tag{10.5}$$

where MEi is the machine capacity (t/h) and MPi is the machine power (kwh).

10.2.2 Mix Design

The mix proportions for the mortar were prepared based on the weight of the materials and in accordance with ASTM C1329. The ratio of water to cement (w/c) was chosen to be 0.48 in order to achieve acceptable flowability following the requirements of ASTM C320, as well as the desired strength. A total of 12 batch samples (i.e. mortar mixes) were cast. The first sample, without any replacement of the binder or ceramic fine aggregate (CFA), acted as the control specimen (0GC0CFA). Table 10.4 presents the mixes prepared with waste ceramic powder (WCP) and CFA. All the designed mixes were de-moulded after 24 hours and then placed inside a water container to cure until the tests were conducted at a mean ambient temperature (AT) of 27 ± 2°C and relative humidity (RH) of 85 ± 5%.

TABLE 10.4
Mix Design for All Mortar Batches (kg/m³)

	Binder		Fine Aggregates	
Mix Label	**OPC**	**WCP**	**RS**	**CFA**
0GC0CFA	550	—	1460	—
10GC0CFA	495	55	1460	—
20GC0CFA	440	110	1460	—
30GC0CFA	395	165	1460	—
40GC0CFA	330	220	1460	—
50GC0CFA	275	275	1460	—
60GC0CFA	220	330	1460	—
0GC25CFA	550	—	1095	365
0GC50CFA	550	—	730	730
0GC75CFA	550	—	365	1095
0GC100CFA	550	—	—	1460
40GC100FCA	330	220	—	1460

TABLE 10.5
Energy Consumption, Greenhouse Gas Emissions and Cost of Production for WCP, OPC, CFA, and RS

Material	Energy Consumption (GJ/ton)	Greenhouse Gas (ton/ton)	Cost (RM/ton)
WCP	1.12	0.045	170
OPC	5.13	0.904	600
CFA	0.111	0.003	10
RS	0.134	0.009	35

10.2.3 Greenhouse Emission, Energy Efficiency, and Cost Analysis

The mortar sustainability and environment benefits were assessed using the total cost of production, greenhouse gas emission and energy consumption of WCP, OPC, CFA, and RS. The estimated values are given in Table 10.5. The results show that the OPC required a huge amount of energy in the manufacturing process compared to WCP, which led to an increase in cost and release of greenhouse gases. OPC presented a 5.13 GJ/ton energy expenditure, which was more than four times the 1.12 GJ/ton calculated for WCP. This difference was related to the higher energy utilization, cost and greenhouse gas emissions associated with OPC, which has a significantly higher greenhouse gas release (0.904 ton/ton) than WCP (0.045 ton/ton). Similar to the release of greenhouse gases, the cost of production for OPC was the highest of all the materials used. This is largely due to the exorbitant amount of energy consumed during the manufacturing process and the extensive effort required to transport the material. The cost of production was 600 RM/ton for OPC compared to 170 RM/ton for WCP. A lower content of OPC in the mortar samples was necessary to achieve a lower level of energy consumption and production cost, as well as improvements in the greenhouse gas emissions and sustainability of the project.

The effects of a WCP replacement for the OPC on the release of greenhouse gases from a blended cement are depicted in Figure 10.1a. The release of greenhouse gases was reduced with an increase in the WCP level in the mortar samples. The density of the gas released was also reduced from 92.9 to 77.5 to 46.8 kg/m^3 with an increase in WCP from 20 to 40 to 60%, respectively. The ceramic-blended cement released less greenhouse gas with any percentage of replacement compared with a standard all-OPC mixture. In the process of manufacturing 1 ton of a blended cement containing 40% WCP, 1 m^3 of greenhouse gas is released, which is a reduction of more than 37% compared to a standard mix. The resulting low level of greenhouse gas emissions associated with the proposed mortar samples prepared using WCP as a binder indicate the strong possibility that a simple and sustainable material can be produced that uses less OPC.

Figure 10.2a displays the impact of the WCP replacement on the cost-effectiveness of the proposed mortar samples compared to OPC as a binder. The utilization of a high content of WCP (i.e. 40%) as a replacement for the OPC saved a considerable amount of cost. This cost calculation by weight was based on the preparation stages of each material (Table 10.5), which directly influenced the final cost of the mortar samples. The cost of the binder was reduced from 380 to 358, 335, 317, 287, and 264 RM/m^3 with the increase in the replacement percentage of WCP for OPC from 0 to 10, 20, 30, 40, 50, and 60%, respectively. The substitution of WCP as a binder in the mortar samples was shown to make a valuable contribution to the generation of sustainable products. Figure 10.14b illustrates the effect on the cost of mortar by replacing RS as the fine aggregate with CFA. A moderate cost savings was realized through this substitution. The price calculation of the materials by weight was based on the preparation stages of each material (Table 10.5), which directly influenced the ultimate cost of the mixtures. The cost of the mortar samples was reduced from more than 380 to 372, 362, 351, and 341 RM/m^3 as the substitution level of CFA for the fine aggregates rose from 0 to 25, 50, 75, and 100%, respectively. Figure 10.14b shows that the use of CFA as a total replacement for RS could significantly contribute to the creation of green products, thus leading to more sustainable development in the field of civil engineering. Figure 10.14c provides a clear comparison of the production cost for the four mixes, which are samples 0GC0CFA, 0GC100CFA, 40GC0CFA, and 40GC100CFA. As can be observed from this figure, the minimum cost of production belongs to sample 40GC100CFA, which was produced using a 40% cement substitute and entirely replacing the standard aggregates with recycled aggregates. A comparison of samples 0GC100CFA and 40GC0CFA shows that any amount of cement substitution results in a notable reduction in cost.

The energy consumption required for the production of each batch of samples was calculated according to the energy consumption and life cycle of the individual materials that were used to manufacture that batch (Figure 10.3). The energy

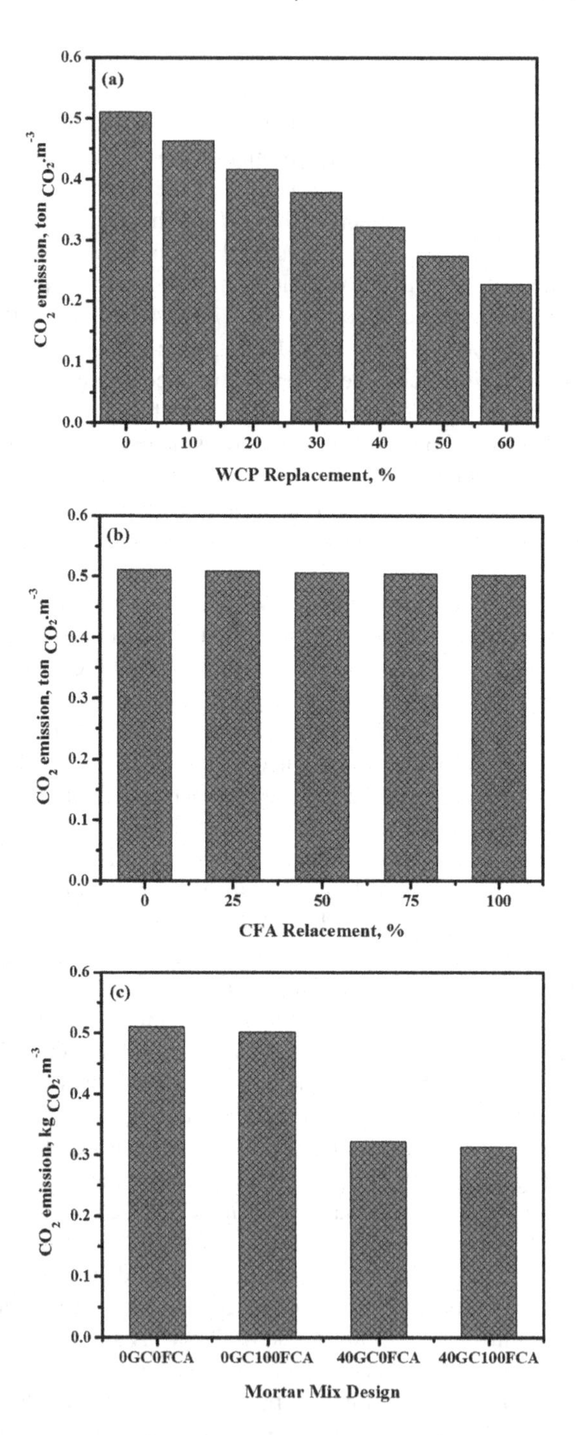

FIGURE 10.1 (a) Greenhouse gas release of WCP replacement, (b) greenhouse gas release of CFA replacement, (c) greenhouse gas release of sample 40GC100CFA.

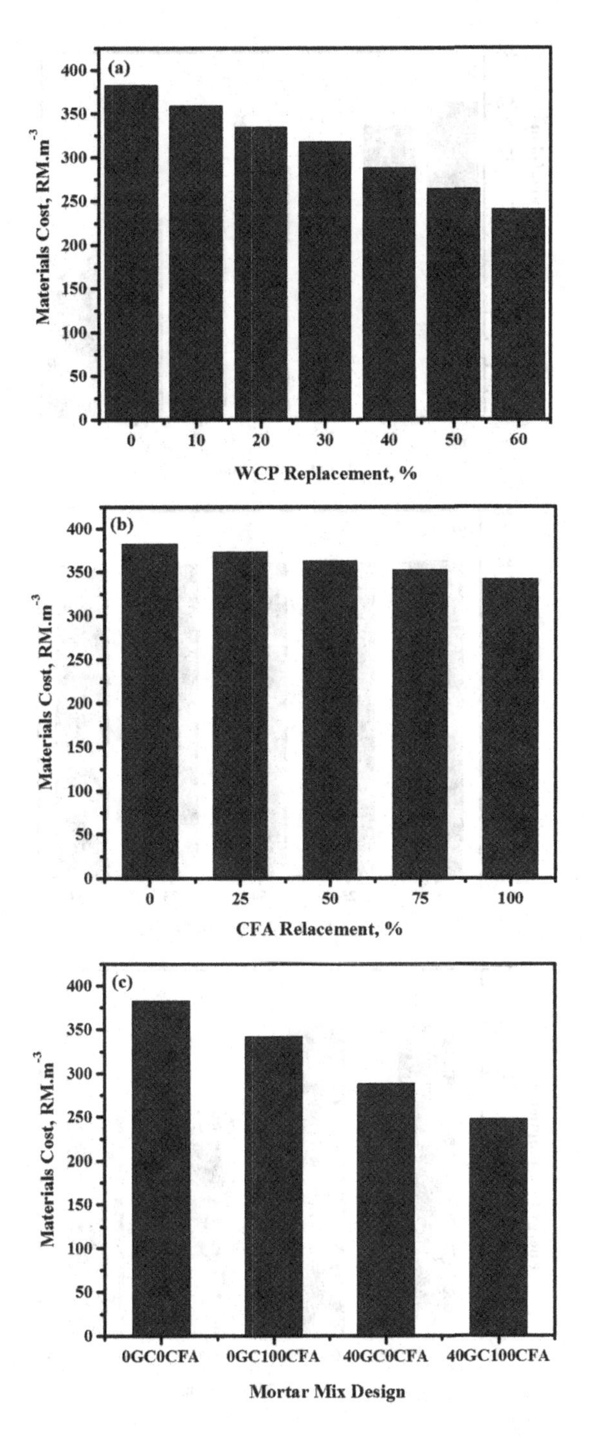

FIGURE 10.2 Effects of different substitution percentages on the cost. (a) WCP, (b) CFA, and (c) WCP and CFA together.

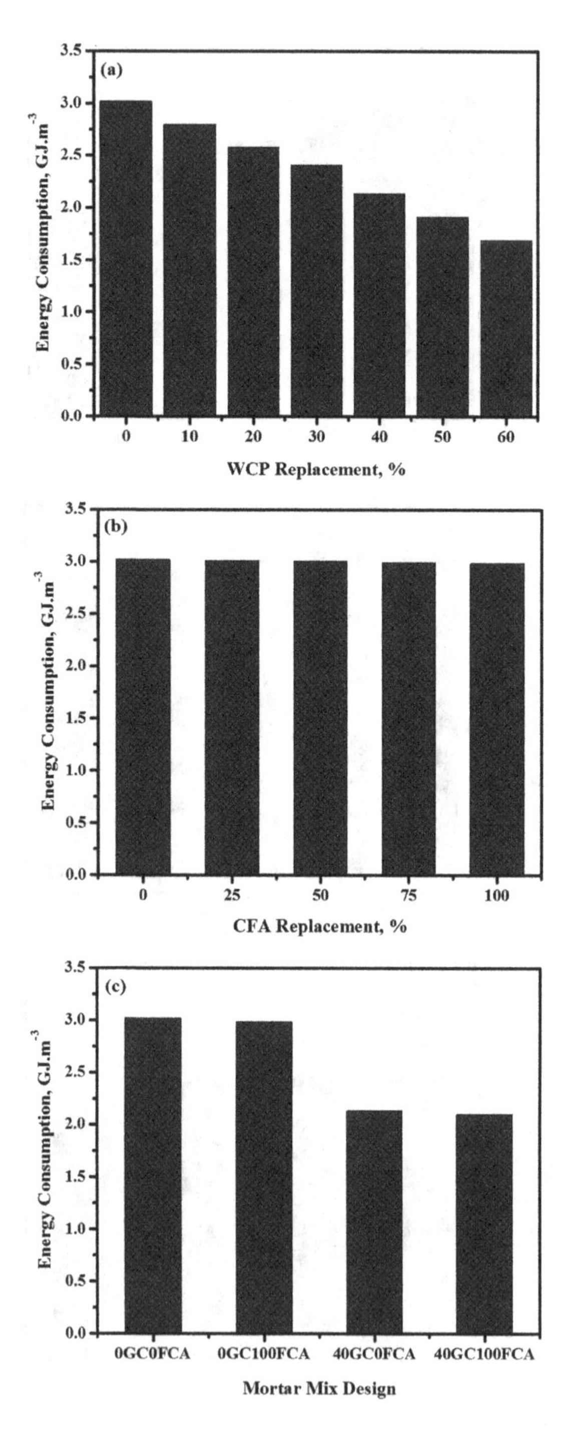

FIGURE 10.3 Effects of various replacement agents on the energy consumption of the mortar samples. (a) WCP, (b) CFA, and (c) WCP and CFA together.

consumption required to produce the binder reduced with the rising level of WCP as a replacement for the OPC. In comparison to the 3.02 GJ/m^3 calculated for a WCP of 0%, the energy utilization was reduced to 2.8, 2.58, 2.41, 2.13, 1.91, and 1.69 GJ/m^3 with an increase in the WCP level to 10, 20, 30, 40, 50, and 60%, respectively. Nonetheless, all the mixes with some percent of WCP consumed less energy during manufacturing than OPC (3.02 GJ/m^3).

In fact, the lower cost of the diesel and electricity used during the preparation of the GC directly influenced the ultimate energy expenditure of the mortar samples. This little outlay, lower greenhouse gas emissions and the low energy expenditure of the WCP were the primary reasons for evaluating the achievability of this sustainable product. From Figure 10.3b, which shows the energy consumption of the mortar samples containing different percentages of CFA as an RS replacement, it can be clearly seen that the replacement of fine aggregates by CFA gave the minimum energy consumption, which was 2.98 GJ/m^3. However, the effect of CFA replacement did not have a significant effect on reducing the energy consumption due to the already low energy consumption associated with RS extraction.

Consequently, it can be concluded that the substitution of OPC with WCP has a significant effect on the reduction of energy consumption due to the large amount of energy consumed in cement manufacturing. However, landfill problems and the preservation of natural resources remain important issues that cannot be eliminated. Replacing WCP and CFA in the same mixture resulted in the smallest energy consumption calculated (2.10 GJ/m^3) compared to the other mixes, which is a step toward the production of sustainable and green construction materials.

10.3 SUSTAINABILITY OF ALKALI-ACTIVATED CONCRETE

10.3.1 Waste Materials Life Cycle

If alkali-activated mortars (AAMs) are to be a viable competing product to OPC-based concretes, they will be required to demonstrate a similar financial cost or lower to the user and/or significant functional, manufacturing or sustainability benefits. In order to be able to compare AAMs with OPC on a sustainability basis, three headline metrics were chosen. In this case, greenhouse emissions, cost and the energy (direct fuel usage and electricity usage) were chosen as three key metrics which are considered to form the main argument for or against the use of AAMs, notwithstanding the fact that other key indicators have a significant role to play such as technical performance, leaching, water usage, hazardous materials content, other environmental emissions of production and the amount of waste volume that can be avoided by utilizing WCP, ground blast furnace slag (GBFS), and FA in alkali-activated or OPC concrete. The three selected metrics are the ones most readily quantified in these early stages of industrial alkali-activated development.

The energy, cost and emission metrics are derived using a life cycle approach. For the purpose of this assessment, this implies the impacts for the production of required feedstocks as well as the manufacture of the binder and any relevant transportation. The importance of this approach is that it allows a valid comparison of the four materials (WCP, GBFS, FA, and OPC); production impacts alone do not give the full picture of the required "embodied" energy and CO_2 in feedstocks. The mixing, laying and curing of the alkali-activated concrete and OPC, and the operational lifetime emissions are not included as they are assumed to be similar for each product. The approach, therefore, may be considered to give a comparable life cycle impact rather than an absolute impact. This is a useful approach for similar products, as it reduces the time required for the assessment.

For each material life cycle, the CO_2 emission, cost, and energy were calculated. The life cycles of WCP, GBFS, and FA are presented in Table 10.6. For all three materials, the costs of collection from the factories are assumed to be zero Malaysian ringgit, as the materials are wastes. The transport distance of each material was also included in life cycle calculation. As shown in Figure 10.4, the GBFS shows the longer transport distance (500 km) compared to 45 and 35 km of FA and WCP, respectively. For cost and diesel consumption, the type of truck engine, volume, speed, and charge of 1 tonne/km were fixed for all types of materials (Table 10.6). According to life cycle of each material, the cost and amount of electricity are calculated depending on capacity, function time and electricity consumption of each machine, as

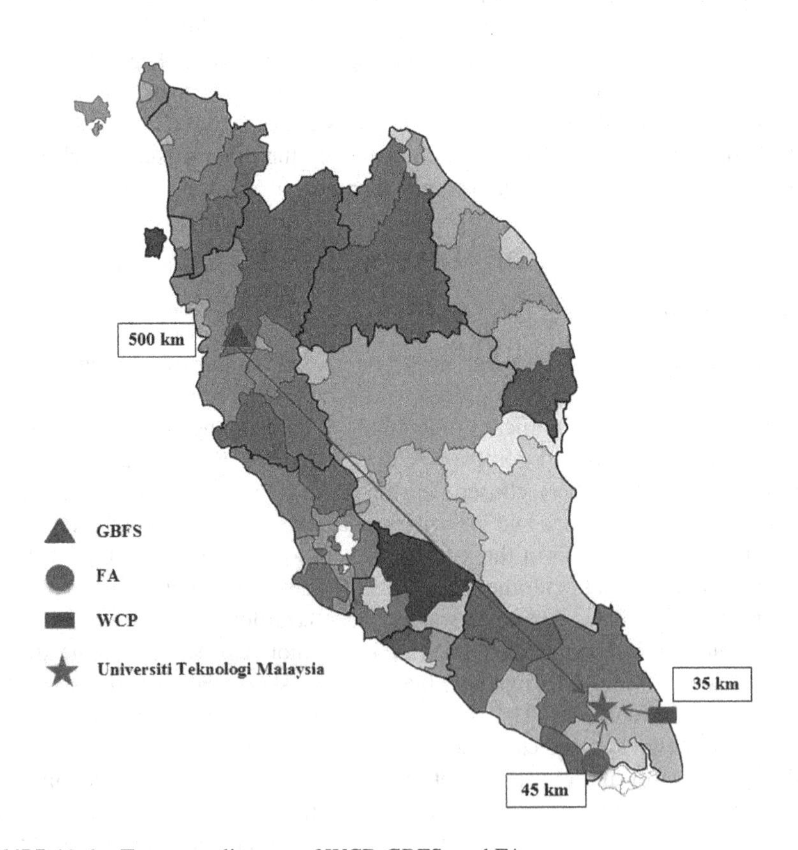

FIGURE 10.4 Transport distance of WCP, GBFS, and FA.

TABLE 10.6
Life Cycle Stages Considered for WCP, GBFS, and FA

Materials	Type	Collection	Transport	Crush	Sieve	Dry	Grind
WCP	Waste	√	√	√	√		√
GBFS	Waste	√	√		√		√
FA	Waste	√	√				

TABLE 10.7
Material and Machine Information

Item	Amount
Truck speed, km/hr	80
Diesel consumption, L/km	0.09
Diesel price, RM/L	2.18
Truck volume, m^3	12
Transport charge of 1 m^3, RM/km	0.75
WCP density, kg/m^3	1780
GBFS density, kg/m^3	1860
FA density, kg/m^3	1350
Crushing machine power, watt	435
Sieving machine power, watt	250
Oven power, watt	1200
Grinding machine power, watt	750
Crushing machine capacity, m^3	0.08
Sieving machine capacity, m^3	0.05
Oven capacity, m^3	0.18
Grinding machine capacity, m^3	0.45
CO_2 emission for 1 L diesel, ton	0.0027
Energy consumption for 1 L diesel, GJ	0.0384
CO_2 emission for 1 kwh electricity, ton	0.00013
Energy consumption for 1 kwh electricity, GJ	0.0036
OPC cost in market including delivery, RM/tonne	600
OPC CO_2 emission, tonne/tonne	0.904
Energy consumption, GJ/tonne	5.13

presented in Table 10.7. The electricity cost for March 2020 was adopted in cost calculation depending on the power of machine by watts as presented in Table 10.8.

Calculation of the total greenhouse gas (GHG) emission, expressed as carbon dioxide equivalent (CO_2eq) per 1000 kg of cement produced, takes into account the collective contribution of CH_4, NO_x, So_x, and CO_2. This also includes the synthetic gases emitted during preparation of the materials, transport of raw materials

TABLE 10.8
Electricity Price According to Rate of Consumption (September 2018)

Tariff (Watt)	Unit (RM/kwh)
1–200	0.218
201–300	0.334
301–600	0.516
601–900	0.546
901+	0.571

and manufacturing in the lab such as crushing, sieving, drying, and grinding. The approach to estimate the total GHG is based on the methodology reported by McLellan et al. (2011) [13] and calculated using Equation 10.6:

$$\text{Total } CO_2 \text{ emission} = \sum_{i=1}^{n} mi\left(\left(di * ei\right) + pi\right) \quad (10.6)$$

where CO_2 emission is the total greenhouse gas emission ($kgCO_2eq$) per tonne of material produced, *mi* is the fraction of component *i*, *di* is the distance transported by a given mode of transport (km), *ei* is the emission factor for the transportation mode ($kgCO_2eq$/(km tonne)), and *pi* is the emissions per unit mass of component *i* produced ($kgCO_2eq$/tonne).

The total cost of the ternary-blended binder is calculated according to the distance transport, total diesel and electricity consumption of each material following the information provided in Tables 10.6, 10.7, and 10.8. Likewise, the total energy for each mixture was calculated depending on diesel and electricity consumption of each type of material including transport and lab treatments. The AAM binder's carbon dioxide emission, cost, and energy consumption are compared to the OPC of mixture achieving 30 MPa strength and required 460 kg/m^3 of OPC.

$$\begin{aligned}\text{Cost of alkali-activated binder} = &\left[\text{wt. WCP} \times \text{cost WCP}\right] \\ &+ \left[\text{wt.GBFS} \times \text{cost GBFS}\right] \\ &+ \left[\text{wt. FA} \times \text{cost FA}\right]\end{aligned} \quad (10.7)$$

10.3.2 Mix Design

The effects of FA replacing GBFS on durability performance of alkali-activated mortars containing a high amount of WCP were studied by preparing five mixtures (Table 10.9). For continuity, all mixtures had equal levels of sodium hydroxide (NH) molarity (4 M) with 50% WCP binder mass for each batch. Each mixture also had equal ratios of binder to fine aggregate (1.0), alkaline solution to binder (0.40), sodium

TABLE 10.9
Mixture Design of Ternary-Blended (WCP, GBFS, and FA) Alkali-Activated Mortars

	Binder (kg/m^3)				Alkaline Solution (kg/m^3)	
Mix	WCP	GBFS	FA	Sand (kg/m^3)	Na_2SiO_3	NaOH
$AAMs_1$	550	550	0	1100	188.6	251.4
$AAMs_2$	550	440	110	1100	188.6	251.4
$AAMs_3$	550	330	220	1100	188.6	251.4
$AAMs_4$	550	220	330	1100	188.6	251.4
$AAMs_5$	550	110	440	1100	188.6	251.4

TABLE 10.10
CO_2 Emission, Energy Consumption and Cost of WCP, GBFS, and FA

Materials	CO_2 Emission (t/t)	Energy (GJ/t)	Cost (RM/t)
WCP	0.045	1.12	170
FA	0.012	0.173	34.35
GBFS	0.152	2.379	449.74

silicate to sodium hydroxide (0.75), and modulus of solution (1.02). GBFS with a 50% content was chosen as the control batch and subsequent mixtures replaced 10, 20, 30 and 40% by mass ratio of GBFS with FA. Sodium hydroxide and sodium silicate were diligently mixed to create a weighed solution which was cooled to room temperature prior to using. In mixing the AAMs, fine aggregates were placed in the mortar mixer machine, before adding binders including FA, GBFS, and WCP. They were mixed in dry conditions for 2 minutes to create a homogeneous mixture before being activated with the alkaline solution. After 4 more minutes of mixing, the final mixture was poured in the moulds as per ASTM C579 [14]. The mixture was poured in two layers into each mould, allowing 15 seconds on the vibration table after each layer to release any air bubbles. Keeping in mind the Malaysian weather, the AAM specimens were then left at a temperature of 27 ± 1.5°C and relative humidity of 75%, for a period of 24 hours, to allow them to cure. Following that time, the specimens were opened and left in the same environment until the day of the test.

10.3.3 Carbon Dioxide Emissions

For the life cycle of WCP, GBFS, and FA, the total cost, CO_2 emission and energy consumption were calculated and are presented in Table 10.10. The results indicated that GBFS required electricity and diesel in preparation stages higher than WCP and FA which led to increasing the cost, carbon dioxide emission, and energy consumption. GBFS presented 2.37 GJ/tonne energy consumption compared to 1.12 and 0.17 GJ/tonne as observed for WCP and FA, respectively. This was also directly proportional to energy consumption, CO_2 emission and increased cost of GBFS and showed higher CO_2 emission (0.152 tonne/tonne) compared to WCP and FA, which were 0.045 and 0.012 tonne/tonne, respectively. Similar to carbon dioxide emission, the cost observed for GBFS was the highest among all the materials, mainly because of the long transport distance affecting an increase in cost of GBFS to 449 RM/tonne compared to 170 and 34 RM/tonne for WCP and FA, respectively. As observed from the results, the reduced content of GBFS in the alkali-activated matrix becomes necessary to achieve the sustainability requirements including lower carbon dioxide emission, cost and energy consumption.

The effects of FA replacing GBFS on carbon dioxide emission of ternary-blended AAMs containing WCP are depicted in Figure 10.5. The carbon dioxide emission tends to drop with the increase in FA content. The emission dropped from 108.3 to 92.9, 77.5, 62.1, and 46.8 kg/m^3 with increasing FA content from 0 to 10, 20, 30, and 40%, respectively. Compared to traditional binder (OPC), the new ternary binder including WCP, GBFS, and FA presented much lower carbon dioxide emission for all alkali-activated matrixes. Compared to 421 kg/m^3 carbon dioxide emission for OPC required to produce 1 m^3, a more than 75% reduction in carbon dioxide emission were achieved for all alkali-activated matrixes. The low carbon dioxide emission results of alkali-activated binder indicate that it is possible to make a simple sustainability comparison on the use of OPC and alkali-activated binder.

10.3.4 Cost-Effectiveness and Energy Efficiency

Figure 10.6 shows the effect of FA replacing GBFS on binder cost calculation of ternary-blended AAMs compared with traditional OPC binder. The use of FA in high volume (40%) as replacement of GBFS also resulted in cost savings. The cost of

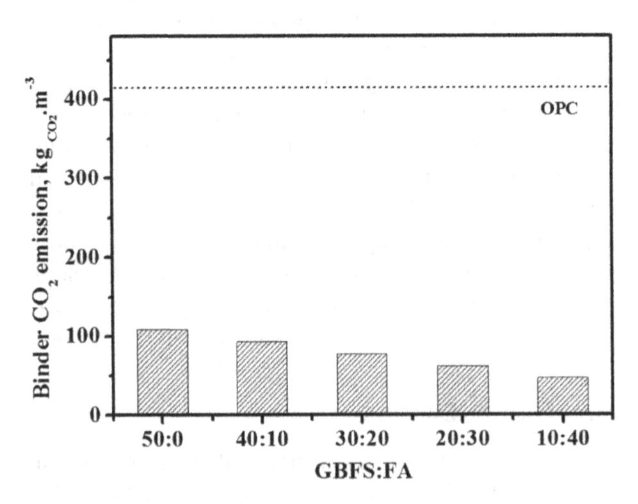

FIGURE 10.5 Carbon dioxide emission of prepared AAMs.

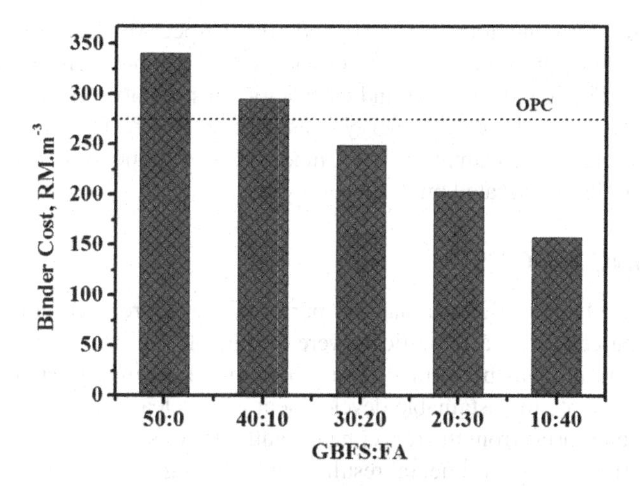

FIGURE 10.6 Effect of FA replacing GBFS on cost of prepared AAMs.

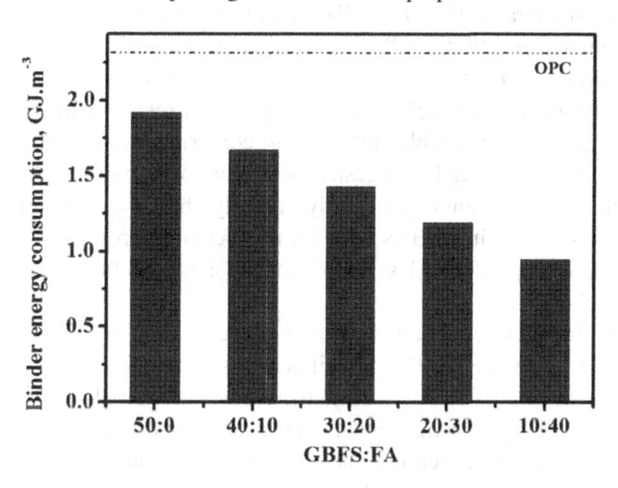

FIGURE 10.7 Energy consumption of prepared AAMs.

materials by weight was based on the life cycle (Table 10.7), which directly affected the final cost of ternary-blended AAM mixtures. It was observed that the binder cost tends to decrease from 340 to 295, 249, 203, and 157 RM/m^3 with an increase in FA content replacing GBFS from 0 to 10, 20, 30, and 40%, respectively. Compared to OPC cost (275 RM/m^3), only mixtures prepared with 20% and higher FA showed a lower cost than traditional binder (OPC). It can be clearly observed that using FA as a replacement for GBFS could contribute as sustainable binders.

Depending on life cycle and energy consumption of each material, the total energy consumption of each alkali-activated matrix was calculated (Figure 10.7). The energy consumption of alkali-activated binder decreased as the FA content increased as the replacement of GBFS in AAMs. Compared with 1.92 GJ/m^3 observed with 0% FA content, the energy consumption dropped to 1.67, 1.43, 1.19, and 0.95 GJ/m^3

with increased FA content to 10, 20, 30, and 40%, respectively. However, all alkali-activated mixtures presented much lower energy consumption compared with OPC (2.36 GJ/m^3). The low diesel cost and electricity consumption during the life cycle of FA directly affected the final energy consumption of AAMs. The low cost, CO_2 emission and energy consumption of FA made it the best option to achieve the sustainability of alkali-activated binder.

10.4 CONCLUSIONS

In the present chapter, the sustainability performance of prepared mortar and concrete utilizing ceramic waste particles were evaluated, and the environmental benefits of the mixture was performed. The utilization of ceramic waste particles was shown to contribute to sustainable development and a cleaner environment by producing a green mortar from the recycling of industrial wastes. Based on the detailed characterizations and experimental results, the following conclusions were drawn:

i. It was demonstrated that the utilization of ceramic waste could be considered environmentally friendly, as the reuse of ceramic waste could significantly reduce CO_2 emissions, save energy, reduce the total consumption of electricity, and reduce fuel consumption, thereby resulting in the worldwide availability of a sustainable and low-cost construction material.
ii. The results of this study suggested new compositions for AAMs that are cheaper and more environmentally friendly than the conventional OPC mortars. By replacing GBFS with FA in AAMs, the resulting product was more efficient, produced less CO_2, reduced costs, and consumed less fuel than OPC mortars.
iii. In order to decrease alkali-activated binder's cost by 42.9% (RM 157) compared to RM 275 with OPC binder, it is proposed to use 40% FA as a replacement of GBFS. This will make the new sustainable product a more suitable alternative to traditional OPC. Additionally, CO_2 emissions will be decreased as the potentially new composition requires less than half the fuel for production than the OPC mortars.
iv. Aside from the positive impacts on the environment, the prepared AAMs also offered a superior product in terms of mechanical and durability properties; something that is likely of interest to many mortar and concrete manufacturers. This suggested substitute for OPC mortars and concrete has a widespread suitability and may also serve to fulfil sustainability aims for companies in the business of sustainable construction.

REFERENCES

1. Mastrucci, A., et al., Life cycle assessment of building stocks from urban to transnational scales: A review. *Renewable and Sustainable Energy Reviews*, 2017. **74**: pp. 316–332.
2. Struble, L. and J. Godfrey. How sustainable is concrete. In International Workshop on Sustainable Development and Concrete Technology, University of Illinois at Urbana-Champaign, USA. 2004: pp. 201–211.

3. Bilodeau, A. and V. M. Malhotra. High-volume fly ash system: The concrete solution for sustainable development. In CANMET/ACI. Séminaire International, 2000. ACI.
4. Huseien, G. F., et al., Geopolymer mortars as sustainable repair material: A comprehensive review. *Renewable and Sustainable Energy Reviews*, 2017. **80**: pp. 54–74.
5. Bashar, I. I., et al., Development of sustainable geopolymer mortar using industrial waste materials. *Materials Today: Proceedings*, 2016. **3**(2): pp. 125–129.
6. Hardjito, D. and B. V. Rangan, Development and properties of low-calcium fly ash-based geopolymer concrete. Curtin University of Technology, Perth, Australia, 2005.
7. Ariffin, M., et al., Effect of ceramic aggregate on high strength multi blended ash geopolymer mortar. *Jurnal Teknologi*, 2015. **77**(16): pp. 33–36.
8. Akbarnezhad, A., et al., Recycling of geopolymer concrete. *Construction and Building Materials*, 2015. **101**: pp. 152–158.
9. Zhu, D. Y., M. Z. Rong, and M. Q. Zhang, Self-healing polymeric materials based on microencapsulated healing agents: From design to preparation. *Progress in Polymer Science*, 2015. **49**: pp. 175–220.
10. He, Z., et al., Facile and cost-effective synthesis of isocyanate microcapsules via polyvinyl alcohol-mediated interfacial polymerization and their application in self-healing materials. *Composites Science and Technology*, 2017. **138**: pp. 15–23.
11. Li, V. C. and E. Herbert, Robust self-healing concrete for sustainable infrastructure. *Journal of Advanced Concrete Technology*, 2012. **10**(6): pp. 207–218.
12. Van Belleghem, B., et al., Quantification of the service life extension and environmental benefit of chloride exposed self-healing concrete. *Materials*, 2016. **10**(1): p. 5.
13. McLellan, B. C., R. P. Williams, J. Lay, A. Van Riessen, G. D. Corder, Costs and carbon emissions for geopolymer pastes in comparison to ordinary portland cement. *Journal of Cleaner Production*, 2011. 19(9–10): pp. 1080–1090.
14. ASTM, C579 Standard test methods for compressive strength of chemical-resistant mortars. In *Grouts, Monolithic Surfacings, and Polymer Concretes*. 2001: ASTM.

11 Alkali-Activated Mortars Containing Ceramic Waste as Repair Material

11.1 INTRODUCTION

Over the years, ordinary Portland cement (OPC) has been extensively used to bind concrete effective for various construction purposes. Several studies [1–4] reported that the concrete shows low durability to the aggressive environments and leads to much deterioration during the life cycle. Wang et al. [5] reported that the restoration and rehabilitation costs which were close to or even exceeded new construction cost. The surfaces of the concrete structures such as sidewalks, pavements, parking decks, bridges, runways, canals, dykes, dams, and spillways progressively deteriorate due to the variety of the physical, chemical, thermal, and biological processes. Actually, the performance of the concrete compositions can greatly be affected by the improper usage of the substances, and physical and chemical conditioning of the environment [2, 6–8]. The immediate consequence is the anticipated need for maintenance and execution of repairs [9, 10]. To overcome these issues, the researchers made dedicated efforts to develop different types of repair materials, with or without OPC, such as emulsified epoxy mortars, sand–epoxy mortars and polymer-modified cement-based mortars. The main aim is to attain the efficient and durable repair materials for the damaged concrete structures. However, the variation in the results obtained by different researchers can be attributed to the difference in the raw materials, specimen geometry, and test methods. For construction purposes, alkali-activated mortars (AAMs) are the newly introduced binders with much higher resistance against severe climates [11, 12]. In the past, intensive efforts have been made to get binders with high performance as sustainable construction materials [13–17].

Recently, there has been an active development in the areas of the alkali-activated pastes, mortars and concrete, which produce substantially lower amounts of carbon dioxide than using OPC, thus much better for the environment. Rather than the high levels of CO_2 created in the production of OPC, the alkaline method has been adopted to make concrete using geopolymers [18–20]. These binders can be produced using wastes like palm oil fuel ash (POFA), metakaolin, fly ash (FA), ceramic tile waste, glass waste, and ground blast furnace slag (GBFS). This development in creating an alternative to OPC has received the attention of professionals in the commercial and academic sectors [21, 22]. The discussions have not only focused on the relatively similar functionality to OPC, but also on the natural features of the materials such as their excellent strength performance, high resistance to sulphuric acid and sulphate attack, ability to withstand heat, the reduced energy consumption, and low level of CO_2 emission [23–26]. Studies on various alumina-silicate materials and their responses as well as microstructural classification activated with various alkaline activators have also been focused on [27–30]. Although, research for the production of ecological pastes/mortars has stepped up in recent times [25, 31], no studies have focused on the development of the ecological mortar mix useful for repair purposes.

Many studies have indicated that the presence of calcium in the FA can significantly affect the resultant hardening characteristics and compressive strength of AAMs [32–34]. Calcium oxide (CaO) generates calcium silicate hydrate (C-S-H) along with the sodium aluminium silicate hydrate (N-A-S-H) gel [35]. The compatible nature of calcium (aluminium) silicate hydrate (C-(A)-S-H) and N-A-S-H gels has a significant influence on the hybrid OPC and alkaline solution-activated aluminium silicate (A-S) systems, generating both products [36, 37]. Earlier studies have used synthetic gels to determine the influence of high pH levels on each gel component. The aqueous aluminate was found to greatly affect C-S-H product formation [38, 39]. Also, the aqueous Ca was shown to modify the N-A-S-H gels and partially replace the sodium (Na) with the Ca to produce (N,C)-A-S-H gels [38, 40]. However, the mechanisms for the formation of such gels and subsequent improvements have not yet been completely understood. To explore the feasibility of achieving green cements for construction purposes, both gels must coexist. Thus, systematic studies on the compatible nature of N-A-S-H and C-A-S-H gels became essential. Some studies have been performed on the materials containing the calcium compounds especially, GBFS [41, 42]. However, most of the earlier research used a high volume of corrosive sodium silicate (NS) and/or sodium hydroxide (NH) to produce the AAMs, which posed health and safety issues to workers during handling [43]. One investigation [44] proposed a simple approach to produce environmentally friendly alkali-activated mortar with improved mechanical properties by overcoming the thermally activated processes and promoting easy management.

The merits of AAMs have made them suitable and sustainable repair material that can be used in different forms such as paste for crack injection, and mortar for section restoration and patch repair [9, 45–48]. Several reports [21, 32, 49] have suggested the utilization of the FA alkali-activated mortar incorporating GBFS as ordinary concrete repairing materials. Alkali-activated mortars were shown to achieve higher shear bond and bending strength than commercial cement-based repairing mortar. The results suggested that it could be a favourable alternative product for the concrete repair work. The term *compatibility* appeared as a popular buzzword in the repair industry, implying the durability of the repairs in general and adequate load-carrying capacity in the case of concrete repairs [50]. Compatibility can be defined as a balance of the physical, chemical, and electrochemical properties and dimensions of a repair material. The existing damaged substrates ensure that the repair can withstand all the stresses induced by the volume changes and chemicals as well as the electrochemical effects without distress and deterioration over a designated period of time. In concrete repair work, the bond compatibility between the repair materials and concrete substrate is one of the important factors that affect the durability and sustainability of the repair work [51, 52]. The coefficient of the thermal expansion is a measure of the length change in a material when it is subjected to a temperature variation. When two materials (repair material and substrate) of different thermal expansion coefficients are joined together and subjected to significant temperature changes, stresses are generated in the composite material. A comprehensive literature survey revealed that no studies have so far been undertaken to evaluate in detail the mechanical properties of such AAMs, especially for the repair of the damaged concrete surfaces. The assessment of the compatibility between the alkali-activated mortars as repair materials and deteriorated concrete substrate has been deficient. Most of the earlier studies have only analyzed the mineralogy and microstructure properties of the AAMs.

Driven by this idea, the present chapter examines the feasibility of achieving a new type of high-performance AAM with the improved mechanical and durable properties to repair damaged concrete surfaces. These newly designed high-performance AAMs were prepared using waste ceramic powder (WCP), GBFS, and FA as the industrial wastes in appropriate proportions. The effects of different ratios of the high-volume WCP replaced by GBFS and FA and activated with low molarity of alkaline activator solution were investigated to determine the flowability and setting time, compressive and bond strength, porosity, resistance to abrasion–erosion, and freeze–thaw cycling as well as thermal compatibility with base concrete of the synthesized AAMs. Furthermore, several tests such as the slant shear bond strength, thermal expansion coefficient and four-point load flexural measurement were conducted to evaluate the bond strength between the AAMs and mortar/concrete substrates.

11.2 MATERIAL CHARACTERIZATIONS

The ceramic wastes was collected from White Horse ceramic manufacturer in Pasir Gudang Johor, Malaysia. During the collection stage, only homogeneous ceramic tile waste was collected (the ceramic tile was of the same thickness with no glassy coating). The ceramic waste was first crushed using a crushing machine then sieved through 600 μm mesh to isolate the large particles. The sieved ceramic was again ground for 6 hours using a Los Angeles abrasion machine to achieve the required particle size according to ASTM 618 of 66% passing 45 μm [53]. Next, the powder was collected and used in the mixing process. Pure GBFS (procured from Ipoh, Malaysia) was used (without any further purification) as a constituent to make the binder free of cement. This slag (off-white in appearance) was distinct from the other auxiliary cementitious substances because it had both cementitious and pozzolanic traits. The GBFS emerged from the hydraulic chemical reaction when water was mixed in. From the X-ray fluorescence (XRF) results, the main oxide compositions were silica and aluminium totalling 84.8% in WCP. The XRF spectra revealed that the GBFS is composed of calcium (51.8%), silicate (30.8%), and alumina (10.9%). The FA (alumina-silicate substance) with a low level of calcium (collected from Tanjung bin power station, Malaysia) was used to make the AAMs. It satisfied ASTM C618 requirements for class F fly ash and appeared grey in colour with 5.2% of calcium,

57.2% of silicate, and 28.8% of aluminium content. The median particle size for the WCP, FA, and GBFS (obtained using the particle size analyzer) was 35, 10, and 12.8 μm, respectively. The Brunauer–Emmett–Teller (BET) surface area of FA and GBFS were calculated to analyze their compressive strength. The specific surface area of WCP and GBFS displayed the value of 12.2 and 13.6 m^2/g, respectively, compared to FA (18.1 m^2/g).

Figure 11.1 shows the XRD patterns of FA and GBFS. The FA revealed pronounced diffraction peaks around $2\theta = 16–30°$, which were allocated to the presence of crystalline silica and alumina compounds. Nonetheless, the occurrences of the other sharp peaks were assigned to the presence of the crystalline phases of the quartz and mullite. As reported by several studies [54–56], the amorphous phases of both WCP and FA play a significant role on the hydration process and gel formulation. The absence of any sharp peak in the XRD pattern of GBFS indeed confirmed their disorder nature. The occurrences of the silica and calcium peak in the pattern were the significant factors towards the GBFS creation. The presence of a high amount of reacting amorphous Si and Ca in the GBFS was advantageous for the AAM synthesis. However, inclusion of the FA was needed to surmount the low level of Al_2O_3 (10.49 wt%) in the GBFS. The scanning electron microscopy (SEM) images of FA and GBFS exhibited that FA consisted of spherical particles with a smooth surface and GBFS was comprised of irregular and angular particles, similar to the one reported earlier [57].

Saturated surface natural sand (siliceous) from the river was utilized as the fine aggregate to prepare the proposed mortars. It was first cleaned using water to lessen the presence of the silts and impurities as per the ASTM C117 standard [58]. This was followed by oven drying at 60°C for 24 hours to remove the moisture before being

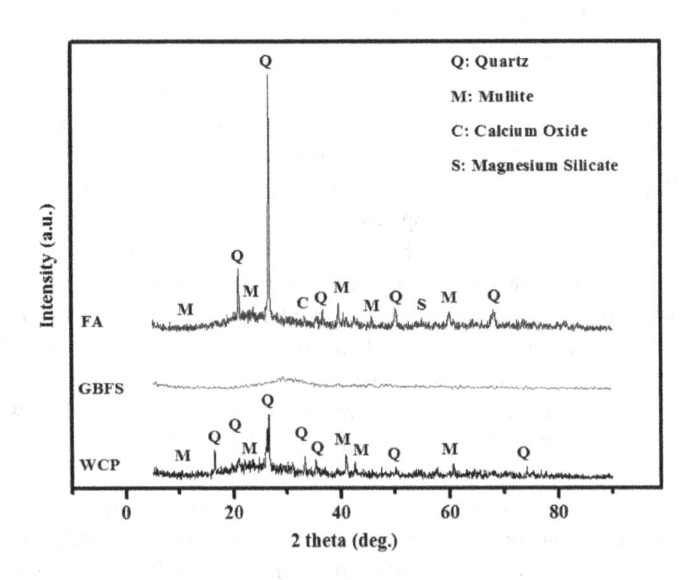

FIGURE 11.1 The XRD patterns of WCP, GBFS, and FA.

graded to match with the ASTM C33-33M specifications [59]. The fineness modulus and specific gravity of the prepared aggregate were 2.9 and 2.6, respectively.

Analytical-grade NH (98% purity) in the form of pellets were dissolved in water to prepare the solution of concentration 4 M (7.4 Na_2O and 92.6% H_2O). An analytical-grade NS blend made of SiO_2 (29.5 wt. %), Na_2O (14.70 wt.%), and H_2O (55.80 wt.%) was utilized. The prepared 2 M NH solution was kept for 24 hours to cool to room temperature. It was then mixed with the NS solution to achieve the ultimate alkaline mixture having a SiO_2:Na_2O ratio of 1.2, wherein the proportion of NS:NH for all alkaline mixtures was maintained at 0.75. The total Na_2O, SiO_2, and H_2O was 10.53, 12.64, and 76.8 (by weight, %), respectively, compared to 20.75, 21.07, and 58.2 of solution prepared for the 14 M of NaOH and a 2.5 of Na_2SiO_3:NaOH ratio as recommended in previous studies (optimum ratios) [54, 60]. Regarding the content of Na_2O, SiO_2, and H_2O, the prepared alkaline solution is environmentally amiable, cheap, consumes less energy, and emits less carbon dioxide.

11.3 DESIGN OF AAM MIXES

For all the AAM mixtures used as repair materials, the ratios of the alkaline solution to binder (S:B) and the binder to fine aggregate (B:A) were 0.40 and 1.0, respectively. As no standard exists for the preparation of AAMs, the ratio of the binder to fine aggregate was selected depending on the trial mixtures which achieved the highest compressive strength at 28 days of age [14, 61]. Two types of industrial waste material (FA and GBFS) were incorporated with WCP to prepare the AAM mix design. The WCP, FA, and GBFS were blended to determine the influence of calcium oxide on the geopolymerization process. The WCP levels were fixed to 50% to 70%. GBFS content was kept in the range of 20% to 50% as the source of CaO. The FA content was kept in the range of 10–30%. The binary blend containing the high-volume WCP (70%) and GBFS (30%) was considered the control sample. Likewise, the content of GBFS was kept to 50% or less because the high volume of GBFS may affect the setting time, reducing it less than 5 minutes. So, the blend containing 70% WCP was prepared as the control sample (Table 11.1). The NH molarity, NS:NH, and modulus of the alkaline solution SiO_2:Na_2O (Ms) were kept constant for all the mixes. Nine substitution levels were implemented to assess the impact of CaO on the geopolymerization process. The CaO content was improved with the increase of GBFS. Conversely, both the SiO_2 and Al_2O_3 content were reduced with the increase in GBFS level depending on the chemical composition (Figure 11.2). The modulus of the alkaline activator solution was calculated using Equation 11.1:

$$Ms = \left[\frac{v}{1+v} \times s\right] / \left[\left(\frac{1}{1+v} \times n1\right) + \left(\frac{v}{1+v} \times n2\right)\right] \tag{11.1}$$

where Ms is the alkaline-activator solution modulus (SiO_2:Na_2O), v is the ratio of the sodium silicate to sodium hydroxide, s is the SiO_2 content from the sodium silicate, $n1$ is the Na_2O content from the sodium hydroxide, and $n2$ is the Na_2O content from the sodium silicate.

TABLE 11.1
Mix Design of the Proposed AAMs (mass%)

	Binder (Mass %)					
No.	WCP	FA	GBFS	Si:Al	Ca:Si	Ca:Al
1	70	0	30	5.09	0.26	1.31
2		10	20	4.62	0.17	0.79
3	60	0	40	4.79	0.37	1.77
4		10	30	4.35	0.27	1.19
5		20	20	4.01	0.18	0.74
6	50	0	50	4.48	0.50	2.24
7		10	40	4.08	0.39	1.59
8		20	30	3.77	0.29	1.09
9		30	20	3.53	0.20	0.70

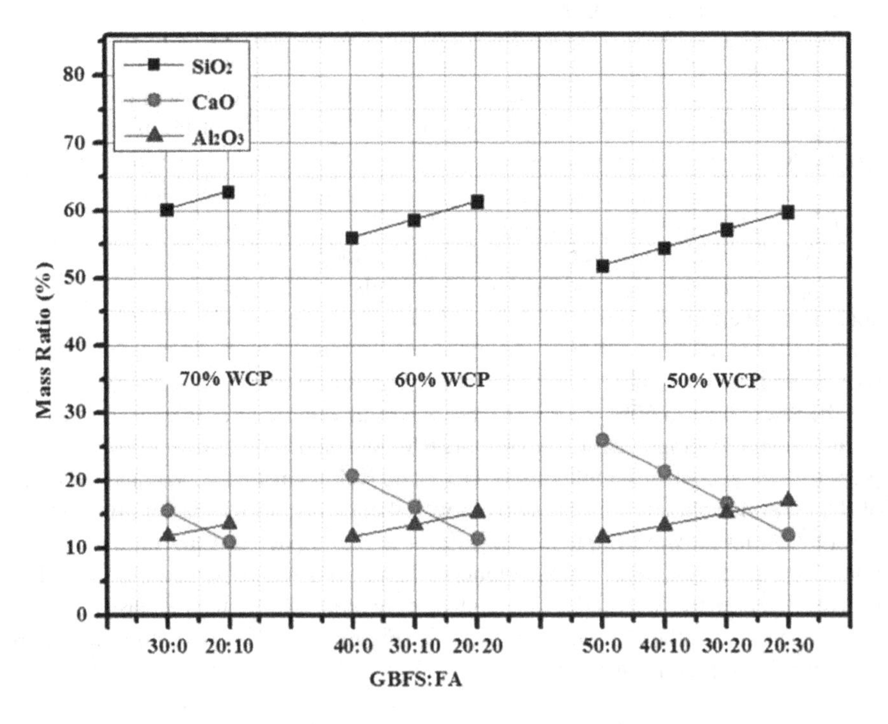

FIGURE 11.2 Effect of GBFS and FA replacement WCP on SiO_2, CaO, and Al_2O_3 content.

The OPC was procured from the Holcim Cement Manufacturing Company following the specified ASTM C150 standard. The OPC was utilized to prepare the high-strength concrete substrates (≥40 MPa) to show the ability of using alkali-activated mortar as high-performance repair materials, where the ratio of the trial mixture of the cement to sand to gravel was 1:1.5:3.0 and the ratio of water to cement (W:C) was

0.48. After 28 days the compressive, splitting tensile and flexural strengths, modulus of elasticity, and porosity of normal concrete substrate (NC) were 43.6 MPa, 4.4 MPa, 5.6 MPa, 28.2 GPa, and 10.2%, respectively.

Following ASTM C109/C109M-16a [62], cube moulds (50 × 50 × 50 mm) were prepared for the compressive strength test. Prism specimens (40 × 40 × 160 mm) were prepared for the flexural strength (FS) test. For the tensile strength (TS) and the slant shear (SS) tests, the cylindrical samples (diameter = 75 mm and depth = 150 mm) were casted. For the shrinkage test, prism specimens of 25 × 25 × 250 mm were prepared. Prior to the casting, the engine oil was applied on the inner surfaces of moulds to make the de-moulding job easier. A homogeneous mixture of NH and NS (by weight) was prepared followed by cooling to room temperature. Next, the AAMs (homogeneous mix of fine aggregate) were synthesized by mixing (via mortar mixer) WCP, FA, and GBFS for 2 minutes under the dry state. The obtained mix was further activated by incorporating the alkali solution. The entire matrix was blended for another 4 minutes using the machine operated at average speed. The prepared fresh mortar was cast in the moulds in two layers, where each layer was consolidated using the vibration table for 15 seconds to eliminate the air voids [63]. Keeping in mind the Singapore weather, the AAM specimens were then left at a temperature of 27 ± 1.5°C and relative humidity of 75% for a period of 24 hours to allow them to cure. Then, the specimens were removed from the moulds and left in the same condition till testing.

11.4 TEST PROCEDURES

11.4.1 Fresh and Strength Tests

The workability of the mortars including the flow diameter, and initial and final setting time were determined by the flow table procedure with modification from ASTM C230/C230M–14 [64] and ASTM C191 [65], respectively. Generally, it is known that the repair materials must possess similar or better strength features than the substrate concrete [2]. For the repair mortar, the ASTM C 109 standard practice was utilized to determine the compressive strength. Samples were tested after 1, 3, 7, 28, 56, and 90 days of curing. The tensile strength of the substrate mortars and the repair component was evaluated using the cylinders of 75 × 150 mm according to the ASTM C496 testing method. The TS of the repair component was evaluated at the curing age of 1, 3, 7, and 28 days. Meanwhile, the tensile strength of the concrete was assessed at the curing age of 28 days. Extra cylinders of the substrate mortar were examined for their tensile strength alongside the slant shear tests performed on the composite cylinders to determine the bond strength of the repair substance. The flexural strength, modulus of the rupture, bend strength and fracture strength are a measure of the mechanical character of the brittle materials. This characterizes the material's ability to resist the deformation under the applied load. The flexural strength test was carried out using the ASTM C78/C78M procedure, where an adequately cured (at age 1, 3, 7, and 28 days) prism specimen of 40 × 40 × 160 mm was used. Three sets of specimens were tested for each of the curing age and their average is reported. The flexural strength was calculated using:

$$\sigma = 3FL/2bd^2 \quad (11.2)$$

where F is the load (force) at the fracture point (N), L is the length of the support span (mm), b is the width (mm), and d is the thickness (mm) of the specimen.

The modulus of elasticity (MOE) test was performed on the sufficiently cured (at age of 28 days) cylindrical specimens of 100 × 200 m following the method documented in ASTM C469/C469M, 2010.

11.4.2 Porosity Test

The porosity of the repair materials has a significant effect on the durability. Generally, the dense, impermeable, highly resistive, or non-conductive repair materials reveal the tendency where the repaired damaged area appears isolated from the adjacent undamaged areas. Consequently, the patched area of concrete shows a large difference in porosity or chloride content from the rest, causing corrosion to remain localized in a limited region. The steel decay rate could accelerate, leading to the early failure either in the scrap or the adjacent concrete. Thus, it is important to make sure that both the concrete substrate and repair component have comparable porosity or density during the selection of a repair material. In this view, the porosity test was performed on the sufficiently cured (at 28 days) cubic (50 × 50 × 50 mm) samples according to the ASTM C642 standard.

11.4.3 Abrasion Resistance Test

The abrasion resistance (AR) test of all AAMs was conducted at the curing ages of 1, 3, 7 and 28 days under dry conditions following the specified Indian Standard (IS 1237-1980) whereby each specimen was weighed correctly by a digital balance. After the initial drying and weighing, the thickness of the specimens was measured at four different points. The grinding path of the abrasion disc testing machine was evenly dispersed with 20 grams of the abrasive powder (sand powder). The AAMs were kept in the holding device of the abrasion machine wherein a load of 300 N was subjected to the specimen. Next, the grinding machine was revolved at 30 rpm. The abrasive powder was constantly fed into the grinding trail to maintain a uniform track distribution related to the test specimen's width. Then, every specimen was abraded for 60 minutes from all the sides and the reading was recorded after each 15 minutes. After the abrasion test, the AAMs were weighed again to determine the weight loss. The thickness of each specimen was also recorded at four points. The degree of the abrasion was estimated from the difference in the measured thickness before and after the testing.

11.4.4 Slant Shear Bond Strength Test

The bond strength between the AAMs and concrete substrate was evaluated via the SS bonding test, where the hardened normal concrete (NC) was diagonally slanted at 30° inclinations from the vertical. According to the ASTM C882, the recommended bond angle of 30° represents the failure stress corresponding to a smooth surface closer to the minimum stress. The concrete substrates were prepared using the aged slant shear concrete cylinders by cutting in half at the 30° and 45° line to the vertical. The saw cut surface of the concrete was used as it was shown to be suitable for the shear bond strength assessment. Half-slanted NC was placed into the cylinder mould, and fresh AAM was poured into the mould. This test was conducted using a compression machine on the specimens after 1, 3, 7, and 28 days of curing. Figure 11.3 displays the step-by-step procedure of the slant shear test.

11.4.5 Freeze–Thaw Cycling Resistance Test

The freeze–thaw cycling resistance test was carried out using the prism-shaped AAM specimens of length 120 mm and cross-sectional area of 40 ×40 × 80 mm following the ASTM C666 standard (–17 to 5°C). Another cubical specimen of 50 × 50 × 50 mm was also tested at the curing age of 28 days. Method A was employed because it was more convenient than method B. After curing for 28 days, all the AAMs were subjected to thawing at room temperature to obtain the ultrasonic pulse velocity (UPV) and the mass. Later, these AAMs were put in a container and submerged inside water, where the water temperature was controlled automatically by the timer to attain every freeze–thaw cycle (a total of 300 cycles). The effectiveness of the AAMs was evaluated depending on the qualitative examination and residual compressive strength development. The variation in the length and dynamic modulus of the number AAMs were monitored after every 50th freeze–thaw cycle for a total of 300 cycles. The durability feature and change in the ultimate length were estimated at the end of freeze–thaw cycling using the relation

$$Df = \frac{PN}{M} \tag{11.3}$$

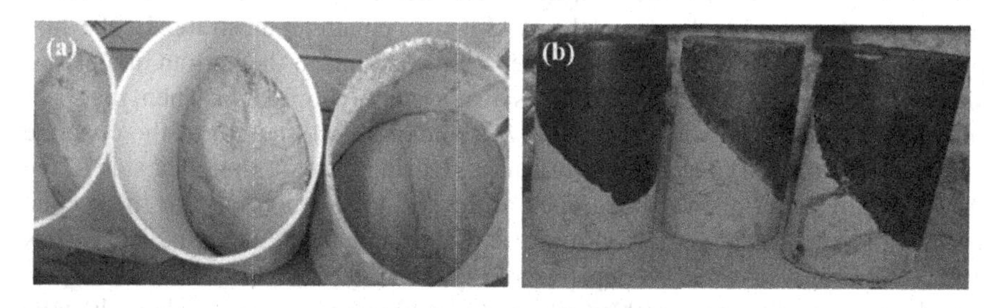

FIGURE 11.3 The preparation steps of the slant shear specimens.

where *Df* is the durability factor of the test specimen, *P* is the relative dynamic modulus of elasticity at *N* cycles, *N* is number cycles at which *P* reached the specified minimum value for discontinuing the test, and *M* is the specified number of cycles at which the exposure was terminated.

11.5 COMPATIBILITY BETWEEN AAM AND CONCRETE SUBSTRATE

11.5.1 Coefficient of Thermal Expansion

The coefficient of thermal expansion (CTE) of the AAMs was measured from the alteration of the length after subjecting them to temperature variation. After connecting the AAM and substrate composite of dissimilar CTE together and subjecting to appreciable temperature variation, stress was produced in the composite. The resultant stress could cause a failure at the composite's interface or within the material of the lower strength. The AAM must have a CTE like the substrate concrete to circumvent the failure at the elevated fluctuating temperature. The CTE is an important attribute of the repairing component subjected to the varied temperature. The thermal consistency test of all the AAMs (acting as repairing material) with the base concrete was carried out by modifying ASTM C884 (intended for thermal compatibility between concrete and an epoxy resin overlay). A concrete block 80 × 100 × 200 mm served as the base for testing. These concrete slabs could sustain ten freeze–thaw cycles needed by the ASTM C666 test method (intended for the concrete resistance evaluation to rapid freeze–thaw cycles called Method A). Various repairing mortars were implemented on the substrate concrete blocks with thickness ranging from 10 to 12 mm with the curing age of 28 days. These AAM-embedded concrete slabs were then subjected to the ten freeze–thaw cycles, wherein the temperature of each cycle varied from 5 to –20°C. At the end of each cycle, each specimen was tested qualitatively for de-bonding together with visual inspection to trace any sign of cracking, scaling, or bond breakage between the base concrete and the repairing mortar [2].

11.5.2 Four-Point Loading Flexural Test

The four-point loading flexural test was performed using two procedures to determine the compatibility between the AAMs and concrete substrate. In the first procedure, the AAM was applied to a depression created at the bottom of a prism-shaped concrete (250 × 100 × 10 mm) as displayed in Figure 11.4a. Then, the specimen was tested identical to ASTM C 78 standard. The strength of the concrete substrate was 43.6 MPa at 28 days of curing. During the test, the filled side of the AAM (repair material) was positioned at the specimen's bottom. For the second procedure, the vertical shear bonding strength between the AAM and NC (100 × 100 × 30 mm) was tested as depicted in Figure 11.4b. The specimens were prepared by casting the prism (100 × 100 × 500 mm) concrete with grade of 40 MPa (C40). Next, these samples were cut from the middle with the required length,

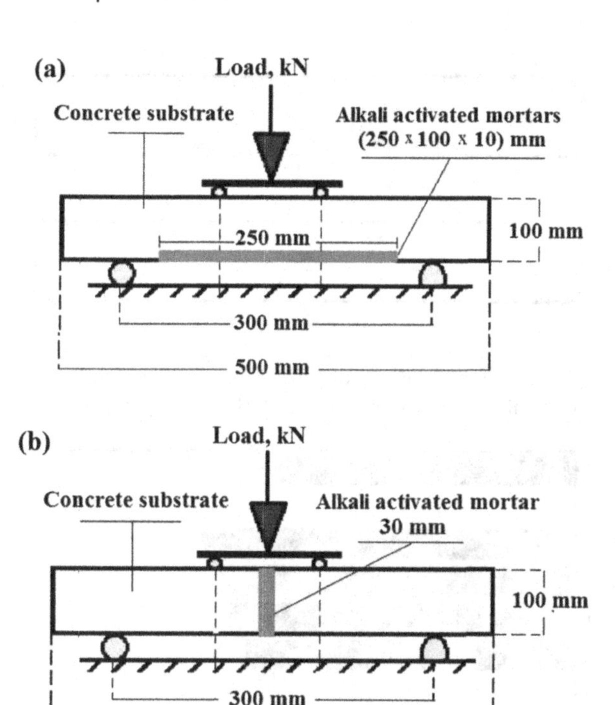

FIGURE 11.4 Diagram displaying the beam compatibility.

width and depth (Figure 11.3) before being placed in the prism moulds to pour the fresh AAMs. They were tested after 28 days of curing using the third point load at the loading rate of 0.2 kN/s. The compatibility or incompatibility of the repair materials with the concrete substrate was assessed via their failure mode. The AAM was declared compatible when the failure passed through the repair material and concrete substrate at the middle third of the beam. Otherwise, the AAM was proven as incompatible with the substrate concrete. The FS of the concrete substrate was analyzed at 58 days of curing equivalent to the repair materials' 28 days of curing testing. The composite beams were tested under simply supported conditions with the two-point loads applied on the top of the beam as shown in Figure 11.3a. The mid-span deflection of the composite beams was measured using a linear variable displacement transducer (LVDT). Figure 11.5 illustrates the failure zone types in the three-point loading and bending stress of the composite beam. The compatibility between AAMs and NC, the coefficient of thermal expansion, four points loading flexure, and bending stress results were compared for cement mortar (OPC) as a reference (cement-to-sand ratio of 1:3, water-to-cement ratio of 0.48, and 34.1 MPa compressive strength at 28 days selected from the trial mixtures) throughout this study.

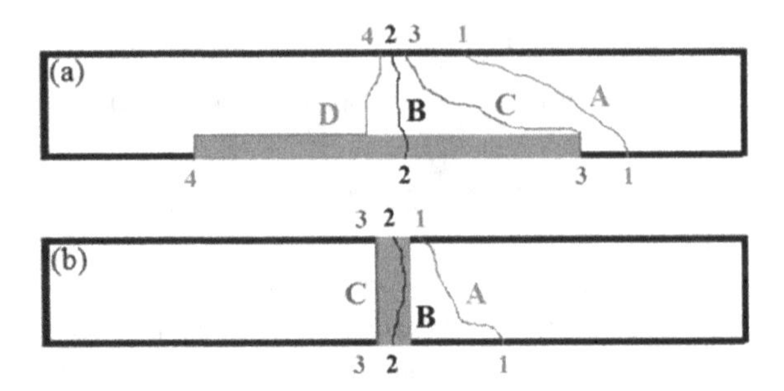

FIGURE 11.5 The compatibility evaluation according to the failure mode: A, B – compatibility; C, D – incompatibility. (a) Composite beam, (b) bonding beam.

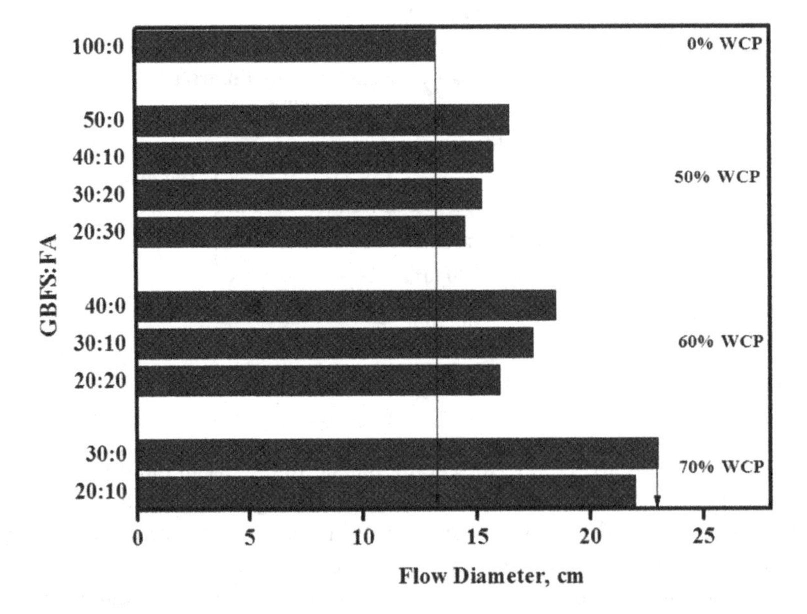

FIGURE 11.6 Flow of AAMs under the effect of high WCP content.

11.6 WORKABILITY PERFORMANCE

Figure 11.6 displays the results of the flow test of the synthesized AAMs with flow values of 16.5, 18.5, and 23 cm when the ceramic content increased from 50 to 60 to 70%, respectively, compared to control sample (AAM_1) which presented a 13.2 cm flow diameter. At each level, when WCP content was fixed and GBFS was replaced by FA, the workability of mortar was reduced. Typically, the flow value decreased with the increase of GBFS and FA content in the mixture. The lower value of flow diameter of ternary-blended AAMs was recorded with the mix AAM_5. The low specific surface area and high particle size distribution of WCP compared to GBFS and FA was attributed to the AAMs' flowability enhancement. Furthermore, it became clear that an increase in FA content decreased the workability of AAMs due to high water adsorption of FA with a porous structure [25, 66, 67] as depicted in Figure 11.2. The other reason to enhance the workability of mortars is to increase the WCP content and reduce the GBFS amount by replacing it with WCP and FA. This affects the rate of chemical reaction [68, 69] and increases the plasticity of the mixture which improves the workability of AAMs.

Figure 11.7 shows the high-volume WCP content-dependent variation in the setting time of AAMs. Mortars with the highest WCP content took the longest time (92 minutes) to set. A sharp decrease in the setting time was observed when the WCP content was reduced and the GBFS content was increased. The effect of FA on the initial and final setting time was clearly observed at each level of high volume WCP. An increase in the FA content significantly influenced the setting time. Both the initial and final setting times increased with the increased level of FA replacing GBFS, i.e. a higher setting time (more than 90 minutes) was recorded with AAM_{10} compared to 8 minutes obtained with control sample (AAM_1). As SiO_2 content increased (64%) and the amount of CaO (10%) decreased in the AAM mixture containing 70% WCP, 20% GBFS, and 10% FA, a delay in the rate of chemical reaction and an increase in setting time were observed. The difference between initial and final setting time also increased with the reduction in GBFS content and increased content of both WCP and FA in the mortar. It also supported the fact that the higher the GBFS content in the mortar, the faster is the setting rate [66, 70]. These results established that WCP and FA, as a part of the ternary-blended binder, was effective in decelerating the setting time of AAMs at ambient condition.

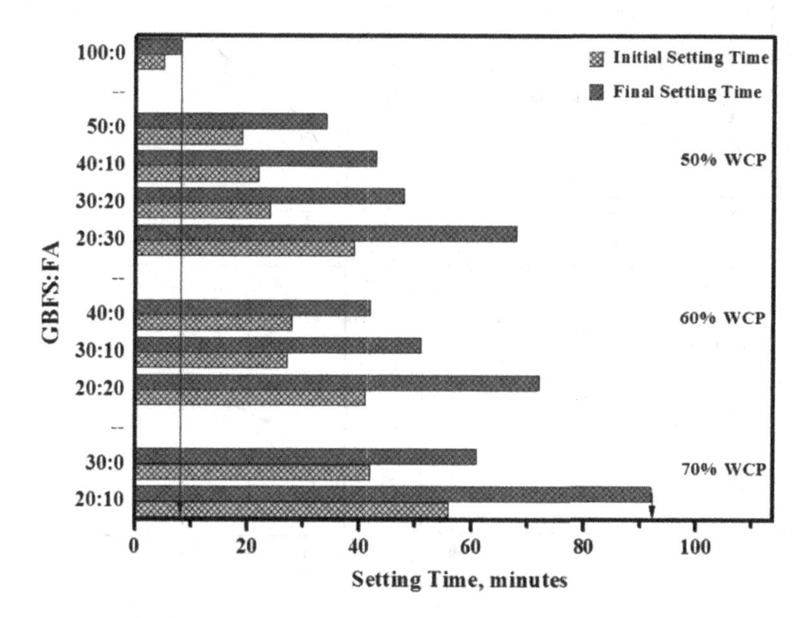

FIGURE 11.7 Effect of GBFS:FA on the setting time of AAMs containing high-volume WCP.

11.7 STRENGTH PERFORMANCE

11.7.1 Compressive Strength

Figure 11.8 shows high-volume content WCP-dependent compressive strength development of alkaliactivated mortars (AAMs). The compressive strength of AAMs was found to vary inversely with the level of increasing WCP content from 50 to 70%, where the strength dropped from 70.1 to 34 MPa at 28 days, respectively. The negative effect of a low content of calcium and a high content of silica was attributed to this drop. This led to the production of fewer C-S-H gels and hence the reduced AAM strength. WCP possesses higher than 70% silica with large particle size (35 μm) compared to GBFS. This influenced the strength development of AAM specimens prepared with high-content WCP. Furthermore, the specimens prepared with high-volume ceramic content achieved 81, 94, and 97% strength at after 28, 56, and 90, respectively days. The influence of high-volume WCP by the percentage of FA replacement of GBFS on the compressive strength of AAMs are depicted in Figure 11.8b–d. An increase in the content of FA increased the amount of silica and Al, thus negatively affecting the calcium content. Subsequently, the reduction in the calcium content from 30 to 10% led to a weakening of the compressive strength from 74.1 to 56.5 MPa, respectively. The occurrence of a low amount of C-S-H product

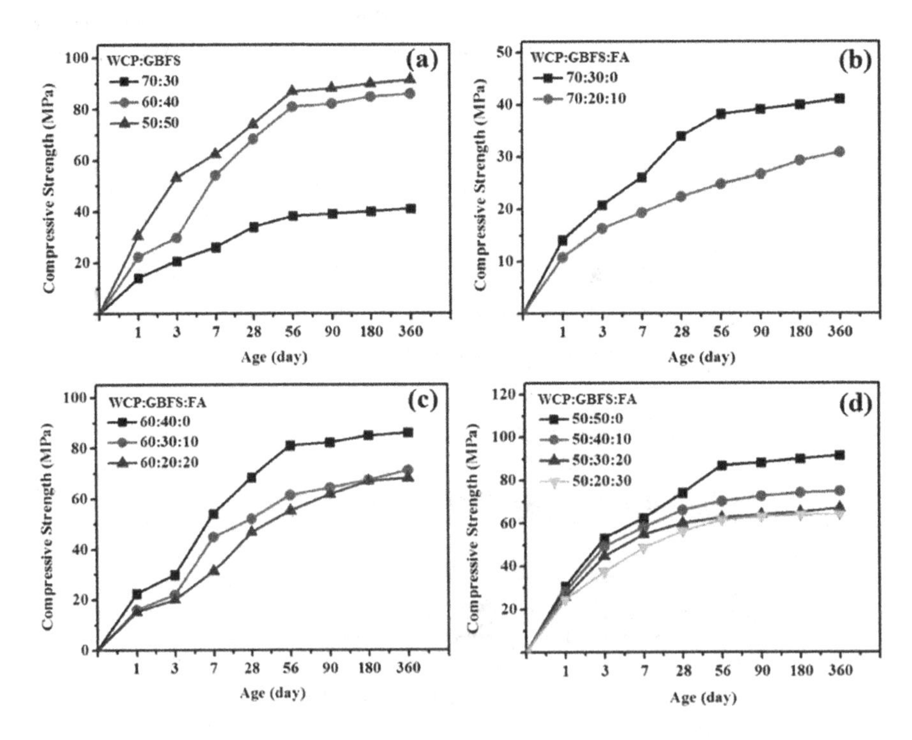

FIGURE 11.8 Effect of FA and GBFS content on the compressive strength of ceramic AAMs.

was ascribed to the low content of calcium and thus played the main role in the loss of AAM strength.

This result was consistent with the findings of Tanakorn et al. [30, 32] and Sanjay et al. [70] on the high-volume GBFS-incorporated mortars. This enhancement in the compressive strength with the increase in the GBFS level was ascribed to the rise of CaO content and reduction of SiO_2 levels in the mortar's matrix. Furthermore, an enhancement in the GBFS content generated a high proportion of CaO to SiO_2 up to 0.95, leading to the formation of the higher amount of C-(A)-S-H gel in the mixtures containing up to 30% of GBFS than the other mixtures [54, 63, 71, 72].

It was established that GBFS is an amorphous and granular substance. It consisted of SiO_2, Al_2O_3, MgO, and high CaO (51.8%), which enables formation of C-S-H gel as the main reaction product with the 1.68 calcium-to-silicate ratio [36]. Upon the incorporation of GBFS, the generated C-S-H gel further improved the strength characteristic compressive strength of AAMs [36, 73, 74]. Kumar et al. [70] demonstrated that at the ambient temperature, the reaction could dominate by the dissolution and precipitation of the C-S-H gel because of the alkali activation of the GBFS. The achieved modification in the setting times and compressive strength was mostly ascribed to the creation of the cementitious C-S-H gel, which in turn enhanced the hardened properties of the AAMs. Influences of the GBFS content on the compressive strength development have been studied by Al-Majidi et al. [69]. The strength of the AAM matrix was found to improve even at early ages due to the increase of the GBFS content. Weiguo et al. [75] investigated the influence of the GBFS content on the AAM matrix and reported an improvement in the compressive and flexural strengths of the mixes with increasing GBFS content. It was acknowledged that the presence of GBFS could accelerate the hydration process and the formation of the C-S-H gel [76]. Furthermore, an increase in the dosage of the slag could strongly accelerate the hydration and enhance the mortars' strength.

The influence of GBFS on gel formation was explained using three basic processes. The first process involved the enhanced compressive strength due to the elevated generation rate of the C-S-H gel originating from the addition of the dissolved Ca on the GBFS surface. It was acknowledged that the higher rate of C-S-H formation in the existence of Ca could result in the water shortage in the mortar matrix and raise its alkaline level, thus allowing elevated dissolution of the existing alumina silicate [70, 77, 78]. The second mechanism could be related to the alkali-activated product of the GBFS that usually predominates the C-A-S-H gel. The subsistence of Al ions led to a higher level of polymerization and notable cross-linking amid the C-S-H chains. In addition, the generation of the N-A-S-H type of gel was considered as the third process to enhance the mortar's strength. The N-A-S-H was the trivial secondary product that coexisted in the composition domain of the primary C-S-H gel category [79]. This could increase the gel compactness by reducing the overall porosity volume and thereby improve the compressive strength of the mortars [80].

Figure 11.9 depicts the effect of SiO_2:Al_2O_3 and CaO:SiO_2 on the development of AAM compressive strength. Specimens prepared with 4.5 or more SiO_2:Al_2O_3 and lower ratio of CaO to SiO_2 (less than 0.20) displayed a lower strength of 22.2 MPa. Furthermore, the compressive strength was enhanced with the decrease of SiO_2:Al_2O_3

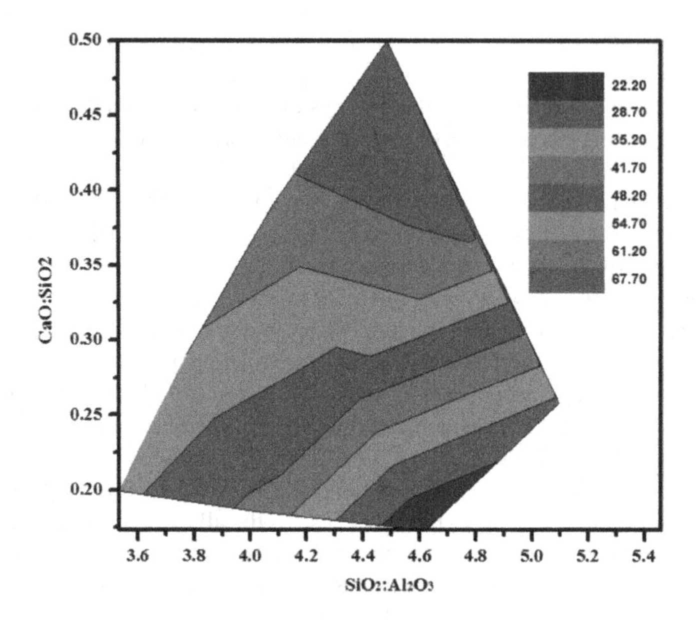

FIGURE 11.9 Effect of varying Ca:Si on the compressive strength of AAMs containing a high volume of WCP.

with a similar ratio of CaO to SiO_2 and recorded values higher than 50 MPa as the silicate-to-aluminium ratio reduced from 4.5 to 3.6. The highest strength (67 MPa) was recorded after 28 days with the GP matrix prepared with a calcium-to-silicate ratio higher than 0.40. The influence on the strength by percentage of SiO_2, Al_2O_3, and CaO are shown in Figure 11.10. The strength of AAM decreased as the percentage of aluminium and silicate increased and calcium content decreased.

11.7.2 XRD

Figure 11.11 depicts the XRD pattern of the AAMs with varied percentages of WCP and GBFS. The intensity of the C-S-H peak decreased as the WCP content increased, especially the peak at 30°. For 70% WCP the observed C-S-H peak at 43° was found to be replaced by a hydrotalcite peak. Calcite peaks at 38.5 and 51° were observed to replace by quartz peak as the WCP content increased above 50%. Five peaks for quartz, located at 17, 21, 28, 38.5, and 51°, were observed at high WCP content, in addition to a weak C-S-H peak at 30°. An increase in the WCP content led to an increase in the non-reacted silicate amount which reduced the C-S-H product. This could be attributed to the reduction of compressive strength from 74.1 to 34.8 MPa at 28 days of age.

Figure 11.12 presents the XRD patterns displaying the effect of FA replacing GBFS on AAM structures containing 50% WCP. The intensity and the number of quartz peaks increased as the FA content increased. The C-S-H peak intensity was reduced as the FA content increased because of the presence of a low amount of

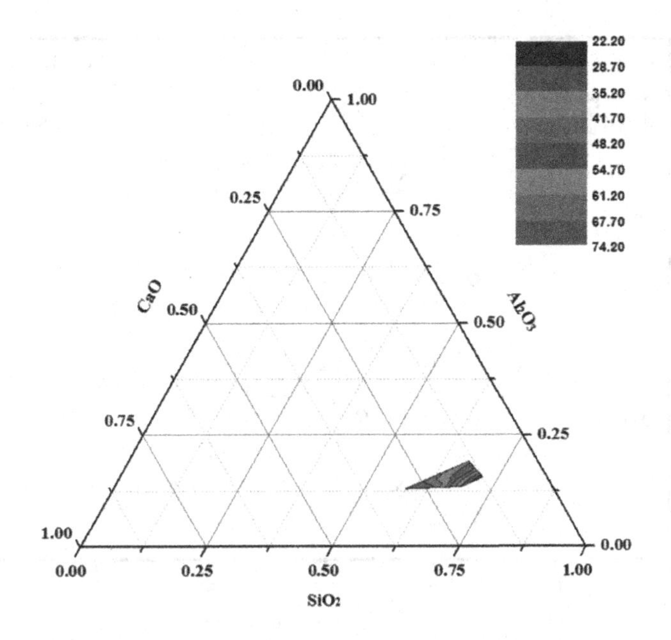

FIGURE 11.10 Effect of varying SiO_2, Al_2O_3, and CaO percentage content on the compressive strength of AAMs containing high-volume FA after 28 days.

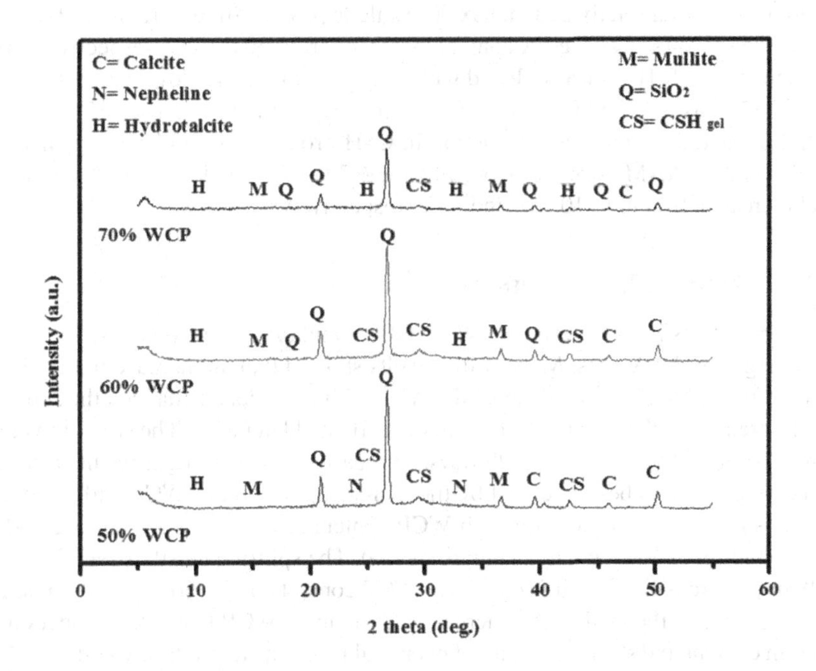

FIGURE 11.11 XRD patterns of AAMs containing high-volume WCP.

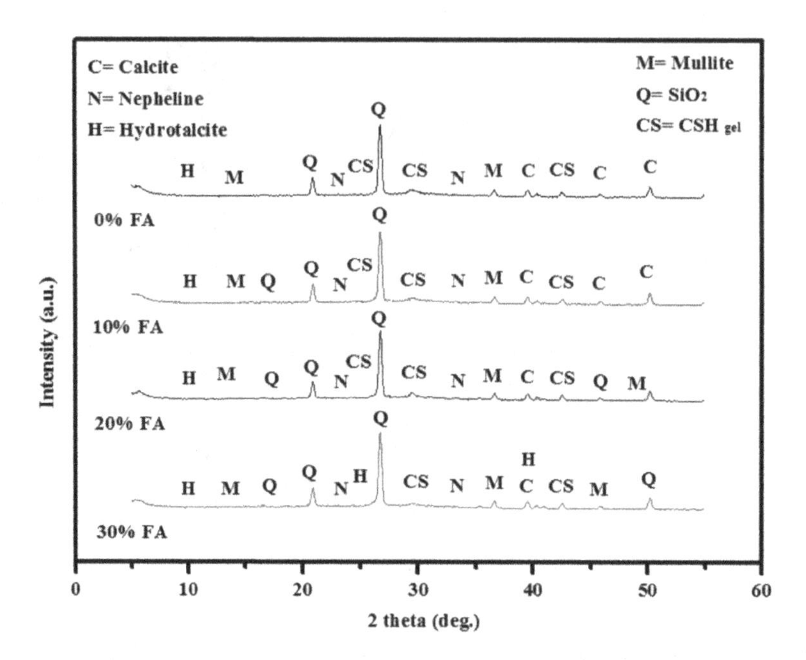

FIGURE 11.12 The XRD patterns showing the effect of FA and GBFS content on the crystalline structures of the AAMs containing 50% WCP.

calcium in the alkali-activated matrix. The calcite peak at 46° was replaced by mullite and then by quartz as the FA content increased from 20 to 30%, respectively. The production of C-S-H gel was reduced with an increase in the FA level where the CSH peak at 27° was replaced by hydrotalcite. An increase in the non-reacted silicate or partially reacted content with reduction in CSH product was evidenced, which led to a drop in the AAMs strength from 74.1 to 66.2 to 60.2 and to 56.5 MPa as the FA level increased from 0 to 10, 20, and 30%, respectively.

11.7.3 Splitting Tensile Strength

Figure 11.13 illustrates the effect of WCP, GBFS, and FA content on the splitting tensile strength of the AAMs. Most of the results showed that an increase in the GBFS content from 20 to 50% in the blended AAMs with FA indeed enhanced the splitting tensile strength of the specimens both at the early and later ages. The strength values were determined at 1, 3, 7, and 28 days. The early tensile strength results after 24 hours were found to be influenced by the increasing content of WCP and presented a lower strength (0.9 MPa) with high WCP content (70%) as compared to 2.4 MPa achieved for 50% WCP content (Figure 11.13a). The splitting tensile strength values at 28 days were 5.4, 5.6, and 2.7 MPa for WCP contents of 50, 60, and 70%, respectively. to replace the GBFS. An increasing content of WCP led to the reduction of calcium content and slowed the rate of chemical reactions to produce C-S-H gel [81].

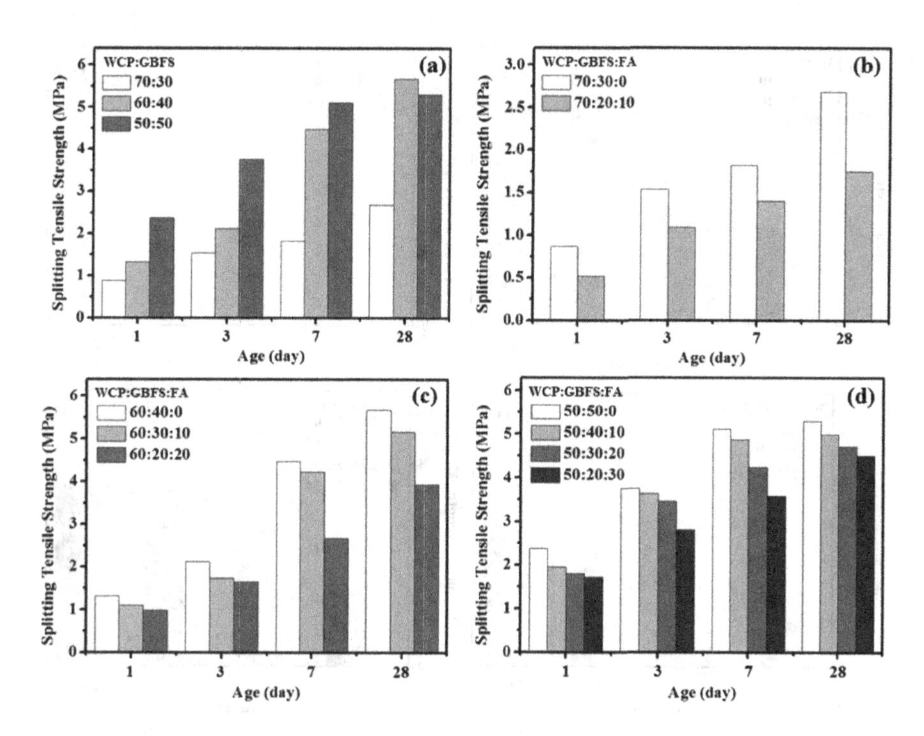

FIGURE 11.13 The effect of varying WCP, FA, and GBFS content on the splitting tensile strength of the AAMs.

The influence of high-volume CWP on the GPMs splitting tensile strength by FA replacing GBFS is shown in Figure 11.13b,c,d with 50, 60, and 70% WCP content, respectively. An inverse relationship was found between the tensile strength of AAM and the content of FA. As the content of FA increased, the strength dropped. The lower value of splitting tensile in this batch was observed for GPM that contained 70% WCP, 20% GBFS, and 10% FA, which revealed a strength of 1.7 MPa after 28 days. Similar results were also reported by Islam et al. [82]. Briefly, the incorporation of GBFS as replacement of FA presented higher splitting tensile strength than concrete substrate.

11.7.4 Flexural Strength and Modulus of Elasticity

Figure 11.14 displays the FS of AAMs containing various percentages of WCP, FA, and GBFS. The influence of increasing contents of GBFS replaced by WCP from 50 to 70% on the flexural strength of AAMs is depicted in Figure 11.14a. The effect of WCP content on the strength at 1, 3, 7, and 28 days was examined. The observed early strength after 24 hours dropped as the content of WCP increased, which were recorded to be 3.8, 3.4, and 1.6 MPa for 50, 60, and 70% of WCP content, respectively. At 28 days, a similar trend was found and the strength again dropped from

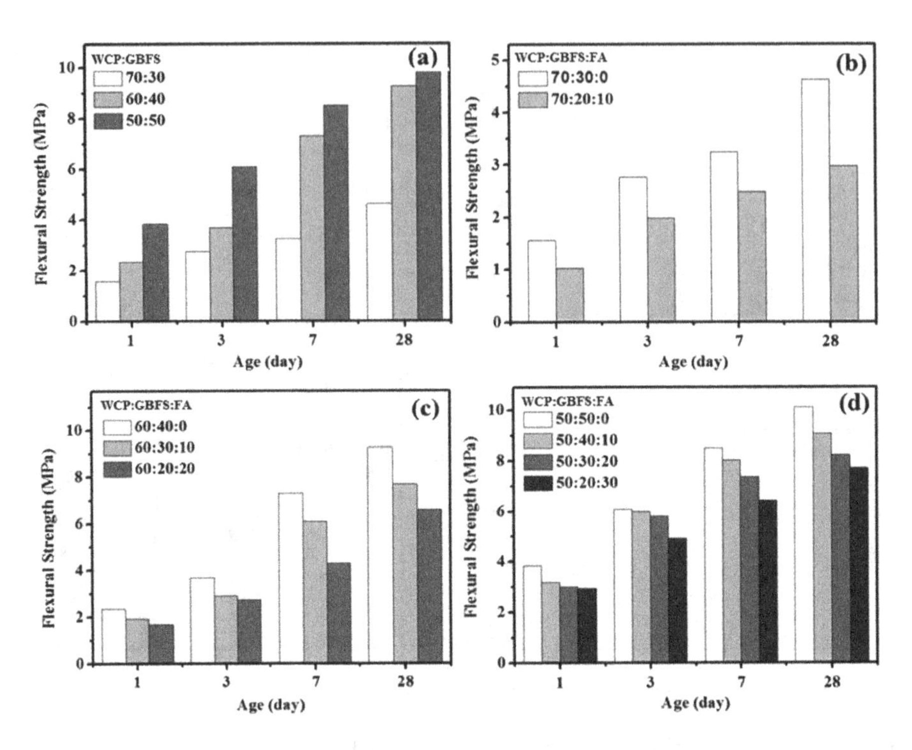

FIGURE 11.14 The effect of the FA-to-GBFS ratios on the FS of the proposed AAMs.

10.1 to 4.6 MPa. Figure 11.14b,c,d shows the effect FA replacement of GBFS at 50, 60, and 70% WCP, respectively. As the content of FA was increased in each level of WCP, the strength dropped. This observed lowest flexural strength (3.0 MPa) after 28 days was achieved for AAMs containing 70% WCP, 20% GBFS, and 10% FA. It is known that the product C-S-H gel is directly proportional to the content of GBFS [83].

Figure 11.15 shows the MOE values of AAMs for varying WCP content (50, 60, and 70%) at 28 days of age. The results showed that as the WCP content replaced GBFS increased, the MOE values decreased. The increase in the WCP percentage replaced GBFS from 50, 60 and 70%, led to a drop in the MOE from 16.3 to 15.8 to 7.4 GPa, respectively, as compared to 19.9 GPa recorded with $AAMs_1$ (100% GBFS). For each high level of WCP content, the effect of FA replacing GBFS was also evaluated. The results showed that an increasing FA content increases the silicate content and reduces the amount of calcium. This in turn led to a drop in the MOE of prepared AAMs. The lowest value of MOE (4.98 GPa) was found for the $AAMs_{10}$ mixture containing 70% WCP, 20% GBFS, and 10% FA. As the WCP and FA content increased, the MOE values of the AAMs remarkably dropped. In addition, the MOE and compressive strength of the AAMs increased, which is consistent with another report [82]. Most of the AAMs revealed MOE values lower than the concrete substrate (NC).

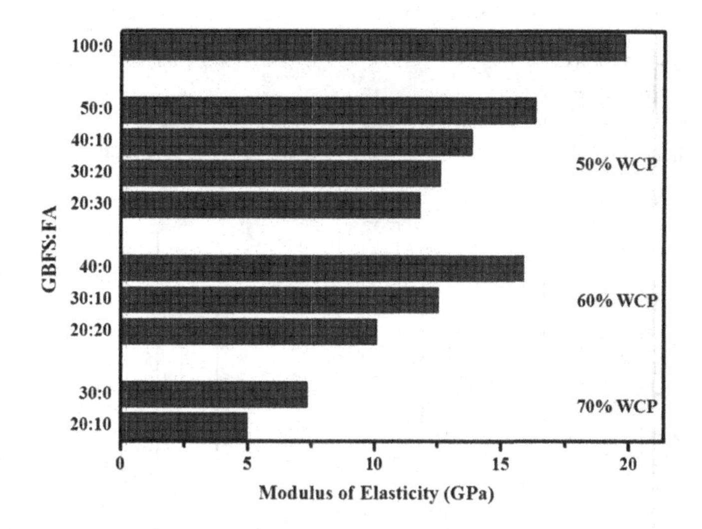

FIGURE 11.15 MOE for AAMs containing high volume of WCP at 28 days of age.

11.8 POROSITY OF AAMS

Figure 11.16 shows the effect of high-volume WCP on the porosity of AAMs. The porosity values were directly proportional to the WCP content, where an increase in the WCP-replaced GBFS from 50 to 60 and 70% led to an increase in the porosity of AAMs from 13.2 to 15.3 and 21.1%, respectively. For 50, 60, and 70% WCP levels, the influence of FA-replaced GBFS on the porosity and water absorption of AAMs was examined. Porosity values of AAMs increased from 13.3 to 16.1, 17.9, and 19.4% with increasing content of FA-replaced GBFS from 0% to 10, 20, and 30%, respectively, with 50% WCP. A similar trend was observed with 60 and 70% WCP containing alkali-activated specimens. As the FA content increased and GBFS content decreased, the porosity level was enhanced. The increasing content of WCP and FA enhanced the non-reacted and partially reacted particles and thereby reduced the C-S-H gel products and made the structure highly porous.

11.9 SURFACE ABRASION RESISTANCE

The influence of high-volume WCP content on the abrasion resistance of AAMs was evaluated and depicted in Figure 11.17. The results were collected at 1, 3, 7, and 28 days showing the abrasion resistance (AR) enhanced with age for all specimens. An inverse relationship was found between abrasion resistance and WCP content. The increase of WCP-replaced GBFS from 50 to 60 and 70% led to an increase in grind depth from 1.74 to 1.96 and 6.04 mm at 1 day of age, and from 1.44 to 1.51 and 4.3 mm after 28 days of age, respectively. A similar trend of results was found with FA-replaced GBFS in high-volume WCP specimens. The early and late abrasion resistance dropped with an increased FA content in 50% WCP specimens from 0 to 10, 20, and 30%, and recorded grind depth of 1.44, 1.46, 1.51, and 1.55 mm

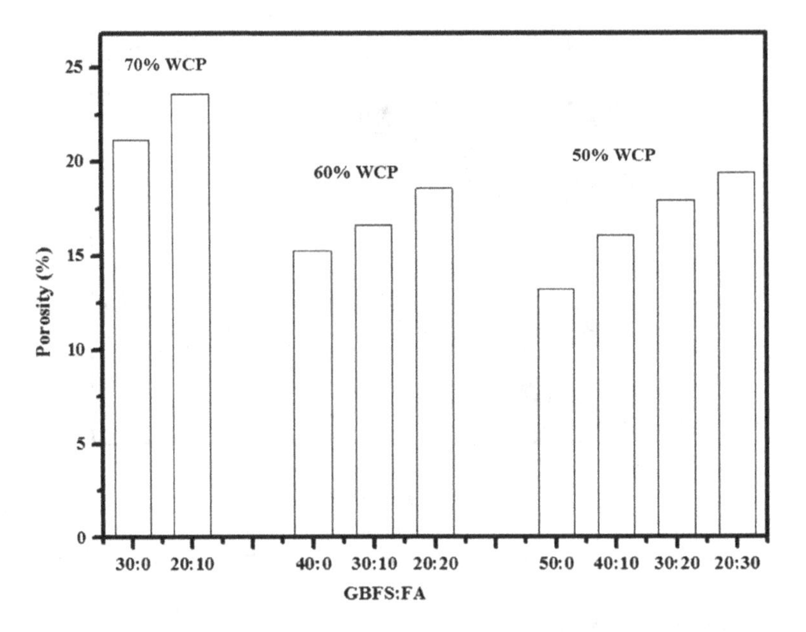

FIGURE 11.16 The effect of changing WCP-to-FA-to-GBFS ratios on the porosity of the AAMs.

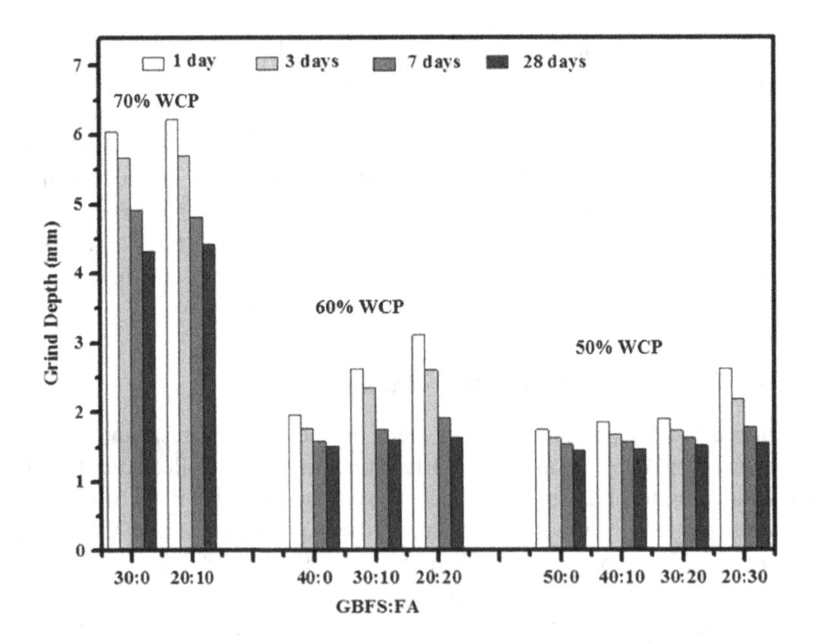

FIGURE 11.17 The effect of varying WCP-to-FA-to-GBFS ratios on the grind depth of the AAMs.

respectively. The results of abrasion resistance were influenced by the reduction of strength and increase in porous structure of alkali-activated specimens with increased WCP content. With the increase in the GBFS content, both the strength and AR were enhanced and the porosity dropped. The AR value was found to be proportional directly to the strength and inversely to the porosity [84]. According to Liu et al. [85], concrete with low porosity, high strength, and strong interfacial bond could enhance the overall concrete abrasion erosion resistance performance. As the porosity of the concrete decreased, the concrete became more impermeable and the AR of the AAM increased. The results of the compressive strength and water absorption were supported by the AR data. Wang et al. [86] demonstrated a reduction in the compressive strength where an increase in the pore volume in the hardened alkali-activated specimens negatively affected AR values.

11.10 FREEZE–THAW RESISTANCE

Effects of WCP replacing GBFS on the residual compressive strength and weight of alkali-activated specimens are illustrated in Figures 11.18 and 11.19. The results indicated an inverse relationship between residual strength and weight with WCP content. The increase in content of WCP from 50 to 60 and 70% drops the strength after 300 freeze–thaw cycles to 56.8, 54.7, and 10.9 MPa, respectively. A similar trend was observed with residual weight. The residual weight reduced from 96.3 and 52.3% with 60 and 70% WCP compared to 96.5% with 50% WCP. The high amount

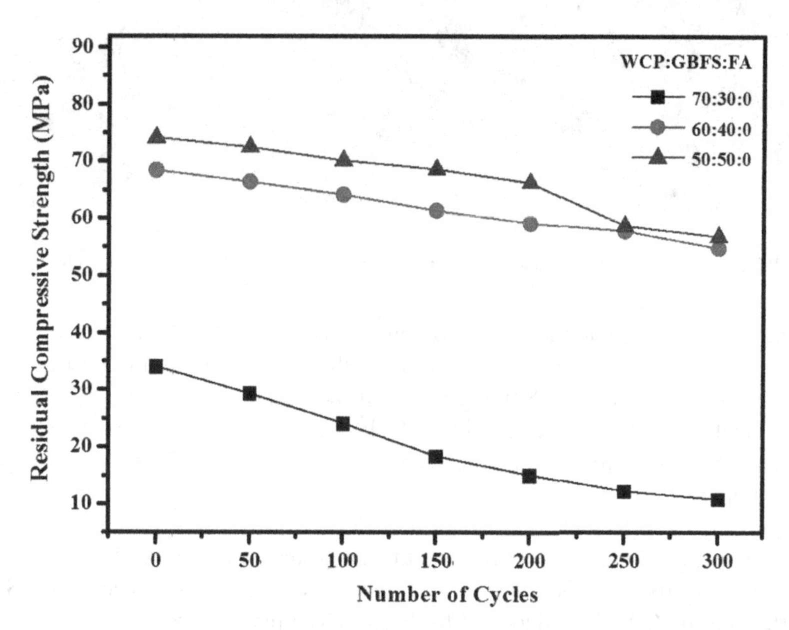

FIGURE 11.18 Effect of high-volume WCP content on residual compressive strength of AAMs.

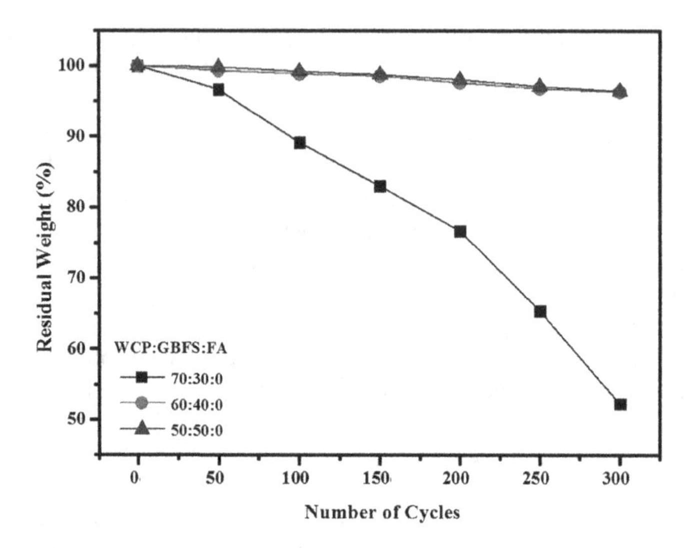

FIGURE 11.19 Effect of high-volume WCP content on residual compressive strength of AAMs.

FIGURE 11.20 Effect of high-volume WCP content on residual weight of AAMs.

of non-reacted silicate led to more a porous structure of 70% WCP as discussed in Section 11.7.1. The number of voids allowed the growth of ice and destroyed the interlock between particles [87]. Figure 11.20 shows the deterioration of AAMs for 50, 60, and 70% of WCP-replaced GBFS. It is clear that low durability and high surface scaling occurred with an increasing content of WCP to 70% as compared to other samples.

Figures 11.21, 11.22, and 11.23 show the effect of FA-replaced GBFS on the residual compressive strength, internal frost crack, and surface scaling of 50% WCP content AAM specimens. It is known that an increased level of FA-replaced GBFS has an effect on the C-S-H product and leads to a more porous structure. The increase in porosity for alkali-activated specimens with increasing FA level allows to growth the ice inside and destroyed the bonding between the particles. The residual strength

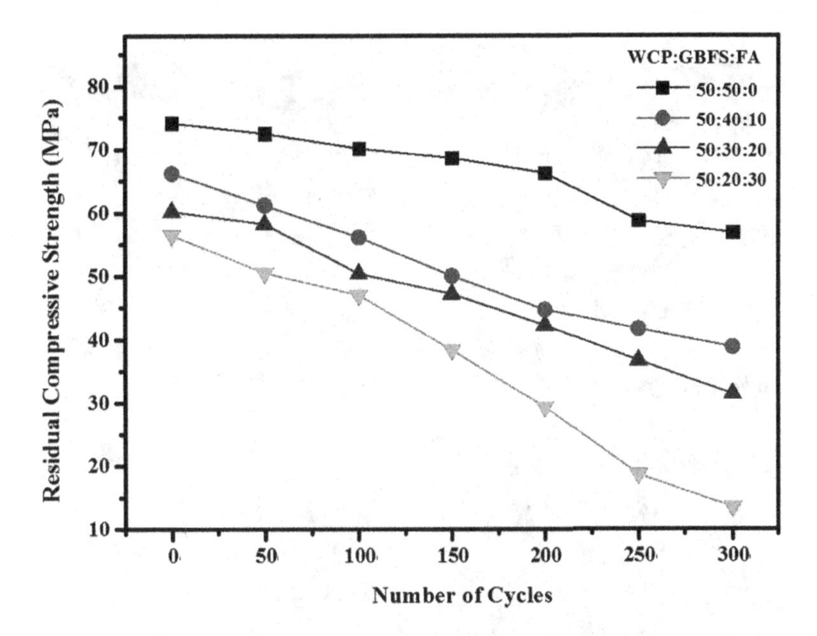

FIGURE 11.21 Effect of FA-replaced GBFS on residual compressive strength of 50% WCP content AAMs.

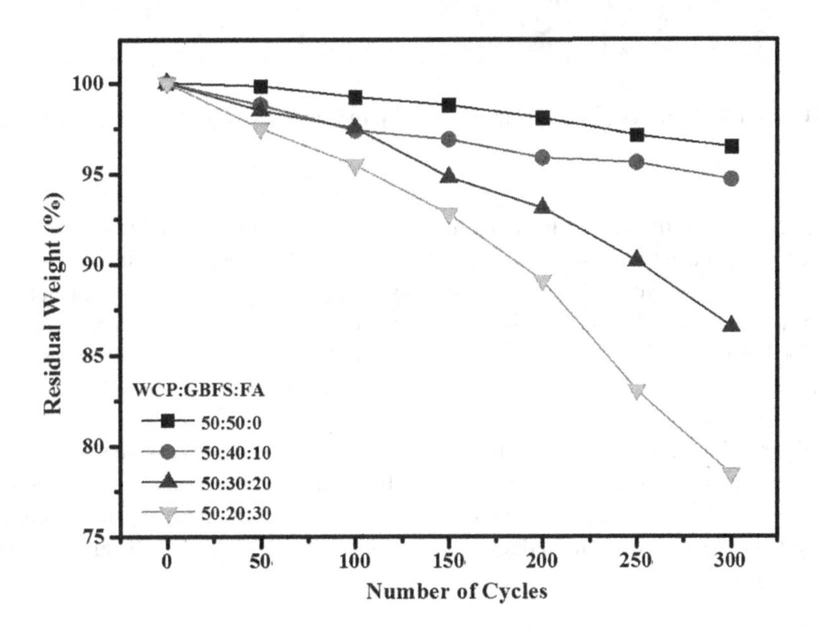

FIGURE 11.22 Effect of FA-replaced GBFS on residual weight of 50% WCP content AAMs.

FIGURE 11.23 Effect of FA-replaced GBFS on the surface texture of 50% WCP-content AAMs.

dropped with an increasing FA content and recorded values of 38.8, 31.4, and 13.5% with 10, 20, and 30 MPa of FA compared to 56.8 MPa was recorded with 0% FA after 300 freeze–thaw cycles. A similar trend was observed for residual weight, where the internal cracks and loss of weight increased with an increasing level of FA-replaced GBFS. Figure 11.23 shows the level of surface scaling, wherein the deterioration was significantly influenced by the FA content. A higher failure was found with AAMs containing 30% FA.

Figure 11.24 depicts the relationship between residual compressive strength and porosity for high-volume WCP content alkali-activated specimens. A linear relation between residual strength and porosity was observed. As the porosity increased, the residual compressive strength decreased (with R^2 value of 0.80 for all samples). The correlation is a follows:

$$Rfc = -4.8894P + 119.91\left(R^2 = 0.8045\right) \tag{11.4}$$

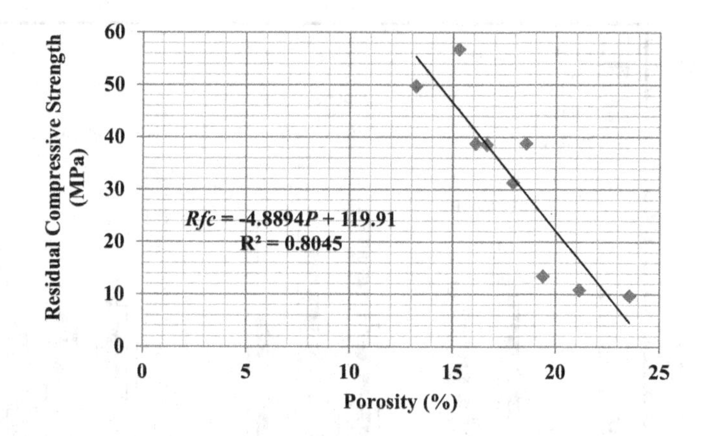

FIGURE 11.24 Correlation of residual compressive strength and porosity of high-volume WCP-content AAMs.

TABLE 11.2
Durability and Length Change of AAMs under Various Freeze–Thaw Cycles

Alkali-Activated Mortars	Number of Cycles	Durability Factor (±1)	Length Change (%)
70% WCP + 30% GBFS	300	78	–0.18
60% WCP + 40% GBFS	300	91	–0.09
50% WCP + 50% GBFS	300	96	–0.07
50% WCP + 40% GBFS + 10% FA	300	89	–0.11
50% WCP + 20% GBFS + 30 % FA	300	81	–0.14

Table 11.2 depicts the durability factor and length change of the studied AAMs under the various freeze–thaw cycles (average of three specimens). The higher durability factor indicated that the material is better to be used when exposed to severe-cold conditions. For instance, the alkali-activated repair material containing 50% WCP and GBFS showed a durability factor over 90%. Conversely, the repair material prepared with a low level GBFS revealed a low durability factor (78%), which was not appropriate for a place (environment) exposed to severe cold.

11.11 BOND STRENGTH PERFORMANCE

11.11.1 Slant Shear Bonding Strength (SSBS)

The results of slant shear bond strength tests for 1, 3, 7, and 28 days at 30° for AAMs containing high-volume WCP are illustrated in Figure 11.25. The bond strengths

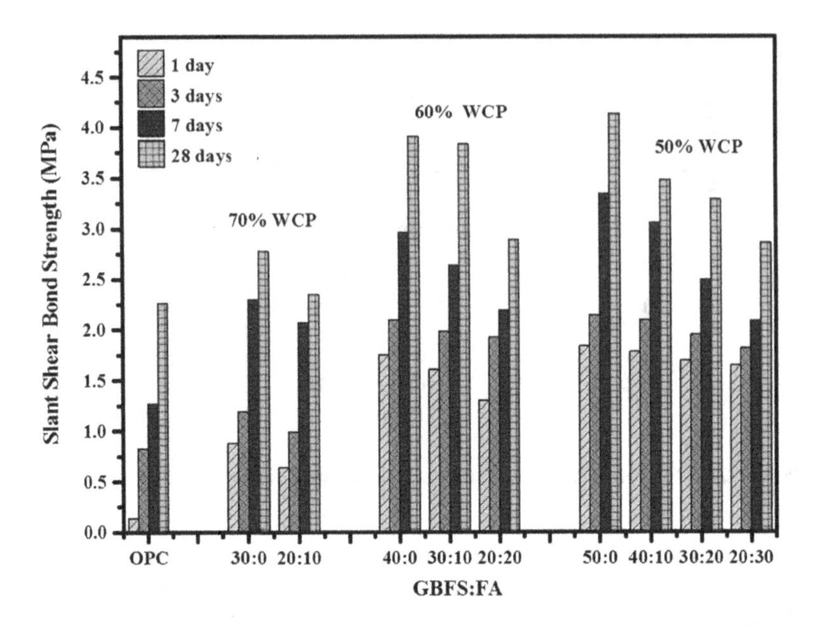

FIGURE 11.25 The 30° slant bond strength of AAM containing high-volume WCP.

were increased with time of aging. However, the bond strength of AAMs dropped with increasing content of WCP replacing GBFS. The bond strength of 50% WCP at 28 days was 4.2 MPa, which dropped from 3.8 to 2.7 MPa with increasing WCP from 60 to 70%, respectively. Similarly, the bond strength of all AAMs was higher than that of normal OPC mortar. It is evident that the AAMs possessed a better bonding character than the cement materials. Figure 11.26 shows the 45° slant shear bond strength results, where the bond strength decreased with an increase in WCP content. Furthermore, the bond strength dropped from 8.9 to 5.2 MPa with an increasing WCP from 50 to 70%, respectively. The effect of FA replacing GBFS in each level of high-volume WCP was evaluated. An increase in the FA content resulted in the reduced bond strength. Meanwhile, all AAMs presented very good bond strength and the failure zone occurred outside the bond zone.

11.11.2 Splitting Tensile Strength/Bond Strength

Figure 11.27 displays the effect of high-volume WCP on bond strength between AAMs and normal mortar (OPC) at 28 days of age evaluated under splitting tensile strength. The bond values revealed an inverse proportionality to the WCP content. An increase in the WCP content from 50 to 70% replacement of GBFS caused a reduction in the bond strength from 2.73 to 1.32 MPa, respectively. The ternary blend containing a high volume of WCP content of 50 and 60% presented excellent bond strength. However, the bond strength of the AAMs dropped at 70% WCP content and displayed a value of 1.32 and 0.79 MPa for FA content of 0 to 10%, respectively.

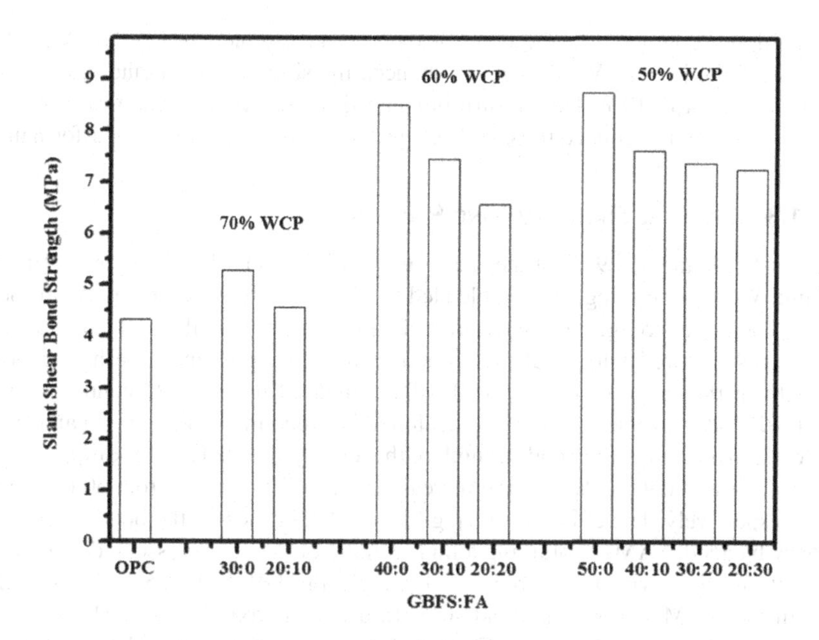

FIGURE 11.26 The 45° slant bond strength of AAM containing high-volume WCP.

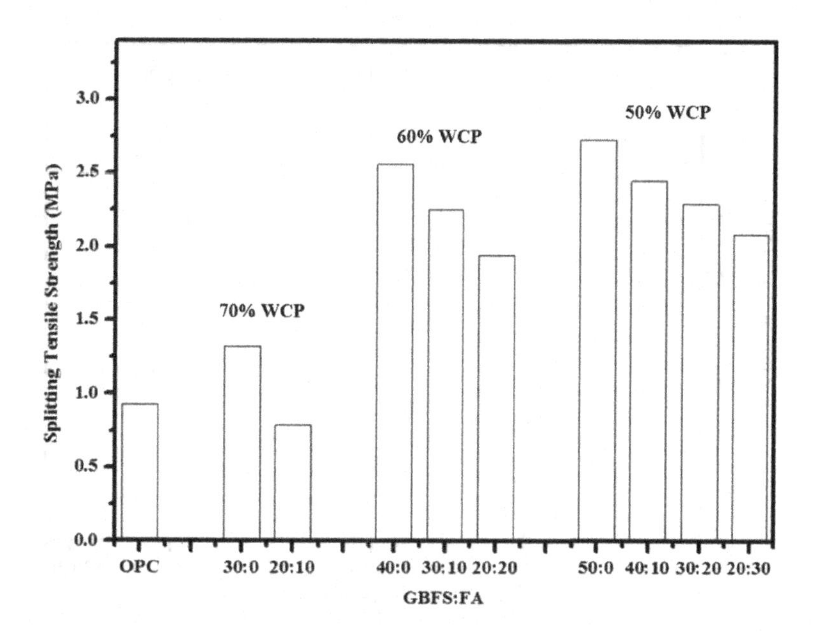

FIGURE 11.27 Splitting tensile strength of high-volume WCP-containing AAM specimens.

It was higher compared to the bond of the control sample (0.93 MPa). A further increase in the level of WCP to 70% enhanced the silicate content and reduced the calcium content to 10%. This in turn influenced the reaction of calcium as well as silicate and thereby reduced the C-S-H gel product and AAM bonding performance.

11.11.3 Flexural Strength/Bond Strength

Figures 11.28 and 11.29 illustrate the results on interfacial bond strength of high-volume WCP containing ternary-blended AAMs and their failure mode, respectively. The bond strength was measured via a four-point load flexural strength test. Specimens repaired with WCP alkali-activated binder presented the higher bond strength in the range of 0.57 to 1.70 MPa compared to the OPC control sample (0.54 MPa). An inverse relationship was found between the bond strength and WCP content. Furthermore, the bond strength of the ternary-blended AAMs dropped from 1.70 to 1.64 and 0.91 MPa with the increase in the WCP content from 50 to 60 and 70%, respectively. The effect of varying FA-to-GBFS ratios on the bond strength of ternary-blended AAMs containing a high volume of WCP also assessed. The bond strength dropped with the increase of FA replacing GBFS. For 50% WCP inclusion in the AAM matrix, the bond strength dropped from 1.70 to 1.31 MPa with increasing FA replacement of GBFS from 0 to 30%, respectively. The lowest bond strength was observed between 0.57 and 0.93 MPa for AAMs containing 70% WCP. Figure 11.28 shows the specimens' recorded bond strength value when the failure pattern occurred in the bond zone. The bond strength between AAMs containing

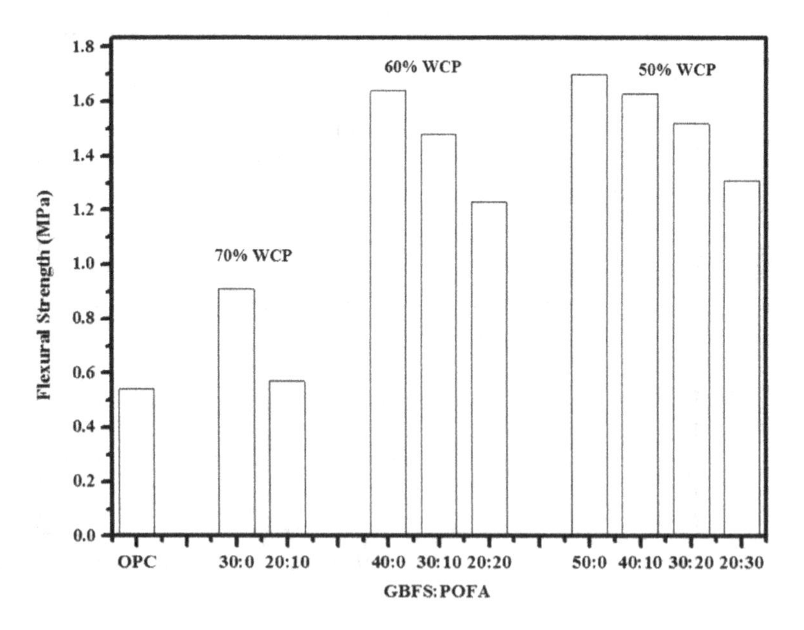

FIGURE 11.28 Four-point load flexural strength for high-volume WCP-containing AAMs as repair materials.

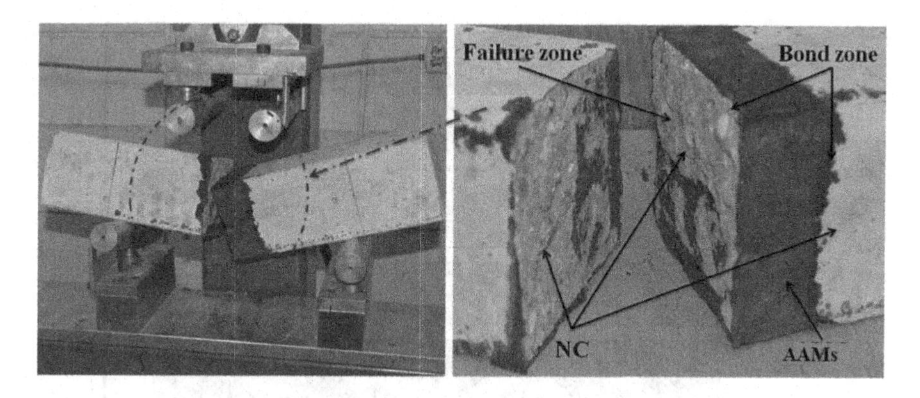

FIGURE 11.29 Failure pattern of high-volume WCP-containing AAMs under flexural strength.

high-volume WCP and NC was evaluated via three points of load flexural strength. The bond strength was found to obey an inverse correlation with WCP and FA contents in the ternary-blend of the alkali-activated matrix. All the AAM matrices exhibited excellent bond strength value as compared to the OPC control, as shown in Figure 11.29.

11.12 COMPATIBILITY BETWEEN AAMS AND CONCRETE SUBSTRATE

11.12.1 Thermal Expansion Coefficient

Figure 11.30 displays the WCP to GBFS to FA ratio dependent de-bonded percentage of the studied AAMs. Mortars containing high volume of GBFS revealed good thermal compatibilities, which virtually remained unchanged even after 25 freeze-thaw cycles. Furthermore, the de-bonded percentage was demonstrated to decrease and the compatibility was enhanced between the AAMs and concrete substrate with the increase of GBFS levels in the AAMs. Interestingly, the de-bonded percentage was dropped from 46 to 24% when the GBFS level was increased from 30 to 50% as WCP replacement, respectively as revealed by the de-bonded patterns. On top, the repair material (AAM) containing 60% of WCP and above displayed an excellent resistance to the thermal expansion and very high compatibility with the concrete substrate compared to the cement mortar (OPC) repair agent.

11.12.2 Four-Point Loading Flexural

Figure 11.31 elucidates the flexural strength of the composite beam (the AAM-embedded concrete substrate) as a function of the WCP-to-FA-to-GBFS ratios. It is known that the rigid materials deflect lower in the flexural test than fragile materials under similar load. In the present composite structure, the FS value was observed to increase with the increase in the GBFS content as WCP replacement. The composite

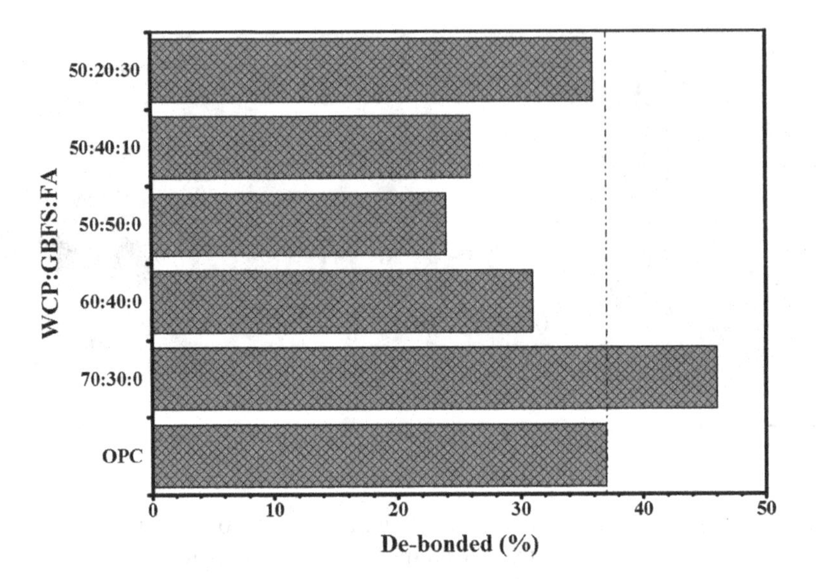

FIGURE 11.30 The WCP:GBFS:FA ratio-dependent de-bonded percentage of the studied AAMs.

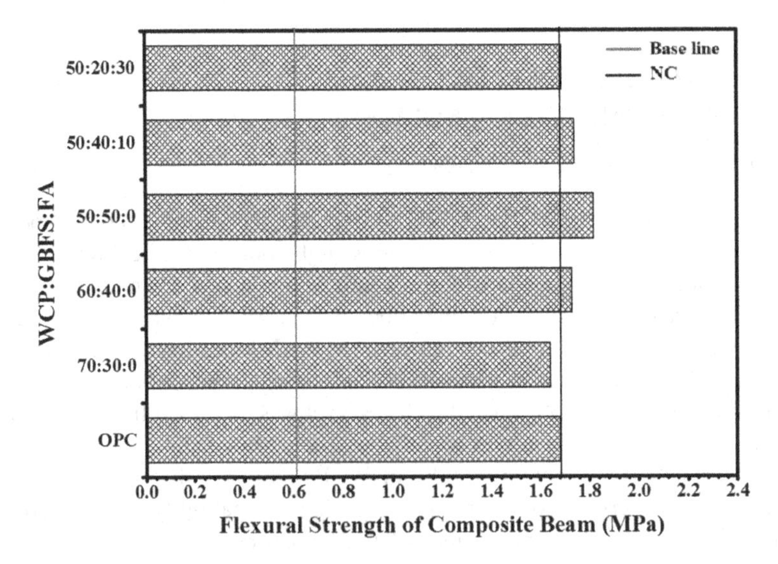

FIGURE 11.31 The FA-to-GBFS ratio-dependent FS of the composite section (the AAMs–concrete substrate).

beams containing 50% of the GBFS and above achieved higher flexural strength than the NC (1.82 MPa). The recorded flexural strengths varied from 1.64 to 1.82 MPa with the increase in the WCP level from 50 to 70%, respectively. Nevertheless, the reduction in the GBFS content to 30% affected the beam resistance and lowered the strength of the AAM below the one possessed by the concrete substrate beam.

TABLE 11.3
GBFS Content–Dependent Results of Four-Point Loading and Failure Zone of the Composite Beam

	Flexural Strength (MPa)			
Mix	Base Line	Concrete Substrate	Composite Beam	Failure Zone
OPC	0.63	1.66	1.67	C
70% WCP	0.63	1.66	1.64	D
60% WCP	0.63	1.66	1.73	A
50% GBFS	0.63	1.66	1.82	A
10% FA	0.63	1.66	1.74	A
30% FA	0.63	1.66	1.69	B

The GBFS levels' influence on the failure mode was also evaluated to determine the compatibility of these materials for the repair work (Table 11.3). Mortars prepared with the high content of WCP (50 and 60%) showed type A failure mode. An increment in the FA content (from 10 and 30%) transformed the failure zone from A to B. The composite beam prepared with cement mortar (OPC) and 70% of the WCP displayed failure zone type C and D, respectively. Furthermore, the mortar prepared with 60% of the WCP and below disclosed a high compatibility advantageous for the repairing work.

11.13 CONCLUSIONS

Due to their several distinct attributes compared to OPC, the alkali-activated mortars arc advantageous for construction purposes especially for repairing damaged concrete. The environmental benefits of AAMs made by recycling various waste materials together with their excellent binding capacity with the OPC motivated the authors to carry out this work. This chapter evaluated the performance of the alkali activated mortars and their cleaner production as repair materials. Depending on the evaluation of the fresh, mechanical attributes and the similarity between the AAMs and concrete substrate (as the repair martial), the following conclusions were drawn:

i. The high performance AAMs with low alkaline solution concentration can be obtained from the waste materials (WCP, GBFS, and FA). Use of GBFS (as waste material) could enhance the Ca^{++} ions concentration in the alkali-activated matrix and substitute the low Na^{+} ions concentration in the geopolymerization development.
ii. The intensity of the geopolymerization can be enhanced via the inclusion of the GBFS, where an increase in the calcium concentration was found to be responsible for the enhanced dissolution and precipitation of the Al_2O_3 and SiO_2.

iii. The replacement of WCP by 50–60% of GBFS in the blended mix resulted in the optimum flowability of the AAMs, where an increase or decrease in the WCP content could diminish the mortars' workability.
iv. The increase in the GBFS content as the replacement of WCP and FA could reduce the initial and final setting times of the AAMs.
v. An increase in the GBFS content could cause the formation of more C-S-H and C-A-S-H gels besides the N-A-S-H gel, which in turn increased the bond strength and improved the microstructures of the AAMs especially after 28 days of age.
vi. Increasing GBFS content could significantly reduce the microstructure porosity, water absorption, and drying shrinkage as well as improve the overall durability of the AAMs, especially to freeze–thaw cycling.
vii. The achievement of an excellent abrasion–erosion resistance and slant shear bond strength of the studied AAMs suggested their potential application as alternative repair materials for the damaged concrete structures.
viii. The results on the compatibility between the AAMs and concrete substrate obtained via the thermal expansion coefficient and four-point loading composite beam for the mortar prepared with the WCP contents of 60% and below revealed their great suitability for repair work.

REFERENCES

1. Jiang, L. and D. Niu, Study of deterioration of concrete exposed to different types of sulfate solutions under drying-wetting cycles. *Construction and Building Materials*, 2016. **117**: pp. 88–98.
2. Mirza, J., et al., Preferred test methods to select suitable surface repair materials in severe climates. *Construction and Building Materials*, 2014. **50**: pp. 692–698.
3. Kumar, G. R. and U. Sharma, Abrasion resistance of concrete containing marginal aggregates. *Construction and Building Materials*, 2014. **66**: pp. 712–722.
4. Huseien, G. F., et al., Effects of ceramic tile powder waste on properties of self-compacted alkali-activated concrete. *Construction and Building Materials*, 2020. **236**: p. 117574.
5. Wang, B., S. Xu, and F. Liu, Evaluation of tensile bonding strength between UHTCC repair materials and concrete substrate. *Construction and Building Materials*, 2016. **112**: pp. 595–606.
6. Huseien, G. F., et al., Synthesis and characterization of self-healing mortar with modified strength. *Jurnal Teknologi*, 2015. **76**(1): pp. 195–200.
7. Sedaghatdoost, A., et al., Influence of recycled concrete aggregates on alkali-activated slag mortar exposed to elevated temperatures. *Journal of Building Engineering*, 2019. **26**: p. 100871.
8. Huseien, G. F., et al., Utilizing spend garnets as sand replacement in alkali-activated mortars containing fly ash and GBFS. *Construction and Building Materials*, 2019. **225**: pp. 132–145.
9. Alanazi, H., et al., Bond strength of PCC pavement repairs using metakaolin-based geopolymer mortar. *Cement and Concrete Composites*, 2016. **65**: pp. 75–82.
10. Huseien, G. F., K. W. Shah, and A. R. M. Sam, Sustainability of nanomaterials based self-healing concrete: An all-inclusive insight. *Journal of Building Engineering*, 2019. 23: pp. 155–171.

11. Ouellet-Plamondon, C. and G. Habert, Life cycle assessment (LCA) of alkali-activated cements and concretes, In *Handbook of Alkali-Activated Cements, Mortars and Concretes*. pp. 663–686. 2015: Elsevier.
12. McLellan, B. C., et al., Costs and carbon emissions for geopolymer pastes in comparison to ordinary Portland cement. *Journal of Cleaner Production*, 2011. **19**(9–10): pp. 1080–1090.
13. Liu, Z., et al., Characteristics of alkali-activated lithium slag at early reaction age. *Journal of Materials in Civil Engineering*, 2019. **31**(12): p. 04019312.
14. Huseien, G. F., et al., Alkali-activated mortars blended with glass bottle waste nano powder: Environmental benefit and sustainability. *Journal of Cleaner Production*, 2019: p. 118636.
15. Wu, Y., et al., Geopolymer, green alkali activated cementitious material: Synthesis, applications and challenges. *Construction and Building Materials*, 2019. **224**: pp. 930–949.
16. Shekhawat, P., G. Sharma, and R. M. Singh, Strength behavior of alkaline activated eggshell powder and flyash geopolymer cured at ambient temperature. *Construction and Building Materials*, 2019. **223**: pp. 1112–1122.
17. Lu, Y., K. W. Shah, and J. Xu, Synthesis, morphologies and building applications of nanostructured polymers. *Polymers*, 2017. **9**(10): p. 506.
18. Li, N., et al., A mixture proportioning method for the development of performance-based alkali-activated slag-based concrete. *Cement and Concrete Composites*, 2018. **93**: pp. 163–174.
19. Huseien, G. F., et al., Geopolymer mortars as sustainable repair material: A comprehensive review. *Renewable and Sustainable Energy Reviews*, 2017. **80**: pp. 54–74.
20. Kubba, Z., et al., Impact of curing temperatures and alkaline activators on compressive strength and porosity of ternary blended geopolymer mortars. *Case Studies in Construction Materials*, 2018. **9**: p. e00205.
21. Huseien, G. F., et al., Evaluation of alkali-activated mortars containing high volume waste ceramic powder and fly ash replacing GBFS. *Construction and Building Materials*, 2019. **210**: pp. 78–92.
22. Phoo-ngernkham, T., et al., Adhesion characterisation of Portland cement concrete and alkali-activated binders. *Advances in Cement Research*, 2018. **31**(2): pp. 69–79.
23. Provis, J. L., A. Palomo, and C. Shi, Advances in understanding alkali-activated materials. *Cement and Concrete Research*, 2015. **78**: pp. 110–125.
24. Li, N., N. Farzadnia, and C. Shi, Microstructural changes in alkali-activated slag mortars induced by accelerated carbonation. *Cement and Concrete Research*, 2017. **100**: pp. 214–226.
25. Huseien, G. F., et al., Effects of POFA replaced with FA on durability properties of GBFS included alkali activated mortars. *Construction and Building Materials*, 2018. **175**: pp. 174–186.
26. Hanjitsuwan, S., et al., Strength development and durability of alkali-activated fly ash mortar with calcium carbide residue as additive. *Construction and Building Materials*, 2018. **162**: pp. 714–723.
27. Li, N., et al., Composition design and performance of alkali-activated cements. *Materials and Structures*, 2017. **50**(3): p. 178.
28. Huseiena, G. F., et al., Potential use coconut milk as alternative to alkali solution for geopolymer production. *Jurnal Teknologi*, 2016. **78**(11): pp. 133–139.
29. Huseien, G. F., et al., Effect of metakaolin replaced granulated blast furnace slag on fresh and early strength properties of geopolymer mortar. *Ain Shams Engineering Journal*, 2016. **9**(4): pp. 1557–1566.

30. Phoo-ngernkham, T., et al., Effects of sodium hydroxide and sodium silicate solutions on compressive and shear bond strengths of FA–GBFS geopolymer. *Construction and Building Materials*, 2015. **91**: pp. 1–8.
31. Huseien, G. F., et al., Waste ceramic powder incorporated alkali activated mortars exposed to elevated temperatures: Performance evaluation. *Construction and Building Materials*, 2018. **187**: pp. 307–317.
32. Phoo-ngernkham, T., et al., High calcium fly ash geopolymer mortar containing Portland cement for use as repair material. *Construction and Building Materials*, 2015. **98**: pp. 482–488.
33. Marinković, S., et al., Environmental assessment of green concretes for structural use. *Journal of Cleaner Production*, 2017. **154**: pp. 633–649.
34. Provis, J. L., Geopolymers and other alkali activated materials: why, how, and what? *Materials and Structures*, 2014. **47**(1–2): pp. 11–25.
35. Chindaprasirt, P., et al., Effect of calcium-rich compounds on setting time and strength development of alkali-activated fly ash cured at ambient temperature. *Case Studies in Construction Materials*, 2018. **9**: p. e00198.
36. Yip, C. K., G. Lukey, and J. Van Deventer, The coexistence of geopolymeric gel and calcium silicate hydrate at the early stage of alkaline activation. *Cement and Concrete Research*, 2005. **35**(9): pp. 1688–1697.
37. Huseien, G. F., et al., The effect of sodium hydroxide molarity and other parameters on water absorption of geopolymer mortars. *Indian Journal of Science and Technology*, 2016. **9**(48): pp. 1–9.
38. Garcia-Lodeiro, I., et al., Compatibility studies between NASH and CASH gels. Study in the ternary diagram $Na_2O–CaO–Al_2O_3–SiO_2–H_2O$. *Cement and Concrete Research*, 2011. **41**(9): pp. 923–931.
39. Jang, J., N. Lee, and H. Lee, Fresh and hardened properties of alkali-activated fly ash/slag pastes with superplasticizers. *Construction and Building Materials*, 2014. **50**: pp. 169–176.
40. Palomo, A., et al., A review on alkaline activation: new analytical perspectives. *Materiales de Construcción*, 2014. **64**(315): p. 022.
41. Kürklü, G., The effect of high temperature on the design of blast furnace slag and coarse fly ash-based geopolymer mortar. *Composites Part B: Engineering*, 2016. **92**: pp. 9–18.
42. Kramar, S., A. Šajna, and V. Ducman, Assessment of alkali activated mortars based on different precursors with regard to their suitability for concrete repair. *Construction and Building Materials*, 2016. **124**: pp. 937–944.
43. Huseien, G. F., et al., Compressive strength and microstructure of assorted wastes incorporated geopolymer mortars: Effect of solution molarity. *Alexandria Engineering Journal*, 2018. **57**(4): pp. 3375–3386.
44. Davidovits, J., Geopolymer cement: A review. In *Technical Papers*, Vol. 21, pp. 1–11. 2013: Geopolymer Institute.
45. Zamanabadi, S. N., et al., Ambient-cured alkali-activated slag paste incorporating micro-silica as repair material: Effects of alkali activator solution on physical and mechanical properties. *Construction and Building Materials*, 2019. **229**: p. 116911.
46. Nunes, V. A., P. H. Borges, and C. Zanotti, Mechanical compatibility and adhesion between alkali-activated repair mortars and Portland cement concrete substrate. *Construction and Building Materials*, 2019. **215**: pp. 569–581.
47. Robayo-Salazar, R., et al., Alkali-activated binary mortar based on natural volcanic pozzolan for repair applications. *Journal of Building Engineering*, 2019. **25**: p. 100785.
48. Geraldo, R. H., et al., Study of alkali-activated mortar used as conventional repair in reinforced concrete. *Construction and Building Materials*, 2018. **165**: pp. 914–919.

49. Wang, J., et al., Effects of fly ash on the properties and microstructure of alkali-activated FA/BFS repairing mortar. *Fuel*, 2019. **256**: p. 115919.
50. Liu, Y., et al., Compatibility of repair materials with substrate low-modulus cement and asphalt mortar (CA mortar). *Construction and Building Materials*, 2016. **126**: pp. 304–312.
51. Pattnaik, R. R. and P. R. Rangaraju, Analysis of compatibility between repair material and substrate concrete using simple beam with third point loading. *Journal of Materials in Civil Engineering*, 2007. **19**(12): pp. 1060–1069.
52. Czarnecki, L., et al., *Polymer Composites for Repairing of Portland Cement Concrete: Compatibility Project*. 1999: NIST.
53. ASTM, Standard specification for coal fly ash and raw or calcined natural pozzolan for use in concrete. 2013: ASTM.
54. Yusuf, M. O., et al., Evolution of alkaline activated ground blast furnace slag–ultrafine palm oil fuel ash based concrete. *Materials & Design*, 2014. **55**: pp. 387–393.
55. Temuujin, J., A. van Riessen, and K. MacKenzie, Preparation and characterisation of fly ash based geopolymer mortars. *Construction and Building Materials*, 2010. **24**(10): pp. 1906–1910.
56. Rickard, W. D., et al., Assessing the suitability of three Australian fly ashes as an aluminosilicate source for geopolymers in high temperature applications. *Materials Science and Engineering: A*, 2011. **528**(9): pp. 3390–3397.
57. Memon, S. A., et al., Development of form-stable composite phase change material by incorporation of dodecyl alcohol into ground granulated blast furnace slag. *Energy and Buildings*, 2013. **62**: pp. 360–367.
58. ASTM, C117 Standard test method for materials finer than 75-µm (no. 200) sieve in mineral aggregates by washing. In *Annual Book of ASTM Standards*. Vol. 4. 2003: ASTM.
59. ASTM, C33 Standard specification for concrete aggregates. In *Annual Book of ASTM Standards*. Vol. 4. 1994: ASTM.
60. Salih, M. A., et al., Development of high strength alkali activated binder using palm oil fuel ash and GGBS at ambient temperature. *Construction and Building Materials*, 2015. **93**: pp. 289–300.
61. Huseiena, G. F., et al., Effect of binder to fine aggregate content on performance of sustainable alkali activated mortars incorporating solid waste materials. *Chemical Engineering*, 2018. **63**: pp. 667–672.
62. ASTM, C109 Standard test method for compressive strength of hydraulic cement mortars (using 2-in. or [50-mm] cube specimens). Vol. 318. 1999: ASTM.
63. Huseien, G. F., et al., Properties of ceramic tile waste based alkali-activated mortars incorporating GBFS and fly ash. *Construction and Building Materials*, 2019. **214**: pp. 355–368.
64. ASTM, Standard specification for flow table for use in tests of hydraulic cement. 2014: ASTM.
65. ASTM, Standard test method for time of setting of hydraulic cement by Vicat needle. 2008: ASTM.
66. Nath, P. and P. K. Sarker, Effect of GGBFS on setting, workability and early strength properties of fly ash geopolymer concrete cured in ambient condition. *Construction and Building Materials*, 2014. **66**: pp. 163–171.
67. Huseien, G. F., et al., Synergism between palm oil fuel ash and slag: Production of environmental-friendly alkali activated mortars with enhanced properties. *Construction and Building Materials*, 2018. **170**: pp. 235–244.
68. Sugama, T., L. Brothers, and T. Van de Putte, Acid-resistant cements for geothermal wells: sodium silicate activated slag/fly ash blends. *Advances in Cement Research*, 2005. **17**(2): pp. 65–75.

69. Al-Majidi, M. H., et al., Development of geopolymer mortar under ambient temperature for in situ applications. *Construction and Building Materials*, 2016. **120**: pp. 198–211.
70. Kumar, S., R. Kumar, and S. Mehrotra, Influence of granulated blast furnace slag on the reaction, structure and properties of fly ash based geopolymer. *Journal of Materials Science*, 2010. **45**(3): pp. 607–615.
71. Puligilla, S. and P. Mondal, Role of slag in microstructural development and hardening of fly ash-slag geopolymer. *Cement and Concrete Research*, 2013. **43**: pp. 70–80.
72. Huseien, G. F., et al., Influence of different curing temperatures and alkali activators on properties of GBFS geopolymer mortars containing fly ash and palm-oil fuel ash. *Construction and Building Materials*, 2016. **125**: pp. 1229–1240.
73. Buchwald, A., H. Hilbig, and C. Kaps, Alkali-activated metakaolin-slag blends—Performance and structure in dependence of their composition. *Journal of Materials Science*, 2007. **42**(9): pp. 3024–3032.
74. Pacheco-Torgal, F., J. Castro-Gomes, and S. Jalali, Investigations on mix design of tungsten mine waste geopolymeric binder. *Construction and Building Materials*, 2008. **22**(9): pp. 1939–1949.
75. Shen, W., et al., Magnesia modification of alkali-activated slag fly ash cement. *Journal of Wuhan University of Technology (Materials Science Edition)*, 2011. **26**(1): pp. 121–125.
76. Yu, R., P. Spiesz, and H. Brouwers, Development of an eco-friendly Ultra-High Performance Concrete (UHPC) with efficient cement and mineral admixtures uses. *Cement and Concrete Composites*, 2015. **55**: pp. 383–394.
77. Khater, H., Effect of calcium on geopolymerization of aluminosilicate wastes. *Journal of Materials in Civil Engineering*, 2011. **24**(1): pp. 92–101.
78. Nath, S. and S. Kumar, Influence of iron making slags on strength and microstructure of fly ash geopolymer. *Construction and Building Materials*, 2013. **38**: pp. 924–930.
79. Myers, R. J., et al., Generalized structural description of calcium–sodium aluminosilicate hydrate gels: The cross-linked substituted tobermorite model. *Langmuir*, 2013. **29**(17): pp. 5294–5306.
80. Li, Z. and S. Liu, Influence of slag as additive on compressive strength of fly ash-based geopolymer. *Journal of Materials in Civil Engineering*, 2007. **19**(6): pp. 470–474.
81. Deb, P. S., P. Nath, and P. K. Sarker, The effects of ground granulated blast-furnace slag blending with fly ash and activator content on the workability and strength properties of geopolymer concrete cured at ambient temperature. *Materials & Design (1980-2015)*, 2014. **62**: pp. 32–39.
82. Islam, A., et al., Engineering properties and carbon footprint of ground granulated blast-furnace slag-palm oil fuel ash-based structural geopolymer concrete. *Construction and Building Materials*, 2015. **101**: pp. 503–521.
83. Song, S. and H. M. Jennings, Pore solution chemistry of alkali-activated ground granulated blast-furnace slag. *Cement and Concrete Research*, 1999. **29**(2): pp. 159–170.
84. Mohebi, R., K. Behfarnia, and M. Shojaei, Abrasion resistance of alkali-activated slag concrete designed by Taguchi method. *Construction and Building Materials*, 2015. **98**: pp. 792–798.
85. Liu, Y.-W., T. Yen, and T.-H. Hsu, Abrasion erosion of concrete by water-borne sand. *Cement and Concrete Research*, 2006. **36**(10): pp. 1814–1820.
86. Wang, S.-D., K. L. Scrivener, and P. Pratt, Factors affecting the strength of alkali-activated slag. *Cement and Concrete Research*, 1994. **24**(6): pp. 1033–1043.
87. Cai, L., H. Wang, and Y. Fu, Freeze–thaw resistance of alkali–slag concrete based on response surface methodology. *Construction and Building Materials*, 2013. **49**: pp. 70–76.

12 Structural Applications of Alkali-Activated Concrete Containing Ceramic Waste

12.1 INTRODUCTION

Research on self-compacting concrete started as far back as 1989 in Japan. Since then research has been evolving towards the development of self-compacting mortar and concrete in various countries. Nowadays self-compacting concrete is one of the most efficient solutions in the world of concrete [1–4]. Self-compacting concrete is a flowable matrix with the ability to resist segregation and fill the mould under gravity without the need of compaction [5–7]. This characteristic allows many benefits including reduction in construction time, the freedom of design work, and improvement in product quality and working environment. Also, this type of concrete brings a positive impact to the environment because it allows the use of additions or fillers from construction and demolition waste [8]. The final strength ranges from normal to high depending on the mix. Compared to conventional vibrated concrete, self-compacting mixes typically require a higher content of fine materials, whereas the maximum nominal size of the coarse aggregate is typically no more than 19 mm [2, 3]. In the cement-free concrete industry, the design of high-performance concrete, improvement of mechanical strength, and self-compacting behaviour are research challenges [9–11].

Alkali-activated mortar and concrete have started to distinguish themselves from standard Portland cement as effective constructional binders without high CO_2 emissions [12–14]. An inorganic polymer material, alkali-activated concrete consists of calcium (Ca) and aluminosilicates (ASs), and is synthesized from alkaline activator solution containing sodium silicate (NS) and sodium hydroxide (NH) and pozzolanic compounds [15, 16]. The production of alkali-activated concrete/mortars is based on the exploitation of a wide range of by-product and agricultural wastes comprising calcium, silica and aluminium (Al) such as, GBFS, palm oil fuel ash (POFA) and fly ash (FA) etc.

Besides being non-recyclable, ceramic waste is also bulky and creates problems for landfill disposal. Hence, for the purposes of natural resource conservation and environmental protection, it is necessary to develop new products with ceramic waste to recycle and to use them in construction projects. The pozzolanic properties of ceramic waste have been confirmed by earlier studies [19–21]. Because they benefit the mechanical and durability properties of concrete, natural pozzolanic wastes were employed as construction material in earlier times [22, 23]. However, industrial wastes are currently the preferred pozzolanic materials because of strict environmental policies [24–26]. The ceramic industry has a significant effect on the environment and relies on landfills to dispose of the massive volumes of its waste. In 2015, ceramic tile production at global level was around 12.4 million square metres [27, 28], of which 10–30% is discarded as waste [19, 29]. Additionally, the greatest proportion of this waste is non-recyclable, which is highly problematic in terms of disposal. To address this issue of extensive amounts of ceramic waste with little usage, researchers have begun to investigate the use of ceramic waste in mortar or concrete instead of cement. This, in turn, is advantageous as it reduces costs and energy consumption, improves the ecological equilibrium, and promotes more careful use of natural resources [30].

Many factors affect the mix design of self-compacting high-performance alkali-activated concrete (SCAAC) including the size and type of coarse and fine aggregate, alkaline activator solution content, concentration of sodium hydroxide, sodium silicate content, and binder chemical composition. In this chapter, we aim at assessing the impact of the binder on performance of SCAAC. Waste ceramic powder (WCP) was collected, ground less than 45 μm, and substituted for GBFS at different ratios to understand the effect on SCAAC's fresh and hardened properties.

12.2 MATERIALS AND MIX DESIGN

Binary blended included WCP and GBFS were used to produce alkai-activated concrete.

The GBFS was used as received from the supplier without any treatment in the lab. Table 12.1 lists the chemical composition of GBFS having density of 2.89 g/cm^3, mean particle size of 12.4 μm, and specific surface area of 4950 cm^2/gm. The GBFS was collected from Ipoh (Malaysia) and used as one of the source materials

TABLE 12.1
X-Ray Fluorescence (XRF) Test of GBFS and WCP

Oxide	GBFS	WCP
SiO_2	30.8	72.8
Al_2O_3	10.9	12.2
Fe_2O_3	0.64	0.56
CaO	51.8	0.01
MgO	4.57	1.0
K_2O	0.36	—
Na_2O	0.45	13.5
SO_3	0.06	—
LOI	0.22	—

to produce cement-free binder. GBFS possesses both cementitious and pozzolanic properties. GBFS develops a specific hydraulic reaction when mixed with water and appears off-white. GBFS meets the requirement of a pozzolanic material because it is comprised of about 90% calcium, silicate, and alumina.

Ceramic waste was collected as waste material from construction industry, which is crushed in the lab using a crushing machine, then sieved (600 μm) before being ground for 6 hours into an average particle size of 45 μm using a Los Angeles machine as shown in Figure 12.1. The main chemical composition of WCP is silica (72.8%), sodium oxide (13.5%), and alumina (12.2%). After grinding, WCP presents as a light grey colour with medium particle size 35 μm.

The X-ray diffraction (XRD) patterns of GBFS and WCP are illustrated in Figure 12.2. The XRD pattern of GBFS verifies its highly amorphous nature because of the absence of any sharp peak. The XRD pattern of WCP revealed pronounced diffraction peaks around 2θ = 16–30°, which are attributed to the crystalline silica and alumina compounds. Nonetheless, the occurrence of other crystalline peaks is ascribed to the presence of crystalline quartz and mullite phases.

Figure 12.3 shows the scanning electron microscopy (SEM) images of GBFS and WCP, which are blended in the self-compacting matrix to achieve the enhanced workability, comprehensive strength, and improved microstructural properties. It is evident that GBFS and WCP are comprised of irregular and angular particles, similar to a previous report [31].

Locally available river sand having a specific gravity of 2.74 and passing 100% through a 2.36 mm sieve was used as fine aggregate in preparation of SCAAC. Crushed stone with a maximum size of 5 mm was used as coarse aggregate. To prepare the alkaline activator solution, sodium hydroxide was added to sodium silicate at a ratio of 1:2.5. To prepare the alkali solution, the concentration of sodium

FIGURE 12.1 WCP preparation stages.

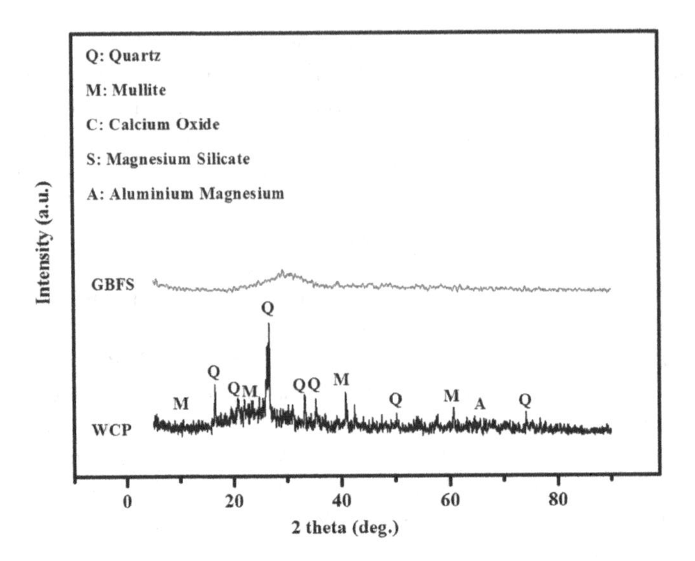

FIGURE 12.2 XRD patterns of GBFS and WCP.

FIGURE 12.3 SEM of GBFS and WCP.

hydroxide was fixed to 4 molarity for all the concrete mixtures. The solution was left for 24 hours before use. The ratio of alkali solution to binder (S:B) was also fixed for all mixtures with a ratio of 0.55 as depicted in Table 12.2.

12.3 SPECIMEN PREPARATION

Each mixture was prepared by weight percentage, being its compounds in dry conditions. Firstly, the aggregates, including river sand and crushed stone, were mixed for 2 minutes in a concrete mixer. After that, the homogeneous-blended binder containing GBFS and WCP was added, and subsequently it was added to the alkaline activator solution, then mix for another 3 minutes until homogeneity. Rheological tests such as the slump flow, J-ring and L-box tests were adopted to measure the

TABLE 12.2
Design of Mixes of SCAAC

Mix Sample	WCP to GBFS	WCP (kg/m^3)	GBFS (kg/m^3)	Total Binder (kg/m^3)	Coarse Aggregate (kg/m^3)	Fine Aggregate (kg/m^3)	NS:NH	NaOH Molarity (M)	S:B
S_1	40:60	180	270	450	900	850	2.5	4	0.55
S_2	50:50	225	225	450	900	850	2.5	4	0.55
S_3	60:40	270	180	450	900	850	2.5	4	0.55
S_4	70:30	315	135	450	900	850	2.5	4	0.55

NS:NH, ratio of sodium silicate to sodium hydroxide.

rheological properties of alkali-activated concrete. Subsequently, casting was done using cubic moulds. The samples were removed from these moulds after 24 hours, then the specimens left to cure in ambient temperature (27°C and 75% relative humidity) until the day of testing. Cubic specimens with dimension of 100 × 100 × 100 mm were used to determine the compressive strength development. Each of the data corresponds to an average of three specimens.

Four alkali-activated reinforced concrete beams were cast to evaluate the effect of WCP content on the beams' flexural strength performance. The first beam (B-S_1) was cast with 40% WCP content as the GBFS replacement and considered as non-self-compacting concrete (control sample). The other three beams were cast with different ratios of WCP content – 50, 60, and 70% – as the GBFS replacement and used to evaluate the effect of WCP on SCAAC performance. The details of the beam design are presented in Table 12.3 and Figure 12.4. Three-point loading was considered to evaluate the flexural strength and failure mood of reinforced alkali-activated concrete after 28 days, as depicted in Figure 12.5.

12.4 FRESH PROPERTIES

Tests to determine the rheological characteristics were performed on the SCAAC including slump flow, T_{50}, and L-box. The slump flow test was done using the Abrams cone in order to evaluate the filling ability of SCAAC to deform under the action of its own weight with restriction, allowing one to visually check the possibility of segregation of the concrete. A range of values for an adequate slump flow

TABLE 12.3
Detail of Alkali-Activated Concrete Beams

No.	Beam ID	Concrete Mix*	Type of Concrete	WCP Content (%)
1	B-S_1	S_1	Normal concrete	40
2	B-S_2	S_2	SCAACs	50
3	B-S_3	S_3	SCAACs	60
4	B-S_4	S_4	SCAACs	70

*Details presented in Table 12.2.

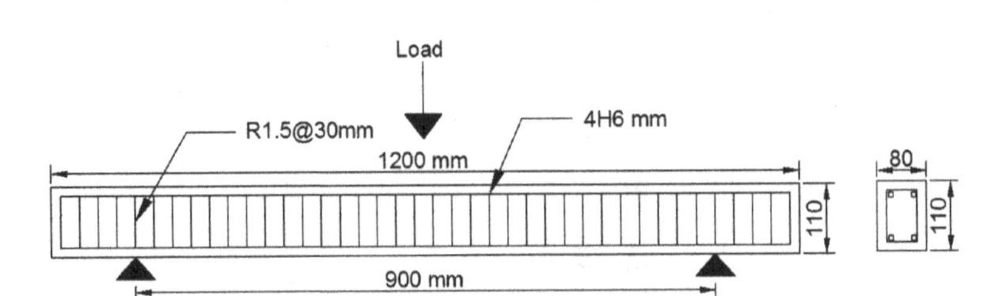

FIGURE 12.4 Detail of designed reinforced concrete beam.

FIGURE 12.5 Testing of reinforced concrete beam.

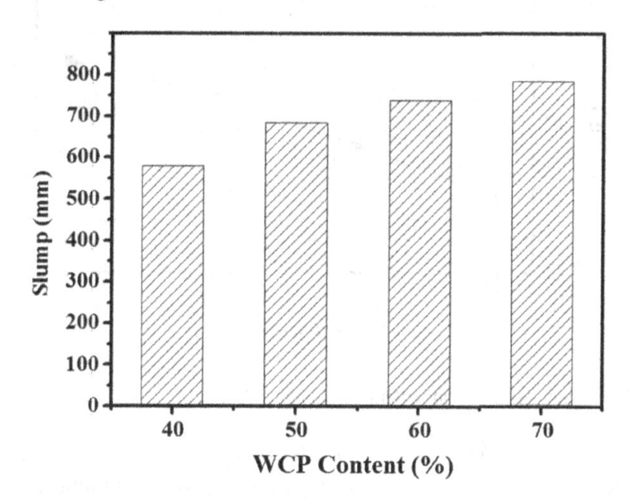

FIGURE 12.6 Effect of WCP content on slump value of SCAACs.

is 600–800 mm [32]. The slump flow time to reach a diameter of 500 mm (T_{50}) for each of the tested mixtures was less than 7 seconds. However, it was observed that the incorporation of ceramic strongly increased the slump flow of the mixtures, as shown in Table 12.4 and Figure 12.6. It was observed that increasing the WCP content led to an increase in the silicate (SiO_2) content and enhanced the workability performance. This observation was also supported by the findings of Majidi et al. [33]. The mixtures with more or equal to 50% WCP content displayed a higher flow compared to the mixtures containing 40% WCP. Also the increase in WCP content improved the passing ability of the mixtures. It was found the ratio of (H_2/H_1) tended to increase from 0.78 to 0.90, 0.96, and 0.98 with increasing WCP content from 40 to 50, 60, and 70%, respectively.

TABLE 12.4
Fresh Properties and Compressive Strength of Prepared Concrete

	Workability			Setting Time		Compressive Strength (MPa)			
Mix Sample	Slump (mm)	L-Box Ratio	T_{50} (sec)	Initial (min)	Final (min)	1 Day	3 Days	7 Days	28 Days
S1	580	0.78	7.0	20	36	32.8	34.4	41.6	56.4
S2	685	0.90	5.0	34	48	28.4	31.2	38.1	47.6
S3	740	0.96	4.0	47	65	21.3	25.8	29.6	42.1
S4	785	0.98	4.0	56	78	16.7	18.2	20.4	31.8

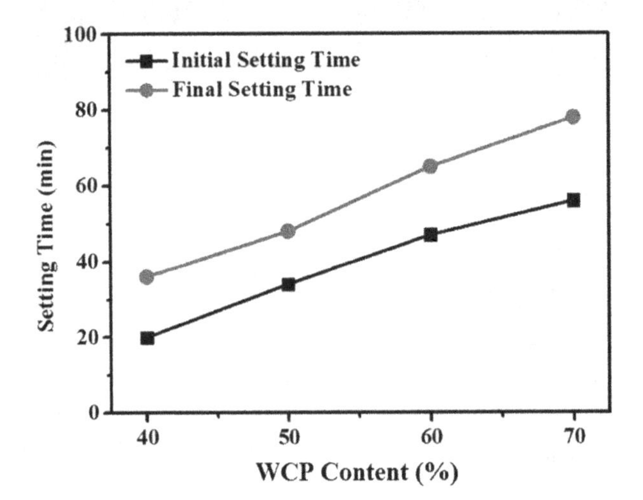

FIGURE 12.7 Effect of WCP on setting time of SCAACs.

Figure 12.7 illustrates the effect of WCP content on stetting time of alkali-activated concrete. It can be clearly observed that as the WCP content increased as the GBFS replacement in the alkali-activated matrix, the initial and final setting time value increased. It was reported [34–36] that when the content of WCP increased from 40 to 50%, the initial and final setting time increased from 20 and 36 minutes to 34 and 48 minutes, respectively. Actually the setting time increased more than 200% with 70% of WCP content compared to a mixture content of only 40% WCP. This increment on setting time is attributed to the reduced amount of calcium oxide with the reduction in GBFS content which led to a slow rate of hydration [37–39].

12.5 COMPRESSIVE STRENGTH

Figure 12.8 shows the effect of WCP as GBFS replacement on strength development of SCAAC. The results showed that the increase in WCP content in the

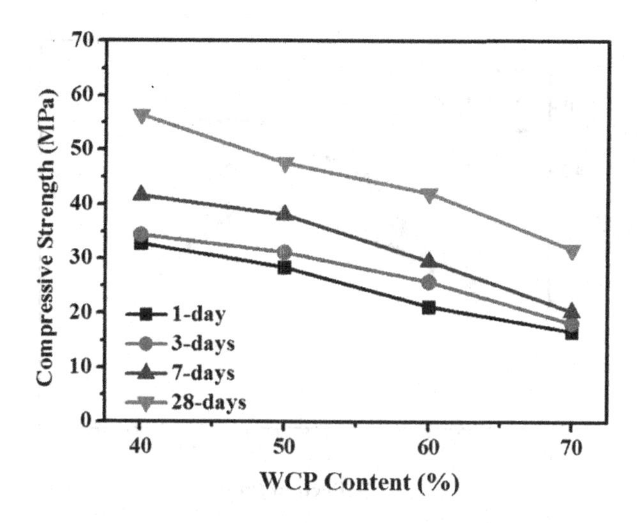

FIGURE 12.8 Compressive strength development of SCAACs containing various WCPs as GBFS replacement.

matrix of alkali-activated negatively affected the compressive strength development. The compressive strength value tended to decrease with an increasing content of WCP for all ages of tested specimens, in a range between 16 and 43% at 28 days of curing age, depending on the percentage of WCP replaced of GBFS. The compressive strength of SCAAC prepared with 50% WCP achieved the highest strength (47.6 MPa) compared to 42.1 and 31.8 MPa with 60 and 70%, respectively. The reduction in compressive strength is considered normal due to the poor adhesion between the paste and aggregate; the increasing WCP content led to a reduction in the calcium oxide and effected the calcium silicate hydrate (C-S-H) [40].

12.6 FLEXURAL BEHAVIOUR

Figure 12.9 shows the load-deflection response for all the tested beams. A significant decrease in the load-carrying capacity and deflection was observed with an increase in the WCP content from 40 to 50, 60, and 70% for beams B-S_1, B-S_2, B-S_3, and B-S_4, respectively. The SCAAC flexural strength results of the tested beams exhibited an almost lower load-deflection response compared to the control beam (B-S_1). The experimental results showed that the crack loadings for the different beams also decreased with increasing WCP content. Compared to beam B-S_1, the cracking load for B-S_2, B-S_3, and B-S_4 decreased by almost 7.4, 27.8, and 49.2 %, respectively. Beam B-S_2 achieved the highest flexural stiffness compared to the other two self-compacting alkali-activated beams (B-S_3 and B-S_4), which may have increased the cracking load. Notably, the increase in WCP content led to an increase in the appearance of cracks, as shown in Table 12.5.

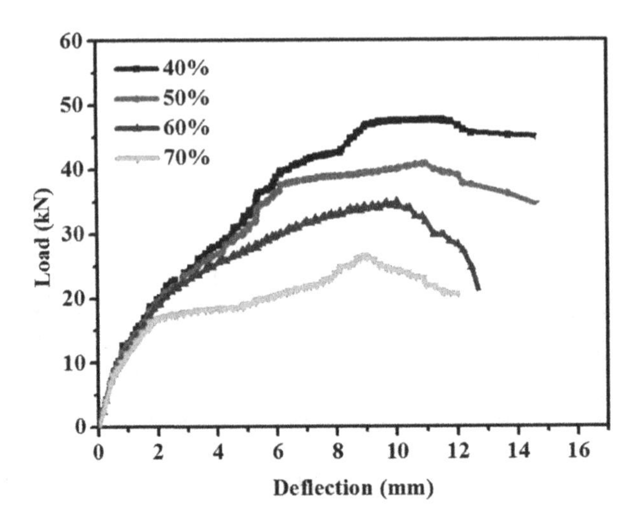

FIGURE 12.9 Load-deflection of reinforcement SCAACs beams prepared with various WCP content.

TABLE 12.5
Cracking, Ultimate Load, and Deflection for Tested Beams

	Cracking Load		Peak Load	
Beam	**Load (kN)**	**Deflection (mm)**	**Load (kN)**	**Deflection (mm)**
B-S_1	20.1	2.06	47.68	11.18
B-S_2	18.6	1.87	40.86	10.97
B-S_3	14.5	1.36	35.11	10.02
B-S_4	10.2	0.83	26.64	9.84

12.7 CONCLUSIONS

Based on the experimental results, the following conclusions can be drawn:

i. The results showed that it is possible to obtain a self-compacting alkali-activated concrete that is environmentally sustainable using WCP as GBFS replacement.
ii. Referring to the fresh properties of alkali-activated concrete, it was observed that with an increased WCP content, the filling and passing ability of SCAAC improved.
iii. Increasing the WCP content causes a decrease in the compressive strength of SCAAC. The compressive strength gradually tends to decrease with an increasing replacement level of GBFS. In the compressive strength, it was observed that at 28 days of curing, the SCAAC it presented up to 43% strength reduction with 70% of substitution of slag by WCP in the mix S_4.

iv. Increasing WCP content from 40% to 50, 60, and 70% reduced the ultimate load of reinforced concrete beams from 47.6 kN to 40.8, 35.1, and 26.6 kN, respectively.

REFERENCES

1. Hassan, I. O., et al., Flow characteristics of ternary blended self-consolidating cement mortars incorporating palm oil fuel ash and pulverised burnt clay. *Construction and Building Materials*, 2014. **64**: pp. 253–260.
2. Krivenko, P., et al., Mechanism of preventing the alkali–aggregate reaction in alkali activated cement concretes. *Cement and Concrete Composites*, 2014. **45**: pp. 157–165.
3. Malek, R., et al., The contribution of class-F fly ash to the strength of cementitious mixtures. *Cement and Concrete Research*, 2005. **35**(6): pp. 1152–1154.
4. Neville, A. M., *Properties of Concrete*. London: Longman, 1995: pp. 1–372.
5. Huseien, G. F., M. Y. Al-Fasih, and H. K. Hamzah, Performance of self-compacting concrete with different sizes of recycled ceramic aggregates. *International Journal of Innovative Research and Creative Technology*, 2015. **1**(3), pp. 264–269.
6. Kamseu, E., et al., Self-compacting geopolymer concretes: Effects of addition of aluminosilicate-rich fines. *Journal of Building Engineering*, 2016. **5**: pp. 211–221.
7. Huseien, G. F., et al., Effects of ceramic tile powder waste on properties of self-compacted alkali-activated concrete. *Construction and Building Materials*, 2020. **236**: p. 117574.
8. Şahmaran, M., H. A. Christianto, and İ. Ö. Yaman, The effect of chemical admixtures and mineral additives on the properties of self-compacting mortars. *Cement and Concrete Composites*, 2006. **28**(5): pp. 432–440.
9. Demie, S., et al. Effects of curing temperature and superplasticizer on workability and compressive strength of self-compacting geopolymer concrete. In *National Postgraduate Conference (NPC), 2011*. 2011: IEEE.
10. Duxson, P., et al., The role of inorganic polymer technology in the development of 'green concrete'. *Cement and Concrete Research*, 2007. **37**(12): pp. 1590–1597.
11. Lachemi, M., et al., Development of cost-effective self-consolidating concrete incorporating fly ash, slag cement, or viscosity-modifying admixtures. *Materials Journal*, 2003. **100**(5): pp. 419–425.
12. Yusuf, M. O., et al., Evolution of alkaline activated ground blast furnace slag–ultrafine palm oil fuel ash based concrete. *Materials & Design*, 2014. **55**: pp. 387–393.
13. Rashad, A. M., Properties of alkali-activated fly ash concrete blended with slag. *Iranian Journal of Materials Science and Engineering*, 2013. **10**(1): pp. 57–64.
14. Huseiena, G. F., et al., Effect of binder to fine aggregate content on performance of sustainable alkali activated mortars incorporating solid waste materials. *Chemical Engineering*, 2018. **63**: pp. 667–672.
15. Patankar, S. V., Y. M. Ghugal, and S. S. Jamkar, Effect of concentration of sodium hydroxide and degree of heat curing on fly ash-based geopolymer mortar. *Indian Journal of Materials Science*, 2014. **2014**: pp. 1–6.
16. Khale, D. and R. Chaudhary, Mechanism of geopolymerization and factors influencing its development: A review. *Journal of Materials Science*, 2007. **42**(3): pp. 729–746.
17. Ranjbar, N., et al., Compressive strength and microstructural analysis of fly ash/palm oil fuel ash based geopolymer mortar under elevated temperatures. *Construction and Building Materials*, 2014. **65**: pp. 114–121.
18. Salih, M. A., et al., Development of high strength alkali activated binder using palm oil fuel ash and GGBS at ambient temperature. *Construction and Building Materials*, 2015. **93**: pp. 289–300.

19. Pacheco-Torgal, F. and S. Jalali, Reusing ceramic wastes in concrete. *Construction and Building Materials*, 2010. **24**(5): pp. 832–838.
20. Huang, B., Q. Dong, and E. G. Burdette, Laboratory evaluation of incorporating waste ceramic materials into Portland cement and asphaltic concrete. *Construction and Building Materials*, 2009. **23**(12): pp. 3451–3456.
21. Pacheco-Torgal, F. and S. Jalali, Compressive strength and durability properties of ceramic wastes based concrete. *Materials and Structures*, 2011. **44**(1): pp. 155–167.
22. Abdollahnejad, Z., et al., Mix design, properties and cost analysis of fly ash-based geopolymer foam. *Construction and Building Materials*, 2015. **80**: pp. 18–30.
23. Nath, P. and P. K. Sarker, Effect of GGBFS on setting, workability and early strength properties of fly ash geopolymer concrete cured in ambient condition. *Construction and Building Materials*, 2014. **66**: pp. 163–171.
24. Huseien, G. F., et al., Effect of metakaolin replaced granulated blast furnace slag on fresh and early strength properties of geopolymer mortar. *Ain Shams Engineering Journal*, 2016. **9**(4): pp. 1557–1566.
25. Roslan, N. H., et al., Performance of steel slag and steel sludge in concrete. *Construction and Building Materials*, 2016. **104**: pp. 16–24.
26. Çevik, A., et al., Effect of nano-silica on the chemical durability and mechanical performance of fly ash based geopolymer concrete. *Ceramics International*, 2018. **44**(11): pp. 12253–12264.
27. Baraldi, L. and MECS-Machinery Economics Studies by ACIMAC, World production and consumption of ceramic tiles. *AMERICA*, 2016. **1**(9.2): p. 7.7.
28. Hussein, A. A., et al., Performance of nanoceramic powder on the chemical and physical properties of bitumen. *Construction and Building Materials*, 2017. **156**: pp. 496–505.
29. Senthamarai, R., P. D. Manoharan, and D. Gobinath, Concrete made from ceramic industry waste: Durability properties. *Construction and Building Materials*, 2011. **25**(5): pp. 2413–2419.
30. Samadi, M., et al., Properties of mortar containing ceramic powder waste as cement replacement. *Jurnal Teknologi*, 2015. **77**(12): pp. 93–97.
31. Memon, S. A., et al., Development of form-stable composite phase change material by incorporation of dodecyl alcohol into ground granulated blast furnace slag. *Energy and Buildings*, 2013. **62**: pp. 360–367.
32. Silva, Y. F., et al., Properties of self-compacting concrete on fresh and hardened with residue of masonry and recycled concrete. *Construction and Building Materials*, 2016. **124**: pp. 639–644.
33. Al-Majidi, M. H., et al., Development of geopolymer mortar under ambient temperature for in situ applications. *Construction and Building Materials*, 2016. **120**: pp. 198–211.
34. Huseien, G. F., et al., Properties of ceramic tile waste based alkali-activated mortars incorporating GBFS and fly ash. *Construction and Building Materials*, 2019. **214**: pp. 355–368.
35. Huseien, G. F., et al., Evaluation of alkali-activated mortars containing high volume waste ceramic powder and fly ash replacing GBFS. *Construction and Building Materials*, 2019. **210**: pp. 78–92.
36. Samadi, M., et al., Waste ceramic as low cost and eco-friendly materials in the production of sustainable mortars. *Journal of Cleaner Production*, 2020: p. 121825.
37. Samadi, M., et al., Influence of glass silica waste nano powder on the mechanical and microstructure properties of alkali-activated mortars. *Nanomaterials*, 2020. **10**(2): p. 324.

38. Phoo-ngernkham, T., et al., High calcium fly ash geopolymer mortar containing Portland cement for use as repair material. *Construction and Building Materials*, 2015. **98**: pp. 482–488.
39. Huseien, G. F. and K. W. Shah, Durability and life cycle evaluation of self-compacting concrete containing fly ash as GBFS replacement with alkali activation. *Construction and Building Materials*, 2020. **235**: p. 117458.
40. Sofi, M., et al., Engineering properties of inorganic polymer concretes (IPCs). *Cement and Concrete Research*, 2007. **37**(2): pp. 251–257.

Index

Printed in the United States
By Bookmasters